Introduction to Reliability Engineering

E. E. Lewis
Department of Mechanical
and Nuclear Engineering
Northwestern University

John Wiley & Sons
New York • Chichester • Brisbane • Toronto • Singapore

Library of Congress Cataloging in Publication Data:

Lewis, E. E. (Elmer Eugene), 1938–
Introduction to reliability engineering.

1. Reliability (Engineering) I. Title.
TA169.L47 1987 620′.00452 86-18891
ISBN 0-471-81199-8

Printed in the United States of America

10 9 8 7 6 5 4 3 2 1

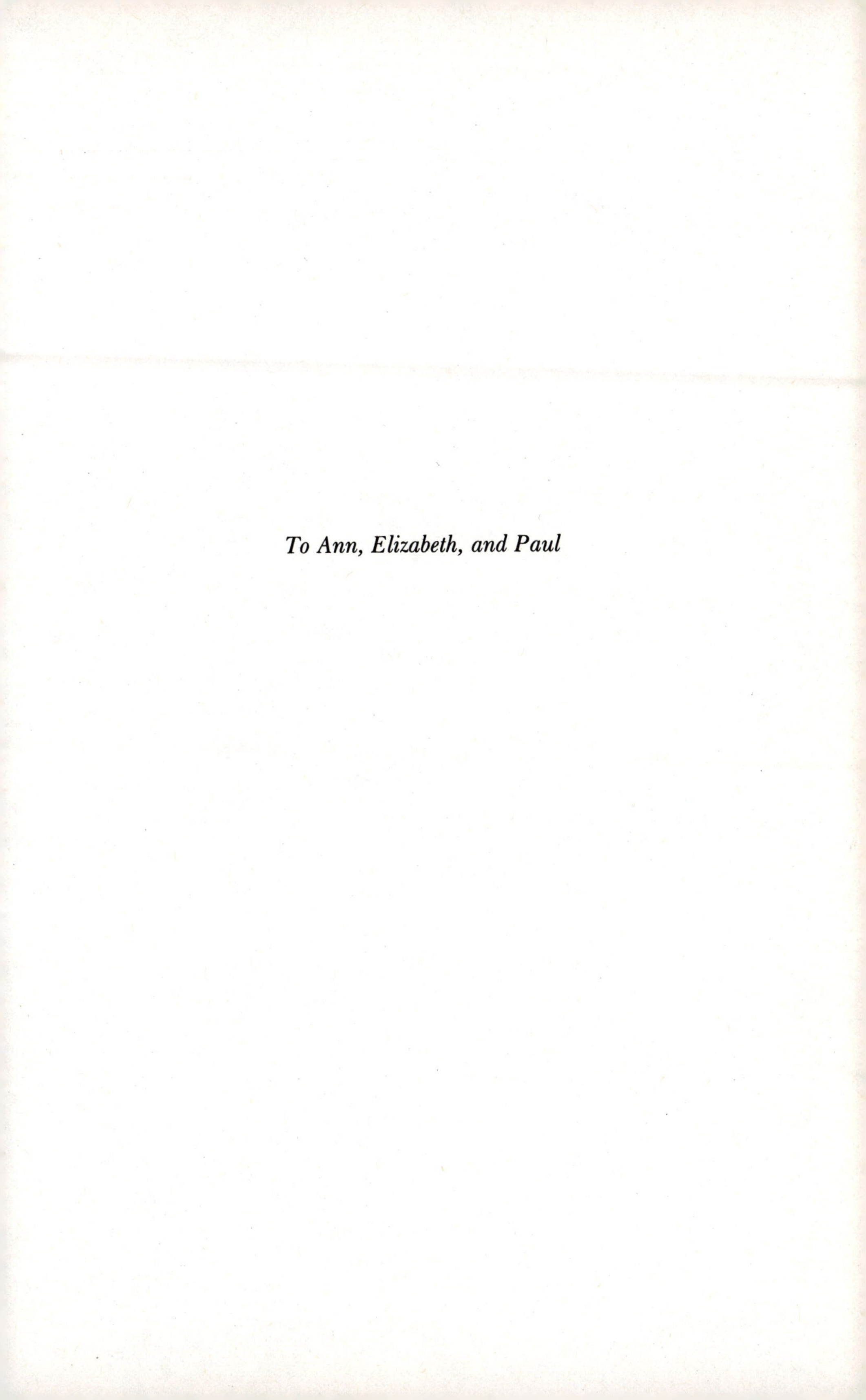

To Ann, Elizabeth, and Paul

About the Author

Elmer E. Lewis is Professor of Mechanical and Nuclear Engineering at Northwestern University. He received a B.S. in engineering physics and an M.S. and Ph.D. in nuclear engineering from the University of Illinois, Urbana. He was a captain in the U.S. Army and then Assistant Professor and Ford Foundation Fellow at MIT before joining Northwestern's faculty in 1968. He has served as Visiting Professor at the University of Stuttgart and as a Guest Scientist at the Nuclear Research Center at Karlsruhe, West Germany; he is a consultant to Los Alamos and Argonne National Laboratories. A fellow of the American Nuclear Society, Professor Lewis is actively involved in research in the areas of reliability modeling, radiation transport, and the physics and safety of nuclear systems. He has lectured widely in the United States and Europe and has authored or co-authored four books and nearly one hundred other research publications.

Preface

Reliability considerations occupy an increasingly important place in engineering practice. Although the details of application differ depending on whether mechanical, electrical, or chemical systems are under analysis, the reliability concepts cut across the specific fields of engineering. In this book I attempt to provide an integrated introduction to the theory and practice of reliability engineering from an interdisciplinary point of view. The book is intended to be useful both to engineering students and to practicing professionals.

The book evolved from a set of lecture notes for a junior–senior level course in reliability engineering at Northwestern University. The course was designed for undergraduate engineering students, most of whom have insufficient time in their schedules for the probability and statistics prerequisites that a more advanced treatment would require. At the same time, I believe that it is advantageous to introduce reliability concepts early in the student's career, with numerous applications and a minimum of advanced mathematics. This approach may motivate students to apply the material to later work in engineering design as well as to pursue the theory in more depth.

The reader is presumed to have completed only the mathematics sequence through ordinary differential equations that is standard to most undergraduate engineering curricula. No prior knowledge of probability or of statistics is assumed; the text is entirely self-contained in this regard. Since the book is focused exclusively on reliability problems, however, it should not be considered as a substitute for a broader course in engineering statistics; the two subjects should complement one another.

The chapters contain nearly 100 solved examples and over 200 exercises, with answers for about half of the exercises given at the end of the book. These are drawn from a variety of engineering fields. The examples, the exercises, and the other material herein, however, can be understood without more advanced knowledge specific to mechanical, civil, electrical, or any of the other engineering disciplines. The physics, chemistry, and basic engineering courses that are common to the first two years of nearly all engineering curricula serve as adequate background.

In organizing the book, I have attempted to provide for some flexibility so that presentation of the material may be adjusted to the students' background and to the number of class hours available. For students with a rudimentary knowledge of probability and statistics, the material in Chapters 2 and 3 may be reviewed rather rapidly. The later sections of Chapter 5 may be eliminated if reliability testing techniques are not to be emphasized. Chapter 6 is essential for those seeking an understanding of the probabilistic basis of safety margins and their relation to reliability considerations; the remainder of the book, however, may be read independently of it. Finally, the Markov processes introduced in Chapter 9 for modeling component failure interactions are not prerequisite to the material covered in Chapter 10.

I am greatly indebted to many for invaluable assistance in preparing this book. I am most appreciative of the patience of numerous students at Northwestern who suffered through the errors and omissions of the early draft manuscripts that were used in lieu of a textbook. A great deal of proofreading and many constructive comments have been provided by friends, colleagues, and students. Among these are M. A. Lawrence, F. Boehm, L. Chastain, J. C. Ovaert, and S. Wu of Northwestern University; H. F. Martz of Los Alamos National Laboratory; R. E. Holtz of Argonne National Laboratory; and C. W. Maynard of the University of Wisconsin. The thoughtful reviews provided by the anonymous referees are also appreciated.

Special thanks are due to Deana Rottner, my secretary, for this is the third book for which she has prepared the manuscript, and without complaint of my illegible scrawl or interminable revisions. Finally, without the continuing support and understanding of my wife, Ann, and of our children, Elizabeth and Paul, the completion of this book would not have been possible.

Evanston, Illinois

Elmer E. Lewis

Contents

"Human error, lack of imagination, and blind ignorance. The practice of engineering is in large measure a continuing struggle to avoid making mistakes for these reasons."

From *The Existential Pleasures of Engineering*

CHAPTER 1

Introduction

1.1 RELIABILITY DEFINED

Reliability considerations are playing an increasing role in virtually all engineering disciplines. As the demands for systems that perform better and cost less increase, there is a concomitant requirement to minimize the probability of failures, whether the failures simply increase costs and inconvenience or gravely threaten the public safety. A substantial body of knowledge has developed both for analyzing the stochastic nature of such failures and for minimizing the probability of their occurrence. In general terms, much of this knowledge cuts across the details of the particular engineering disciplines in which it is applied, thus making a book of this nature feasible. Within this general framework, however, there is a rich variety of contexts in which reliability considerations appear. Indeed, deeper insight into failures and their prevention is to be gained by comparing and contrasting the reliability characteristics of systems of differing characteristics: computers, electromechanical machinery, energy conversion systems, chemical and materials processing plants, and structures, to name a few.

In the broadest sense, reliability is associated with dependability, with successful operation, and with the absence of breakdowns or failures. It is necessary for engineering analysis, however, to define reliability quantitatively as a probability. Thus reliability is defined as the probability that a component, device, equipment, or system will perform its intended function for a specified period of time under a given set of conditions. This definition is succinct, making it necessary to elaborate on the terms used in it.

The term *reliability* may be applied to almost any object, which is the reason that the terms *system, device, equipment,* and *component* are all used in the definition. Each of these terms, however, has somewhat different connotations. For example, the reliability of a system may often include considerations of operator errors, whereas that of a component normally would not. Similarly, systems and equipment are often thought of in terms of configurations of devices, components, or parts. In a general probabilistic approach to reliability, it matters little what the object of analysis is called, for in specific contexts the meaning is clear. For the most part we shall refer to the reliability of the system, with the understanding that any of the other terms may be substituted. In addition, if there is need to dissect the subject of reliability analysis, we shall consider the system as a set of interacting components working together as an integrated whole. Where more than one level of division is required, a number of different hierarchies may be used; for example, system, subsystem, component, and part.

A system is said to fail when it ceases to perform its intended function. When there is total cessation of function—an engine stops running, a structure collapses, a piece of communication equipment goes dead—the system has clearly failed. Often, however, it is necessary to define failure quantitatively in order to take into account the more subtle forms of failure; through deterioration or instability of function. Thus a motor that is no longer capable of delivering a specified torque, a structure that exceeds a specified deflection, or an amplifier that falls below a stipulated gain has failed. Intermittent operation or excessive drift in electronic equipment and the machine tool production of out-of-tolerance parts are other failures.

The way in which time is specified in the definition of reliability may also vary considerably, depending on the nature of the system under consideration. For example, in an intermittently operated system one must specify whether calendar time or the number of hours of operation is to be used. If the operation is cyclic, such as that of a switch, time is likely to be cast in terms of the number of operations. If reliability is to be specified in terms of calendar time, it may also be necessary to specify the frequency of starts and stops, and the ratio of operating to total time.

Finally, specifying the conditions under which a system is to operate in many situations requires extensive analysis. Such conditions may be divided roughly into the principal design loads and the environmental effects. Design loads, for example, might be the weight that a structure must support, the electrical load on a generator, the rate of information transfer on a telecommunication system, or the impact loading on a landing gear. Infrequent but severe loading such as those caused by floods or earthquakes may also be included in the definition of conditions. The environmental conditions, which also may be viewed as loadings, must also be taken into consideration. Temperature extremes, dust, salt, and humidity are examples of environmental loadings.

In addition to the reliability itself, other quantities are used to characterize the reliability of a system. The mean time to failure and failure rate are examples, and in the case of repairable systems, so also are the availability and mean time to repair. The definition of these and other terms will be introduced as needed.

Reliability is defined positively, in terms of a system performing its intended function, and no distinction is made between failures. In reality there is a great deal of concern not only with the probability of failure but also with the potential consequences of different modes of failure. In particular, attention must be given to failures that present severe safety problems, not just economic loss or inconvenience. The disciplines of system safety analysis and risk assessment must often go hand in hand with reliability considerations. Thus the manufacturer of a home appliance must be concerned with reliability because frequent failures and resulting customer dissatisfaction are major concerns. But, in addition, careful scrutiny must also be given to insure that the failures that do occur do not create a safety hazard such as shock or electrocution. For other systems, such as an aircraft engine, there is likely to be much less distinction between reliability and safety considerations, for most failures may be assumed to be synonymous with the creation of safety hazards. In either case, safety considerations frequently enter into the discussions of reliability in the chapters that follow.

1.2 PERFORMANCE AND RELIABILITY

Much of engineering endeavor is concerned with designing and building products for improved performance. We strive for lighter and therefore faster aircraft, for thermodynamically more efficient energy conversion devices, for faster computers and for larger, longer-lasting structures. The pursuit of such objectives, however, often requires designs incorporating features that more often than not may tend to be less reliable than older, lower-performance systems. The trade-offs between performance and reliability are often subtle; they often involve loading, system complexity, and new materials and concepts. Thus any product with both improved performance and reliability is a significant advance in engineering design.

Load is most often used in the mechanical sense of the stress on a structure. But here we interpret it more generally so that it also may be the thermal load caused by high temperature, the electrical load on a generator, or even the information load on a telecommunications system. Whatever the nature of the load on a system or its components may be, performance is frequently improved through increased loading. Thus, by decreasing the weight of an aircraft, we increase the stress levels in its structure; by going to higher—thermodynamically more efficient—temperatures we are forced to operate materials under conditions in which there are heat-induced losses of strength and more rapid corrosion. By allowing for ever-increasing flows of information in communications systems, we approach the limits in frequency at which switching or other digital circuits may operate. Such approaches to the physical limits of systems or their components result in an increased number of failures unless appropriate countermeasures are taken. Thus specifications for a purer material, tighter dimensional tolerance, and a host of other measures are required to reduce uncertainty in the performance limits, and thereby permit one to operate close to these limits without increasing the probability of exceeding them.

The performance of a system is often increased at the expense of increased complexity, the complexity usually being measured by the number of required components. Once again, reliability will be decreased unless compensating measures are taken, for it may be shown that if nothing else is changed, reliability decreases with each added component. In these situations reliability can only be maintained if component reliability is increased or if component redundancy is built into the system.

Probably the greatest improvements in performance have come through the introduction of entirely new materials or devices to achieve a particular goal. For, in contrast to the trade-off faced with increased loading or complexity, more fundamental advances may have the potential for both improved performance and greater reliability. Certainly, the history of technology is a study of such advances; the replacement of wood by metals in machinery and structures, the replacement of piston with jet aircraft engines, and the replacement of vacuum tubes with solid-state electronics all led to fundamental advances in both performance and reliability.

Even with major advances in technology, however, reliability may be a severe problem, particularly during the early stages of introducing a new technological advance. The engineering community must proceed through a learning experience to reduce the uncertainties in the limits in loading on the new device, to understand its susceptibilities to adverse environments, and to perfect the procedures for fabrication, manufacture, and construction.

1.3 RELIABILITY REQUIREMENTS

As we have just indicated, at any stage of technological development trade-offs must often be made between reliability and performance; similarly, trade-offs are often required between reliability and cost. How such trade-offs are made, and the criteria on which they are based, is deeply imbedded in the essence of engineering practice. For the criteria and considerations are as varied as the uses to which technology is put. Some examples illustrate this point.

As a first example, consider a race car. If one looks at the history of automobile racing at the Indianapolis 500, from year to year, one finds that the performance is continually improving, if measured as the average speed of the qualifying cars. At the same time, the reliability of these cars, measured as the probability that they will finish the race, remains uniformly low at less than 50%.* This should not be surprising, for in this situation performance is everything, and one must tolerate a high probability of breakdown if there is to be any chance of winning the race.

At the opposite extreme is the design of a commercial airliner, where mechanical breakdown could well result in a catastrophic accident. In this case reliability is the overriding design consideration; degraded speed, payload, and fuel economy are accepted in order to maintain a very small probability of catastrophic failure.

An intermediate example might be in the design of a military aircraft, for here the trade-off to be achieved between reliability and performance is more equally balanced. Reducing reliability may again be expected to increase the incidence of fatal accidents. Nevertheless, if the performance of the aircraft is not sufficiently high, the number of losses in combat may negate the mission of the aircraft, with a concomitant loss of life.

In contrast to these life or death implications, reliability of many products may be viewed in more routine economic terms. The design of a piece of machinery, for example, may involve trade-offs between the increased capital costs entailed if high reliability is to be achieved, and the increased costs of repair and of lost production that will be incurred from lower reliability. Even here, more subtle issues come into play. For consumer products, the higher initial price that may be required for a more reliable item must be carefully weighed against the purchaser's annoyance with the possible failure of a less reliable item and the cost of replacement or repair.

1.4 SYSTEM LIFE CYCLE AND RELIABILITY

In the foregoing discussion we have referred primarily to the relationships of reliability to design. In reality, important reliability considerations appear throughout the entire life cycle of a system. We must first discuss the life cycles of a system. Broadly speaking, the life cycle may be divided into the following categories:

1. Definition and conceptual design.
2. Detailed design and development.
3. Manufacture, construction, or both.
4. Operation.

The reliability activities taking place during each of these four periods of time may be quite different, depending on the nature of the system. Those for a mass-produced consumer product differ greatly, for instance, from those for single and few-of-a-kind systems, such as a large structure

* R. D. Haviland, *Engineering Reliability and Long Life Design,* Van Nostrand, New York, 1964, p. 114.

or a chemical processing plant. It is nevertheless useful to examine reliability from conceptual design through operation.

Definition and Conceptual Design

In the project definition the objectives of the system are set forth in the form of one or more functional requirements. Thus, for a generator the power output is specified, for a computer the speed and memory, for a structure the load-bearing capacity, and so on. In addition, the environment in which the system is to function must also be determined: the range of temperature and humidity, the concentrations of dust or other contaminants, as well as the severity of mechanical impact, vibration, power surges, or other loadings. Finally, the service life to which the system is to be designed must be specified.

From such requirements a conceptual design is formulated that in broad form outlines how the system is to function and provides the general plan for its construction. Will an engine be a diesel or gasoline-fueled, and will it be air- or water-cooled? Will a bridge be reinforced concrete or steel, and what is the architectural form that it shall take? Will a steam generator be once-through or U tube, and so on.

From the functional requirements comes the definition of failure, and thus also of reliability. Reliability requirements may then be set, and the trade-offs between reliability, cost, and functional requirements may be examined as the design proceeds into the detailed phase.

Detailed Design and Development

The conceptual design must be converted into a detailed set of drawings and specifications from which the system can be built. During this phase, maintenance requirements and procedures are also likely to take shape in some detail. As the design proceeds, experiment, testing, and analysis are likely to be required to choose between alternatives, to solve problems, and to predict the performance of subsystems or components.

Reliability considerations should permeate this stage of the design: in setting safety factors and design margins; in eliminating unnecessary complexities; in translating system reliability criteria into reliability requirements for subsystems, components, and parts; and in setting time intervals for inspection, maintenance, and replacement of parts subject to wear. It is at this stage that the detailed examination of potential failure mechanisms and modes is most beneficial, for often they may be eliminated or mitigated without nearly the expense that would be incurred were they to require major redesign or retrofitting at a later time. It is also during the detailed-design stage that the experience of the designer and the use of standards and codes of good practice are most beneficial in eliminating potential safety hazards from the system. The designer must work toward a fail-safe and foolproof system.

In the later stages of the design process, prototypes are built and the first reliability tests may be performed. Although the number of prototypes available and the time deadlines for the beginning of production may not allow enough units of the final design to be tested to failure to gain statistically meaningful predictions of reliability, the failures that do occur are valuable in that they add to the understanding of the failure mechanisms and thus provide a basis for improving reliability by modifying the design or through maintenance procedures. Even with a single prototype the process of test-fix-test-fix, in which failures are analyzed and design refinements made, can be used to increase substantially the reliability of the final product. In the design of few-of-a-kind systems, such as large structures, power plants, or chemical processing facilities, the data gained from models, pilot plants, or subsystem prototypes provide a valuable understanding of failure modes and may suggest refinements that will increase system reliability.

Manufacture/Construction

Historically, reliability considerations during the manufacture of a system are most closely related to the practices of quality control. If the reliability inherent in the design is to be achieved in the built system, it must not be compromised by substandard components or materials, missing parts, improper assembly, or other defects in the manufacturing processes. Reliability in manufacture is monitored and controlled by the methods of statistical quality control. The materials and components entering the construction process must be monitored by carefully planned sampling procedures, and stringent process control is required to identify and eliminate problems in manufacture.

Reliability testing on manufactured items is exceedingly important. Through such testing the manufacturing procedure can be modified to eliminate shortcomings that result in weak units subject to early failure. Moreover, more units are likely to be available for testing than when only prototypes were available. Thus statistical methods can be used more successfully in making quantitative estimates of reliability. Equally important, the manufactured units provide a truer picture of the reliability achieved in the product as it is delivered to the user than do carefully constructed prototypes.

Verification of end product reliability by testing to failure is not possible in large one-of-a-kind systems. Moreover, since such systems are likely to be constructed under field conditions that are more variable than those found in a manufacturing plant, there is more cause for concern that the system reliability may be compromised. Very stringent acceptance criteria on components, careful supervision and control of the construction process, and often an elaborate set of proofs or acceptance tests are necessary in such situations.

Operations

Reliability considerations do not end with the completion of manufacture, or even after carefully designed packaging has been provided to avoid both damage in shipment and deterioration in storage. Therefore, during the life of a system, careful attention must be given to maintenance and operation and to the feedback of field information for improvements in design and manufacture.

The data collected on field failures are particularly valuable because they are likely to provide the only estimates of reliability that incorporate the loadings, environmental effects, and imperfect maintenance found in practice. On both component and system levels such a data base is invaluable for predicting the reliability of future designs and for improving design.

The early collection of field failure data is important because it may reveal unanticipated environmental loadings, shortcomings in field maintenance, user abuse, or other factors that were not or could not be fully anticipated by the designers. Such information may be used to improve reliability of the operating system by revising operational practices or maintenance schedules. In some instances replacement parts will need to be designed. If the reliability problems are severe enough or have major safety implications, the product will have to be recalled for modifications or the system shut down for retrofitting.

1.5 PREVIEW

In the chapters that follow we must first introduce a number of probability concepts in order to treat reliability in a quantitative manner. With the concepts developed in Chapters 2 and 3, we are able to formulate reliability problems quantitatively. In Chapter 4 we investigate reliability and its relationship to failure rates and other phenomena where the primary variable is time. Then, in Chapter 5, we turn from probability to statistics to examine how the reliability or related characteristics can be estimated from test data or from statistical information gathered from product failures in the field.

In Chapter 6 we examine the relationship of reliability to the loading on a system and the capacity of the system to withstand the various types of loads. This entails, among other things, an exposition of the probabilistic treatment of safety factors, design margins, and of wearin and wearout effects.

In Chapter 7 we discuss the approaches by which redundant configurations may be used to improve reliability, and also some of the potential pitfalls involved therein. Then, in Chapter 8, the effects of both preventive and corrective maintenance are examined, and the concepts of maintainability and availability are introduced. In Chapter 9 Markov processes are introduced in order to examine more closely the interaction of component failures.

In Chapter 10 we depart from the traditional definition of reliability to examine the prevention of particular modes of failure that may lead to serious accidents. Fault trees and some of the related methodologies that are employed in such system safety engineering are discussed.

Bibliography

Arsenault, J. E., and J. A. Roberts (eds.), *Reliability and Maintainability of Electronic Systems,* Computer Science Press, Potomac, MD, 1980.

Green, A. E., and A. J. Bourne, *Reliability Technology,* Wiley, New York, 1972.

Haviland, R. D., *Engineering Reliability and Long Life Design,* Van Nostrand, New York, 1964.

Ireson, W. G. (ed.), *Reliability Handbook,* McGraw-Hill, New York, 1966.

Kapur, K. C., and L. R. Lamberson, *Reliability in Engineering Design,* Wiley, New York, 1977.

McCormick, N. J., *Reliability and Risk Analysis,* Academic Press, New York, 1981.

Reliability Guidebook, Asian Productivity Organization, Tokyo, Japan, 1972.

CHAPTER 2

Probability and Sampling

2.1 INTRODUCTION

Fundamental to all reliability considerations is an understanding of probability, for reliability is defined as just the *probability* that a system will not fail under some specified set of circumstances. In this chapter we define probability and discuss the logic by which probabilities can be combined and manipulated. We then examine sampling techniques by which the results of tests or experiments can be used to estimate probabilities. Although quite elementary, the notions presented will be shown to have immediate applicability to a variety of reliability considerations ranging from the relationship of the reliability of a system to its components to the common acceptance criteria used in quality control.

2.2 PROBABILITY CONCEPTS

We shall denote the probability of an event, say a failure, X, as $P\{X\}$. This probability has the following interpretation. Suppose that we perform an experiment in which we test a large number of items, for example, light bulbs. The probability that a light bulb fails the test is just the relative frequency with which failure occurs when a very large number of bulbs are tested. Thus, if N is the number of bulbs tested and n is the number of failures, we may define the probability formally as

$$P\{X\} = \lim_{N \to \infty} \frac{n}{N}. \tag{2.1}$$

Equation 2.1 is an empirical definition of probability. In some situations symmetry or other theoretical arguments also may be used to define probability. For example, one often assumes that the probability of a coin flip resulting in "heads" is 1/2. Closer to reliability considerations, if one has two pieces of equipment, A and B, that are chosen from a lot of equipment of the same design and manufacture, one may assume that the probability that A fails before B is 1/2. If the hypothesis is doubted in either case, one must verify that the coin is true or that the pieces of equipment are identical by performing a large number of tests to which Eq. 2.1 may be applied.

Probability Axioms

Clearly, the probability must satisfy

$$0 \leqslant P\{X\} \leqslant 1. \tag{2.2}$$

Now suppose that we denote the event not X by $\tilde{X}$. In our light-bulb example, where X indicates failure, $\tilde{X}$ then indicates that the light bulb passes the test. Obviously, the probability of passing the test, $P\{\tilde{X}\}$, must satisfy

$$P\{\tilde{X}\} = 1 - P\{X\}. \tag{2.3}$$

Equations 2.2 and 2.3 constitute two of the three axioms of probability theory. Before stating the third axiom we must discuss combinations of events.

We denote by $X \cap Y$ the event that both X and Y take place. Then, clearly, $X \cap Y = Y \cap X$. The probability that both X and Y take place is denoted by $P\{X \cap Y\}$. The combined event $X \cap Y$ may be understood by the use of a Venn diagram shown in Fig. 2.1*a*. The area of the square is equal to one. The circular areas indicated as X and Y are, respectively, the probabilities $P\{X\}$ and $P\{Y\}$. The probability of both X and Y occurring, $P\{X \cap Y\}$, is indicated by the cross-hatched area. For this reason $X \cap Y$ is referred to as the intersection of X and Y, or simply as X *and* Y.

Suppose that one event, say X, is dependent on the second event, Y. We define the conditional probability of event X, given event Y as $P\{X|Y\}$. The third axiom of probability theory is

$$P\{X \cap Y\} = P\{X|Y\}P\{Y\}. \tag{2.4}$$

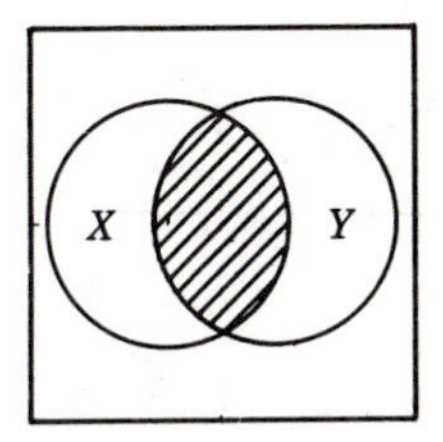

(a) $X \cap Y$ *(b)* $X \cup Y$

FIGURE 2.1 Venn diagrams for the intersection and union of two events.

That is, the probability that both X and Y will occur is just the probability that Y occurs times the conditional probability that X occurs, given the occurrence of Y. Provided that the probability that Y occurs is greater than zero, Eq. 2.4 may be written as a definition of the conditional probability:

$$P\{X|Y\} = \frac{P\{X \cap Y\}}{P\{Y\}}. \tag{2.5}$$

Note that we can reverse the ordering of events X and Y, by considering the probability $P\{X \cap Y\}$ in terms of the conditional probability of Y, given the occurrence of X. Then, instead of Eq. 2.4, we have

$$P\{X \cap Y\} = P\{Y|X\}P\{X\}. \tag{2.6}$$

An important property that we will sometimes assume is that two or more events, say X and Y, are mutually independent. For events to be independent, the probability of one occurring cannot depend on the fact that the other is either occurring or not occurring. Thus

$$P\{X|Y\} = P\{X\} \tag{2.7}$$

if X and Y are independent, and Eq. 2.4 becomes

$$P\{X \cap Y\} = P\{X\}P\{Y\}. \tag{2.8}$$

This is the definition of independence, that the probability of two events both occurring is just the product of the probabilities of each of the events occurring. Situations also arise in which events are mutually exclusive. That is, if X occurs, then Y cannot, and conversely. Thus $P\{X|Y\} = 0$ and $P\{Y|X\} = 0$; or more simply, for mutually exclusive events

$$P\{X \cap Y\} = 0. \tag{2.9}$$

With the three probability axioms and the definitions of independence in hand, we may now consider the situation where either X or Y or both may occur. This is referred to as the union of X and Y or simply $X \cup Y$. The probability $P\{X \cup Y\}$ is most easily conceptualized from the Venn diagram shown in Fig. 2.1*b*, where the union of X and Y is just the area of the overlapping circles indicated by cross hatching. From the cross-hatched area it is clear that

$$P\{X \cup Y\} = P\{X\} + P\{Y\} - P\{X \cap Y\}. \tag{2.10}$$

If we may assume that the events X and Y are independent of one another, we may insert Eq. 2.8 to obtain

$$P\{X \cup Y\} = P\{X\} + P\{Y\} - P\{X\}P\{Y\}. \tag{2.11}$$

Conversely, for mutually exclusive events, Eqs. 2.9 and 2.10 yield

$$P\{X \cup Y\} = P\{X\} + P\{Y\}. \tag{2.12}$$

EXAMPLE 2.1

Two circuit breakers of the same design each have a failure-to-open-on-demand probability of 0.02. The breakers are placed in series so that both must fail to open in order for the circuit breaker system to fail. What is the probability of system failure (*a*) if the failures are independent, and (*b*) if the probability of a second failure is 0.1, given the failure of the first? (*c*) In part *a* what is the probability of one or more breaker failures on demand? (*d*) In part *b* what is the probability of one or more failures on demand?

Solution $X \equiv$ failure of first circuit breaker
$Y \equiv$ failure of second circuit breaker
$P\{X\} = P\{Y\} = 0.02$ ◀

(*a*) $P\{X \cap Y\} = P\{X\}P\{Y\} = 0.0004.$
(*b*) $P\{Y|X\} = 0.1$
$P\{X \cap Y\} = P\{Y|X\}P\{X\} = 0.1 \times 0.02 = 0.002.$ ◀
(*c*) $P\{X \cup Y\} = P\{X\} + P\{Y\} - P\{X\}P\{Y\}$
$= 0.02 + 0.02 - (0.02)^2 = 0.0396.$ ◀
(*d*) $P\{X \cup Y\} = P\{X\} + P\{Y\} - P\{Y|X\}P\{X\}$
$= 0.02 + 0.02 - 0.1 \times 0.02 = 0.038.$ ◀

Combinations of Events

The foregoing equations state the axioms of probability and provide us with the means of combining two events. The procedures for combining events may be extended to three or more events, and the relationships may again be presented graphically as Venn diagrams. For example, in Fig. 2.2*a* and *b* are shown, respectively, the intersection of *X*, *Y*, and *Z*, $X \cap Y \cap Z$; and the union of *X*, *Y*, and *Z*, $X \cup Y \cup Z$. The probabilities $P\{X \cap Y \cap Z\}$ and $P\{X \cup Y \cup Z\}$ may again be interpreted as the cross-hatched areas.

The following observations are often useful in dealing with combinations of two or more events. Whenever we have a probability of a union of events, it may be reduced to an expression involving only the probabilities of the individual events and their intersection. Equation 2.10 is an example of this. Similarly, probabilities of more complicated combinations involving

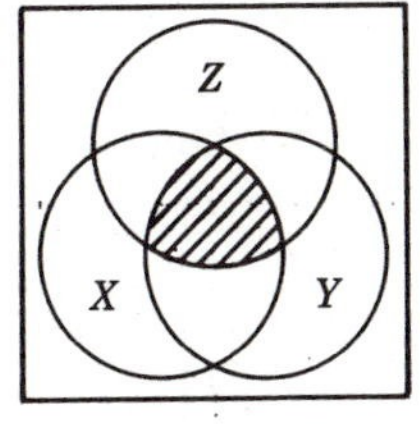

(*a*) $X \cap Y \cap Z$

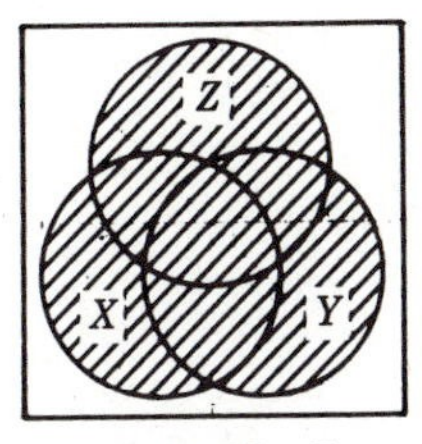

(*b*) $X \cup Y \cup Z$

FIGURE 2.2 Venn diagrams for the intersection and union of three events.

unions and intersections may be reduced to expressions involving only probabilities of intersections. The intersections of events, however, may be eliminated only by expressing them in terms of conditional probabilities, as in Eq. 2.6, or if the independence may be assumed, they may be expressed in terms of the probabilities of individual events as in Eq. 2.8.

The treatment of combinations of events is streamlined by using the rules of Boolean algebra listed in Table 2.1. If two combinations of events are equal according to these rules, their probabilities are equal. Thus, since according to Rule 1a, $X \cap Y = Y \cap X$, we also have $P\{X \cap Y\} = P\{Y \cap X\}$. The communicative and associative rules are obvious. The remaining rules may be verified from a Venn diagram. For example, in Fig. 2.3*a* and *b*, respectively, we show the distributive laws for $X \cap (Y \cup Z)$ and $X \cup (Y \cap Z)$. Note that in Table 2.1, ϕ is used to represent the null event for which $P\{\phi\} = 0$, and I is used to represent the universal event for which $P\{I\} = 1$.

We may illustrate how the combinations involving more than two events may be expressed in terms only of the intersections of events. Consider, for

TABLE 2.1 Rules of Boolean Algebra[a]

Mathematical symbolism	Designation
(1a) $X \cap Y = Y \cap X$	Commutative law
(1b) $X \cup Y = Y \cup X$	
(2a) $X \cap (Y \cap Z) = (X \cap Y) \cap Z$	Associative law
(2b) $X \cup (Y \cup Z) = (X \cup Y) \cup Z$	
(3a) $X \cap (Y \cup Z) = (X \cap Y) \cup (X \cap Z)$	Distributive law
(3b) $X \cup (Y \cap Z) = (X \cup Y) \cap (X \cup Z)$	
(4a) $X \cap X = X$	Idempotent law
(4b) $X \cup X = X$	
(5a) $X \cap (X \cup Y) = X$	Law of absorption
(5b) $X \cup (X \cap Y) = X$	
(6a) $X \cap \tilde{X} = \phi^b$	Complementation
(6b) $X \cup \tilde{X} = I^b$	
(6c) $\widetilde{(\tilde{X})} = X$	
(7a) $\widetilde{(X \cap Y)} = \tilde{X} \cup \tilde{Y}$	de Morgan's theorem
(7b) $\widetilde{(X \cup Y)} = \tilde{X} \cap \tilde{Y}$	
(8a) $\phi \cap X = \phi$	Operations with I
(8b) $\phi \cup X = X$	
(8c) $I \cap X = X$	
(8d) $I \cup X = I$	
(9a) $X \cup (\tilde{X} \cap Y) = X \cup Y$	These relationships are unnamed.
(9b) $\tilde{X} \cap (X \cup \tilde{Y}) = \tilde{X} \cap \tilde{Y} = \widetilde{(X \cup Y)}$	

[a] Adapted from H. R. Roberts, W. E. Vesley, D. F. Haastand, and F. F. Goldberg, *Fault tree Handbook*, NUREG-0492, U.S. Nuclear Regulatory Commission, 1981.
[b] ϕ = null set; I = universal set.

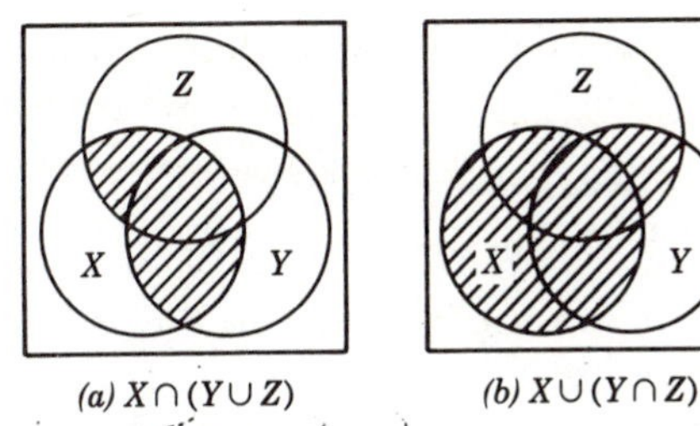

FIGURE 2.3 Venn diagrams for combinations of three events.

instance, the probability $P\{X \cap (Y \cup Z)\}$. We use Rule 3a to write

$$P\{X \cap (Y \cup Z)\} = P\{(X \cap Y) \cup (X \cap Z)\}, \tag{2.13}$$

but this is just the union of the two composites $X \cap Y$ and $X \cap Z$. Therefore, by Eq. 2.10,

$$P\{X \cap (Y \cup Z)\} = P\{X \cap Y\} + P\{X \cap Z\} - P\{(X \cap Y) \cap (X \cap Z)\}, \tag{2.14}$$

but the associative Rules 2a and b allow us to eliminate the parentheses from the last expression by first writing

$$(X \cap Y) \cap (X \cap Z) = (Y \cap X) \cap (X \cap Z), \tag{2.15}$$

and then using law 4a to obtain

$$Y \cap (X \cap X) \cap Z = Y \cap X \cap Z = X \cap Y \cap Z. \tag{2.16}$$

Using this intermediate result, we then have

$$P\{X \cap (Y \cup Z)\} = P\{X \cap Y\} + P\{X \cap Z\} - P\{X \cap Y \cap Z\}. \tag{2.17}$$

Thus we have written the probability of the combination of events entering in terms of the probability of intersections.

In reliability analysis one frequently encounters relationships involving the union of three or more events. These probabilities too can be expanded in terms of the intersection of events. For example,

$$P\{X \cup Y \cup Z\} = P\{X \cup (Y \cup Z)\} \tag{2.18}$$

by the associative law. Thus we have the union of events X and $(Y \cup Z)$. Then we may use Eq. 2.10 to obtain

$$P\{X \cup Y \cup Z\} = P\{X\} + P\{Y \cup Z\} - P\{X \cap (Y \cap Z)\}. \tag{2.19}$$

The second term on the right is expanded as in Eq. 2.10 to obtain

$$P\{Y \cup Z\} = P\{Y\} + P\{Z\} - P\{Y \cap Z\}. \tag{2.20}$$

We have already obtained the last term in Eq. 2.10, the intersection of events

Z and $(Y \cup Z)$, in Eq. 2.17. Thus, by combining Eqs. 2.19, 2.20, and 2.17, we obtain

$$P\{X \cup Y \cup Z\} = P\{X\} + P\{Y\} + P\{Z\} - P\{X \cap Y\} - P\{X \cap Z\} - P\{Y \cap Z\} + P\{X \cap Y \cap Z\}. \tag{2.21}$$

Thus we may express the probability of event $X \cup Y \cup Z$ entirely in terms of the probabilities of intersections of events.

If the intersections in expressions such as Eq. 2.21 are to be eliminated, they must be expressed in terms of conditional probabilities. If the events are independent, the conditional probabilities, in turn, may be expressed as products of the probabilities of individual events. Thus Eq. 2.4 may be used to express $P\{X \cap Y\}$ as a conditional probability, and if X and Y are independent,

$$P\{X \cap Y\} = P\{X\}P\{Y\}. \tag{2.22}$$

The expressions for $P\{X \cap Z\}$ and $P\{Y \cap Z\}$ may be treated analogously. Similarly, the intersection of three or more events is treated in the following way. First, consider $Y \cap Z$ as a composite event:

$$P\{X \cap Y \cap Z\} = P\{X \cap (Y \cap Z)\}. \tag{2.23}$$

Then, from the definition of conditional probability in Eq. 2.4, we have

$$P\{X \cap Y \cap Z\} = P\{X|Y \cap Z\}P\{Y \cap Z\}. \tag{2.24}$$

Then we again apply Eq. 2.4 to $P\{Y \cap Z\}$,

$$P\{X \cap Y \cap Z\} = P\{X|Y \cap Z\}P\{Y|Z\}P\{Z\}, \tag{2.25}$$

which may be used in Eq. 2.21. A most important case occurs when X, Y, and Z are independent. Then $P\{X|Y \cap Z\} \rightarrow P\{X\}$ and $P\{Y|Z\} \rightarrow P\{Y\}$, and we have simply

$$P\{X \cap Y \cap Z\} = P\{X\}P\{Y\}P\{Z\}. \tag{2.26}$$

Thus, if X, Y, and Z are independent, the probability of $X \cup Y \cup Z$ reduces to an expression involving only the probabilities of the individual events X, Y, and Z:

$$P\{X \cup Y \cup Z\} = P\{X\} + P\{Y\} + P\{Z\} - P\{X\}P\{Y\} - P\{X\}P\{Z\} - P\{Y\}P\{Z\} + P\{X\}P\{Y\}P\{Z\}. \tag{2.27}$$

In dealing with repeated failures and multicomponent systems in the following chapters, we shall frequently need to evaluate the probability of several events, say $X_1 \cup X_2 \cup \cdots \cup X_n$, where each of the events is independent of all the others. For n independent events the intersection is obtained simply by generalizing Eq. 2.26 to

$$P\{X_1 \cap X_2 \cap X_3 \cap \cdots \cap X_n\} = P\{X_1\}P\{X_2\}P\{X_3\} \cdots P\{X_n\}. \tag{2.28}$$

For the union of n independent events, $P\{X_1 \cup X_2 \cup \cdots \cup X_n\}$ may be obtained most simply in the following manner. Recalling that $\tilde{X}_i$ signifies not X_i, we observe that

$$P\{X_1 \cup X_2 \cup \cdots \cup X_n\} + P\{\tilde{X}_1 \cap \tilde{X}_2 \cap \cdots \cap \tilde{X}_n\} = 1. \tag{2.29}$$

In this expression the term on the far left is just the probability that one or more of the events X will occur. Since the only other possible result is that none of the events occur, the two probabilities must sum to one. For example, if X_i signifies the failure of the ith out of n engines that are tested, the first term in Eq. 2.29 is just the probability that one or more failures will occur, and the second term is the probability that no failures will occur.

Now assume that the events are mutually independent. The probability that no event will occur is then just

$$P\{\tilde{X}_1 \cap \tilde{X}_2 \cap \cdots \cap \tilde{X}_n\} = P\{\tilde{X}_1\}P\{\tilde{X}_2\} \cdots P\{\tilde{X}_n\}. \tag{2.30}$$

Then, using Eq. 2.3, we have

$$P\{\tilde{X}_1 \cap \tilde{X}_2 \cap \cdots \cap \tilde{X}_n\} = [1 - P\{X_1\}][1 - P\{X_2\}] \cdots [1 - P\{X_n\}]. \tag{2.31}$$

Finally, substituting this result into Eq. 2.29 yields the probability that one or more events will occur:

$$P\{X_1 \cup X_2 \cup \cdots \cup X_n\} = 1 - \prod_{n'=1}^{n} [1 - P\{X_{n'}\}]. \tag{2.32}$$

The same reasoning may be used to prove that

$$P\{\tilde{X}_1 \cup \tilde{X}_2 \cup \tilde{X}_3 \cup \cdots \cup \tilde{X}_n\} = 1 - \prod_{n'=1}^{n} [1 - P\{\tilde{X}_{n'}\}]. \tag{2.33}$$

EXAMPLE 2.2

A critical seam in an aircraft wing must be reworked if any one of the 28 identical rivets is found to be defective. Quality control inspections find that 18% of the seams must be reworked. (*a*) Assuming that the defects are independent, what is the probability that a rivet will be defective? (*b*) To what value must this probability be reduced if the rework rate is to be reduced below 5%?

Solution (*a*) Let X_i represent the failure of the ith rivet in Eq. 2.31. Then, since $P\{X_1\} = P\{X_2\} = \cdots P\{X_{28}\}$,

$$0.18 = P\{X_1 \cup X_2 \cup \cdots \cup X_{28}\} = 1 - [1 - P\{X_1\}]^{28}$$

$$P\{X_1\} = 1 - (0.82)^{1/28} = 0.0071. \blacktriangleleft$$

(*b*) Since $0.05 = 1 - [1 - P\{X_1\}]^{28}$,

$$P\{X_1\} = 1 - (0.95)^{1/28} = 0.0018. \blacktriangleleft$$

Independent Failures

The foregoing relationships may be used to demonstrate two important reliability phenomena that reappear and are treated in greater depth in subsequent chapters. These phenomena concern the effect of repeated demand or loading on a system and the number of components contained in a system.

Repeated Demand

Suppose that each time a demand is made on a system, the probability of success is r, giving a corresponding probability of failure of

$$p = 1 - r. \tag{2.34}$$

The term *demand* here is quite general; it may be the switching of an electric relay, the opening of a valve, the start of an engine, or even the stress on a bridge as a truck passes over it. Whatever the application, there are two salient points. First, we must be able to count or at least infer the number of demands; and second, the probability of success with each demand must be independent of the number of previous demands.

We define the reliability R_n as the probability that the system will still be operational after n demands. Let X_n signify the event of success in the nth demand. Then, if the probabilities of success at each demand are mutually independent, R_n is given by Eq. 2.28 as

$$R_n = P\{X_1\}P\{X_2\} \cdots P\{X_n\}, \tag{2.35}$$

or since $P\{X_n\} = r$ for all n,

$$R_n = r^n. \tag{2.36}$$

Then, using Eq. 2.34, we obtain

$$R_n = (1 - p)^n. \tag{2.37}$$

We may put this result in a more useful approximate form. First, note that the exponential of

$$\ln R_n = \ln(1 - p)^n = n \ln(1 - p) \tag{2.38}$$

is

$$R_n = \exp[n \ln(1 - p)]. \tag{2.39}$$

If the probability for failure on demand is small, we may make the approximation

$$\ln(1 - p) \simeq -p \tag{2.40}$$

for $p \ll 1$, yielding

$$R_n \simeq e^{-np}. \tag{2.41}$$

Since $p \ll 1$ is nearly always a good approximation, we see that the reliability decays exponentially with the number of demands. If the rate at which demands are made on the system is roughly constant, we may write

$$t \approx n\,\Delta t, \tag{2.42}$$

where Δt is the averge time interval between demands and t is the total time elapsed. We may then calculate the reliability $R(t)$, defined as the probability that the system will still be operational at time t, as

$$R(t) \simeq e^{-\lambda t}, \tag{2.43}$$

where λ is called the failure rate and is given here by

$$\lambda = p/\Delta t. \tag{2.44}$$

In succeeding chapters departures from these approximations will be examined, and the factors that influence λ and p will be investigated.

EXAMPLE 2.3

A manufacturer of relief valves knows that by the end of 100 operations 1% of the valves will have failed. If we assume that the failures are random and independent, (*a*) what is the failure probability per demand? (*b*) What is the reliability for a design life in 500 demands? (*c*) If the average user operates the valve three times per week, what is the probability of failure during the first year?

Solution (*a*) From Eq. 2.41 we have

$$e^{-100p} = 0.99;$$

Therefore

$$p = -\frac{1}{100}\ln 0.99 \simeq 10^{-4}. \blacktriangleleft$$

(*b*) Again from Eq. 2.41

$$R_{500} = e^{-500p} = \exp[-500 \times 10^{-4}] = e^{-0.05} \simeq 0.95. \blacktriangleleft$$

(*c*) The average time between operations is

$$\Delta t = \frac{1}{3 \times 52} = 0.0064 \text{ (year)}.$$

Hence from Eq. 2.44

$$\lambda = \frac{10^{-4}}{0.0064} = 0.0156.$$

Thus

$$R(t) = \exp(-0.0156t)$$

and

$$1 - R(1) = 1 - \exp(-0.0156 \times 1) = 0.0155. \blacktriangleleft$$

Multiple Components

If there is no redundancy in a system, any one of the system's components that fails results in system failure. If we define X_m as the probability that the mth component operates successfully for some specified period of time, the reliability of a system with M components is then the probability

$$R_M = P\{X_1 \cap X_2 \cap \cdots \cap X_M\}. \tag{2.45}$$

Suppose that we assume that the success (or failure) of each component is independent of that of the others. Then Eq. 2.28 holds. If we also assume that all the components have the same probability of success, $P\{X_m\} = r_0$, the system reliability is

$$R_M = r_0^M. \tag{2.46}$$

Then, defining the component failure probability as

$$p_0 = 1 - r_0, \tag{2.47}$$

we obtain

$$R_M = (1 - p_0)^M. \tag{2.48}$$

With the assumption that $p_0 \ll 1$, we may again apply the approximation of Eq. 2.40 to obtain

$$R_M = e^{-Mp_0} \tag{2.49}$$

From this idealized relationship we see that reliability tends to deteriorate as the system complexity, measured by M, the number of components, increases. In later chapters the repercussions of this relationship are examined and the effects of redundant components studied.

EXAMPLE 2.4

A manufacturer of circuits knows that 5% of his circuit boards fail in proof testing due to independent diode failures. The failure of any diode causes board failure. (*a*) If there are 100 diodes on a board, what is the probability of any one diode's failing? (*b*) If the size of boards is increased to contain 500 diodes, what percent of the new boards will fail in proof testing? (*c*) What must the failure probability per diode be if the 5% failure rate is to be maintained for the 500 diode boards?

Solution (*a*) $R_{100} = 1 - 0.05 = e^{-100p}$, $p = -\dfrac{1}{100}\ln(0.95) = 0.5 \times 10^{-3}$. ◀

(*b*) $1 - R_{500} = 1 - e^{-500p} = 1 - \exp[-500 \times 0.5 \times 10^{-3}] = 0.22 = 22\%$. ◀

(*c*) $R_{500} = 0.95 = e^{-500p'}$, $p' = -\dfrac{1}{500}\ln(0.95) = 0.1 \times 10^{-3}$. ◀

2.3 DISCRETE RANDOM VARIABLES

Frequently, in reliability considerations, we need to know the probability that a specific number of events will occur, or we need to determine the average number of events that are likely to take place. For example, suppose that we have a computer with N memory chips and we need to know the probability that none of them, that one of them, that two of them, and so on, will fail during the first year of service. Or suppose that there is a probability p that a Christmas tree light bulb will fail during the first 100 hours of service. Then, on a string of 25 lights, what is the probability that there will be n ($0 \leq n \leq 25$) failures during this 100-hr period? To answer such reliability questions, we need to introduce the properties of discrete random variables. We do this first in general terms, before treating two of the most important discrete probability distributions.

Properties of Discrete Variables

A discrete random variable is a quantity that can be equal to any one of a number of discrete values $x_0, x_1, x_2, \ldots, x_n, \ldots, x_N$. We refer to such a variable with the bold-faced character $\mathbf{x}$, and denote by x_n the values to which it may be equal. In many cases these values are integers so that $x_n = n$. By *random* variables we mean that there is associated with each x_n a probability $f(x_n)$ that $\mathbf{x} = x_n$. We denote this probability as

$$f(x_n) = P\{\mathbf{x} = x_n\}. \tag{2.50}$$

We shall, for example, often be concerned with counting numbers of failures (or of successes). Thus we may let $\mathbf{x}$ signify the number n of failures in N tests. Then $f(0)$ is the probability that there will be no failure, $f(1)$ the probability of one failure, and so on. The probabilities of all the possible outcomes must add to one

$$\sum_n f(x_n) = 1, \tag{2.51}$$

where the sum is taken over all possible values of x_n.

The function $f(x_n)$ is referred to as the *probability mass function* (PMF) of the discrete random variable $\mathbf{x}$. A second important function of the random variable is the *cumulative distribution function* (CDF) defined by

$$F(x_n) = P\{\mathbf{x} \leq x_n\}, \tag{2.52}$$

the probability that the value of $\mathbf{x}$ will be less than or equal to the value x_n. Clearly, it is just the sum of probabilities:

$$F(x_n) = \sum_{n'=0}^{n} f(x_{n'}). \tag{2.53}$$

Closely related is the *complementary cumulative distribution function* (CCDF), defined by the probability that $\mathbf{x} > x_n$:

$$\tilde{F}(x_n) = P\{\mathbf{x} > x_n\}. \tag{2.54}$$

It is related to PMF by

$$\tilde{F}(x_n) = 1 - F(x_n) = \sum_{n'=n+1}^{N} f(x_{n'}), \tag{2.55}$$

where x_N is the largest value for which $f(x_n) > 0$.

It is often convenient to display discrete random variables as bar graphs of the PMF. Thus, if we have, for example,

$$f(0) = 0,\ \ f(1) = \tfrac{1}{16},\ \ f(2) = \tfrac{1}{4},\ \ f(3) = \tfrac{3}{8},\ \ f(4) = \tfrac{1}{4},\ \ f(5) = \tfrac{1}{16},$$

the PMF may be plotted as in Fig. 2.4*a*. Similarly, from Eq. 2.53 the bar graph for the CDF appears as in Fig. 2.4*b*.

Several important properties of the random variable $\mathbf{x}$ are defined in terms of the probability mass function $f(x_n)$. The *mean* value, μ, of $\mathbf{x}$ is

$$\mu = \sum_n x_n f(x_n), \tag{2.56}$$

and the *variance* of $\mathbf{x}$ is

$$\sigma^2 = \sum_n (x_n - \mu)^2 f(x_n), \tag{2.57}$$

which may be reduced to

$$\sigma^2 = \sum_n x_n^2 f(x_n) - \mu^2. \tag{2.58}$$

The mean is a measure of the expected value or central tendency of $\mathbf{x}$ when a very large sampling is made of the random variable, whereas the variance is a measure of the scatter or dispersion of the individual values of x_n about μ. It is also sometimes useful to talk about the most probable value of $\mathbf{x}$:

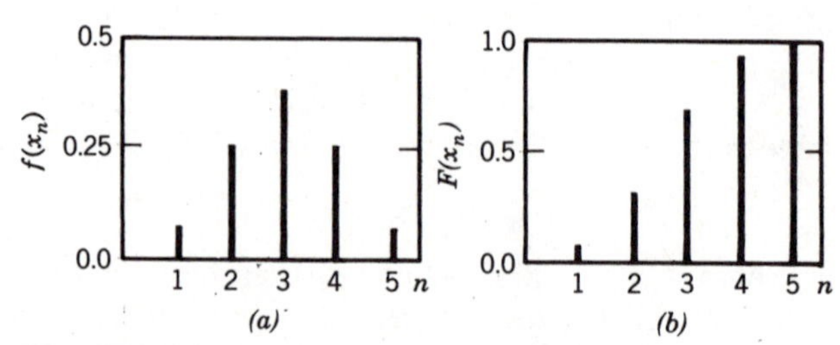

FIGURE 2.4 Discrete probability distribution: (*a*) probability mass function (PMF), (*b*) corresponding cumulative distribution function (CDF).

the value of x_n for which the largest value of $f(x_n)$ occurs, assuming that there is only one largest value. Finally, the median value is defined as that value $\mathbf{x} = x_n$ for which the probability of obtaining a smaller value is 1/2:

$$\sum_{n' \geq n} f(x_{n'}) = \tfrac{1}{2}. \tag{2.59}$$

and consequently,

$$\sum_{n' \leq n} f(x_{n'}) = \tfrac{1}{2}. \tag{2.60}$$

EXAMPLE 2.5

A discrete probability distribution is given by

$$f(x_n) = An \qquad n = 0, 1, 2, 3, 4, 5$$

(*a*) Determine A.
(*b*) What is the probability that $\mathbf{x} \leq 3$?
(*c*) What is μ?
(*d*) What is σ?

Solution (*a*) From Eq. 2.51,

$$1 = \sum_{n=0}^{5} An = A(0 + 1 + 2 + 3 + 4 + 5) = 15A$$

$$A = \frac{1}{15}. \blacktriangleleft$$

(*b*) From Eq. 2.52 and 2.53,

$$P\{\mathbf{x} \leq 3\} = F(3) = \sum_{n=0}^{3} \frac{n}{15} = \frac{1}{15}(0 + 1 + 2 + 3) = \frac{2}{5}. \blacktriangleleft$$

(*c*) From Eq. 2.56,

$$\mu = \sum_{n=0}^{5} n\,\frac{n}{15} = \frac{1}{15}(0 + 1 + 4 + 9 + 16 + 25) = \frac{11}{3}. \blacktriangleleft$$

(*d*) Using Eq. 2.58, we first calculate

$$\sum_{n=0}^{5} x_n^2 f(x_n) = \sum_{n=0}^{5} \frac{1}{15} n^3 = \frac{1}{15}(0 + 1 + 8 + 27 + 64 + 125) = 15.$$

to obtain for the variance

$$\sigma^2 = 15 - \mu^2 = 15 - \left(\frac{11}{3}\right)^2 = 1.555$$

$$\sigma = 1.247. \blacktriangleleft$$

The idea of the expected value is an important one. In general, if there is a function $g(x_n)$ of the random variable $\mathbf{x}$, the *expected value* $E\{g\}$ is defined for a discrete random variable as

$$E\{g\} = \sum_n g(x_n)f(x_n). \tag{2.61}$$

Thus the mean and variance given by Eqs. 2.56 and 2.57 may be written as

$$\mu = E\{x\} \tag{2.62}$$

$$\sigma^2 = E\{(x - \mu)^2\} \tag{2.63}$$

or as in Eq. 2.58,

$$\sigma^2 = E\{x^2\} - \mu^2. \tag{2.64}$$

The quantity $\sigma = \sqrt{\sigma^2}$ is referred to as the standard error or *standard deviation* of the distribution. The notion of expected value is also applicable to the continuous random variables discussed in the following chapter.

The Binomial Distribution

The binomial distribution is the most widely used discrete distribution in reliability considerations. To derive it, suppose that p is the probability of failure for some piece of equipment in a specified test and

$$q = 1 - p \tag{2.65}$$

is the corresponding success (i.e., nonfailure) probability. If such tests are truly independent of one another, they are referred to as Bernoulli trials.

We wish to derive the probability

$$f(n) = P\{\mathbf{n} = n|N, p\} \tag{2.66}$$

that in N independent tests there are n failures. To arrive at this probability, we first consider the example of the test of two units of identical design and construction. The tests must be independent in the sense that success or failure in one test does not depend on the result of the other. There are four possible outcomes, each with an associated probability: qq is the probability that neither unit fails, pq the probability that only the first unit fails, qp the probability that only the second unit fails, and pp the probability that both units fail. Since these are the only possible outcomes of the test, the sum of the probabilities must equal one. Indeed,

$$p^2 + 2pq + q^2 = (p + q)^2 = 1, \tag{2.67}$$

and by the definition of Eq. 2.66

$$f(0) = q^2, \quad f(1) = 2qp, \quad f(2) = p^2. \tag{2.68}$$

In a similar manner the probability of n independent failures may also be covered for situations in which a larger number of units undergo testing. For example, with $N = 3$ the probability that all three units fail independently is obtained by multiplying the failure probabilities of the individual units together. Since the units are identical, the probability that none of the three fails is qqq. There are now three ways in which the test can result in one unit failing: the first fails, pqq; the second fails, qpq; or the third fails, qqp. There are also three combinations that lead to two units failing: units 1 and 2 fail, ppq; units 1 and 3 fail, pqp; or units 2 and 3 fail, qpp. Finally, the probability of all three units failing is ppp.

In the three-unit test the probabilities for the eight possible outcomes must again add to one. This is indeed the case, for by combining the eight terms into four we have

$$q^3 + 3q^2p + 3qp^2 + p^3 = (q + p)^3 = 1. \tag{2.69}$$

The probabilities of the test resulting in 0, 1, 2, or 3 failures are just the successive terms on the left:

$$f(0) = q^3, \quad f(1) = 3q^2p, \quad f(2) = 3qp^2, \quad f(3) = p^3. \tag{2.70}$$

The foregoing process may be systematized for tests of any number of units. For N units Eq. 2.70 generalizes to

$$C_0^N q^N + C_1^N pq^{N-1} + C_2^N p^2q^{N-2} + \cdots + C_{N-1}^N p^{N-1}q + C_N^N p^N = (q + p)^N = 1, \tag{2.71}$$

since $q = 1 - p$. For this expression to hold, it may be shown that the C_n^N must be the binomial coefficients. These are given by

$$C_n^N = \frac{N!}{(N - n)!n!}. \tag{2.72}$$

A convenient way to tabulate these coefficients is in the form of Pascal's triangle; this is shown in Table 2.2. Just as in the case of $N = 2$ or 3, the $N + 1$ terms on the left-hand side of Eq. 2.71 are the probabilities that

TABLE 2.2 Pascal's Triangle

								1									$N = 0$
							1		1								$N = 1$
						1		2		1							$N = 2$
					1		3		3		1						$N = 3$
				1		4		6		4		1					$N = 4$
			1		5		10		10		5		1				$N = 5$
		1		6		15		20		15		6		1			$N = 6$
	1		7		21		35		35		21		7		1		$N = 7$
1		8		28		56		70		56		28		8		1	$N = 8$

there will be 0, 1, 2, . . . , N failures. Thus the PMF for the binomial distribution is

$$f(n) = C_n^N p^n (1 - p)^{N-n}, \qquad n = 0, 1, \ldots, N. \tag{2.73}$$

That the condition Eq. 2.51 is satisfied follows from Eq. 2.71. The CDF corresponding to $f(n)$ is

$$F(n) = \sum_{n'=0}^{n} C_{n'}^N p^{n'} (1 - p)^{N-n'}. \tag{2.74}$$

The mean number of failures is determined by substituting Eq. 2.73 into Eq. 2.56. After some algebra, the result may be shown to be

$$\mu = Np. \tag{2.75}$$

The variance is most directly determined from Eq. 2.64. First, using Eq. 2.71, we may show with some algebra that the definition of expected value in Eq. 2.61 yields

$$E\{n^2\} = N(N - 1)p^2 + Np. \tag{2.76}$$

Consequently, inserting this expression and Eq. 2.75 into 2.64 yields

$$\sigma^2 = Np(1 - p). \tag{2.77}$$

EXAMPLE 2.6

Ten compressors with a failure probability $p = 0.1$ are tested. (*a*) What is the expected number of failures $E\{n\}$? (*b*) What is σ^2? (*c*) What is the probability that none will fail? (*d*) What is the probability that two or more will fail?

Solution (*a*) $E\{n\} = \mu = Np = 10 \times 0.1 = 1.$ ◀

(*b*) $\sigma^2 = Np(1 - p) = 10 \times 0.1(1 - 0.1) = 0.9.$ ◀

(*c*) $P\{\mathbf{n} = 0|10, p\} = f(0) = C_0^{10} p^0 (1 - p)^{10} = 1 \times 1 \times (1 - 0.1)^{10} = 0.349.$ ◀

(*d*) $P\{\mathbf{n} \geq 2|10, p\} = 1 - f(0) - f(1) = 1 - C_0^{10} p^0 (1 - p)^{10} - C_1^{10} p^1 (1 - p)^9$
$= 1 - (1 - 0.1)^{10} - 10 \times 0.1 \times (1 - 0.1)^9 = 0.264.$ ◀

The Poisson Distribution

Situations in which the probability of failure p becomes very small, but the number of tested units N is large, are frequently encountered. It then becomes cumbersome to evaluate the large factorials appearing in the binomial distribution. For this, as well as for a variety of situations discussed in later chapters, the Poisson distribution is employed.

The Poisson distribution may be shown to result from taking the limit of the binomial distribution as $p \to 0$ and $N \to \infty$, with the product Np

remaining constant. To obtain the distribution, we first multiply Eq. 2.73 by N^n/N^n and rearrange the factors to yield

$$f(n) = \frac{N!}{(N - n)!n!N^n}\left(\frac{Np}{1 - p}\right)^n (1 - p)^N. \tag{2.78}$$

Then writing

$$(1 - p)^N = \exp[N \ln(1 - p)], \tag{2.79}$$

we have

$$f(n) = \frac{N!}{(N - n)!N^n}\frac{1}{n!}\left(\frac{Np}{1 - p}\right)^n \exp[N \ln(1 - p)]. \tag{2.80}$$

If we now assume that $p \ll 1$, so that we may write $\ln(1 - p) \simeq -p$, it follows that

$$f(n) \simeq \frac{N!}{(N - n)!N^n}\frac{1}{n!}\left(\frac{Np}{1 - p}\right)^n e^{-Np}. \tag{2.81}$$

Finally, we take the limit $N \rightarrow \infty$ and $p \rightarrow 0$ with the stipulation that

$$\mu \equiv Np \tag{2.82}$$

remains constant. In doing this, we note that for any finite value of n,

$$\lim_{N\rightarrow\infty} \frac{N!}{(N - n)!N^n} = \lim_{N\rightarrow\infty} \frac{(N - n + 1)(N - n + 2) \cdots (N - 1)N}{N^n} = 1. \tag{2.83}$$

Thus the result is the PMF for the Poisson distribution,

$$f(n) = \frac{\mu^n}{n!} e^{-\mu}. \tag{2.84}$$

Unlike the binomial distribution, the Poisson distribution can be expressed in terms of a single parameter μ. Thus $f(n)$ may be written as the probability

$$P\{\mathbf{n} = n|\mu\} = \frac{\mu^n e^{-\mu}}{n!}, \qquad n = 0, 1, 2, \ldots. \tag{2.85}$$

The condition Eq. 2.51 must, of course, be satisfied. This may be verified by first recalling the power series expansion for the exponential function:

$$e^{\mu} = \sum_{n=0}^{\infty} \frac{\mu^n}{n!}. \tag{2.86}$$

Thus we have

$$\sum_{n=0}^{\infty} f(n) = \sum_{n=0}^{\infty} \frac{\mu^n}{n!} e^{-\mu} = e^{\mu}e^{-\mu} = 1. \tag{2.87}$$

In the foregoing equations we have chosen μ to represent Np because it may be shown to be the mean of the Poisson distribution. With some algebra the Poisson distribution may be shown to yield

$$E\{n\} = \sum_{n=0}^{\infty} n \frac{\mu^n}{n!} e^{-\mu} = \mu. \tag{2.88}$$

In the Poisson distribution the variance is equal to the mean. First determining that

$$E\{n^2\} = \sum_{n=0}^{\infty} \frac{n^2 \mu^n}{n!} e^{-\mu} = \mu(\mu + 1), \tag{2.89}$$

we may use Eq. 2.58 to write

$$\sigma^2 = \mu. \tag{2.90}$$

EXAMPLE 2.7

Do the preceding 10-compressor example approximating the binomial distribution by a Poisson distribution. Compare the results.

Solution (*a*) $\mu = Np = 1$. ◀
(*b*) $\sigma^2 = \mu = 1$ ◀ (0.9 for binomial).
(*c*) $P\{\mathbf{n} = 0|\mu = 1\} = e^{-\mu} = 0.3678$ ◀ (0.3874 for binomial).
(*d*) $P\{\mathbf{n} \geq 2|\mu = 1\} = 1 - f(0) - f(1) = 1 - 2e^{-\mu} = 0.2642$ ◀ (0.2639 for binomial).

The Poisson distribution is useful in several other types of reliability problems. We shall return to it in Chapter 4.

2.4 ATTRIBUTE SAMPLING

The discussions in the preceding section illustrate how the binomial and Poisson distributions can be determined given the parameter p, which often denotes a probability of failure. In reliability engineering and the associated discipline of quality assurance, however, one rarely has the luxury of knowing p, *a priori*. Rather, one often is faced with the problem of estimating a failure probability, a mean number of failures, or some other quantity approximately from a limited amount of data. One cannot, for instance, test all items by running them for their design life to determine their failure probability exactly. Not only would the test be too lengthy and costly, but the test in many cases would create unacceptable wear on the items that did survive.

The problem of estimating characteristics from samples of limited size is a fundamental problem of statistical inference and is treated in textbooks on statistical analysis.* In what follows we only touch on some of the rudimentary computational and graphical methods in order to illustrate their use in the estimate of failure probabilities from test data. In this chapter we deal only with the attribute sampling associated with discrete distributions. By attribute sampling it is understood that the outcome of each test or experiment is a simple pass or fail. The sampling of continuous distributions is discussed in later chapters.

Sampling Distribution

Suppose that we want to estimate the failure probability p of a system and also gain some idea of the precision of the estimate. Our experiment consists of testing N units for failure, with the assumption that the N units are representative of a much larger—perhaps infinite—number of units of the same design and manufacture. If there are n failures, the failure probability, defined by Eq. 2.1, may be estimated by

$$\hat{p} = \frac{n}{N}. \tag{2.91}$$

We use the caret ˆ to indicate that $\hat{p}$ is an estimate, rather than the true value of p. It is referred to as a point estimate of p.

The difficulty, of course, is that if the test is repeated, a different value of n, and therefore a different estimate $\hat{p}$, is likely to be obtained. The number of failures is a random variable that obeys the binomial distribution discussed in the preceding section. Thus $\hat{p}$ is also a random variable; its distribution may be obtained by substituting $n = N\hat{p}$ into Eq. 2.73:

$$f(\hat{p}) = C^N_{\hat{p}N} p^{\hat{p}N}(1 - p)^{(N-\hat{p}N)}, \qquad \hat{p} \equiv 0, \frac{1}{N}, \frac{2}{N}, \ldots, 1. \tag{2.92}$$

This discrete probability mass function is referred to as the sampling distribution. It tells us that the probability of obtaining a particular value of $\hat{p}$ from our test is just $f(\hat{p})$, given that the true value is p. From the sampling distribution, we may show that Eq. 2.91 is an unbiased estimator: an estimator is said to be unbiased if when the sampling is repeated many times, the mean value of the estimator, the mean, taken over all the samples, yields the true value. Thus, for $\hat{p}$ to be unbiased, we must have

$$E\{\hat{p}\} \equiv \mu_{\hat{p}} \equiv \sum_{\hat{p}} \hat{p} f(\hat{p}) = p. \tag{2.93}$$

This may easily be shown to be the case, as follows.

* See for example, L. L. Lapin, *Probability and Statistics in Modern Engineering*, Brooks/Cole, Belmont, CA, 1983.

EXAMPLE 2.8

(*a*) Show that $\hat{p}$ is an unbiased estimator of p.
(*b*) Find the variance of $f(\hat{p})$.

Solution (*a*) Combining Eqs. 2.92 and 2.93 yields

$$\mu_{\hat{p}} = \sum_{\hat{p}} \hat{p} C^N_{\hat{p}N} p^{\hat{p}N}(1 - p)^{(N-\hat{p}N)}.$$

Then substituting $\hat{p} = n/N$, we have

$$\mu_{\hat{p}} = \frac{1}{N} \sum_{n} n C^N_n p^n (1 - p)^{(N-n)}.$$

Comparing this expression to Eqs. 2.56, 2.62, and 2.73, we see that the sum is equal to the expected value of n given by Eq. 2.75 as $E\{n\} = Np$; thus $\mu_{\hat{p}} = p$. ◀
(*b*) Inserting Eq. 2.92 into Eq. 2.58, we obtain

$$\sigma^2_{\hat{p}} = \sum_{\hat{p}} \hat{p}^2 C^N_{\hat{p}N} p^{\hat{p}N}(1 - p)^{(N-\hat{p}N)} - \mu^2_{\hat{p}}.$$

Then we substitute $\hat{p} = n/N$ and $\mu_{\hat{p}} = p$:

$$\sigma^2_{\hat{p}} = \frac{1}{N^2} \sum_{n} n^2 C^N_n (1 - p)^{(N-n)} - p^2.$$

Comparing the first term to Eqs. 2.61 and 2.73, we see that it is just $E\{n^2\}/N^2$. Thus, from Eq. 2.76, we have

$$\sigma^2_{\hat{p}} = \frac{1}{N^2} [N(N - 1)p^2 + Np] - p^2$$

and

$$\sigma^2_{\hat{p}} = \frac{1}{N} p(1 - p) \quad ◀$$

In addition to using the sampling distribution to prove that the estimate $\hat{p} = n/N$ is unbiased, we may gain some idea of how precise the estimate is for a given sample size N. By plotting $f(\hat{p})$ as in Fig. 2.5, for several different values of N, we see—not surprisingly—that the larger the sample size becomes, the smaller the probability that a value of $\hat{p}$ with a large error will be obtained. For the example shown in Fig. 2.5, in which the true value is $p = 0.25$, the probability that the estimate $\hat{p}$ will be in error by more than 0.10 is about 50% for $N = 10$, about 20% for $N = 20$, and only about 10% when $N = 40$.

Unfortunately, we do not know beforehand what the value of p is. If we did, we would not be interested in using the estimator n/N to obtain an approximate value. Therefore, we would like to estimate the precision of $\hat{p}$ without knowing the exact value of p. For this we must introduce the somewhat more subtle notion of confidence limits.

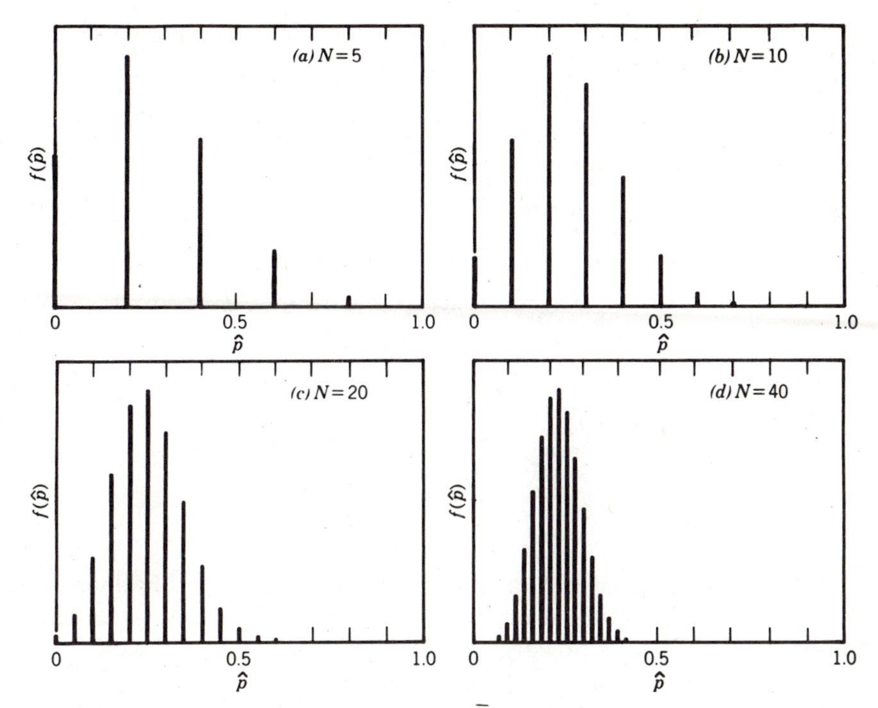

FIGURE 2.5 Probability mass function for binomial sampling where $\bar{p} = 0.25$.

Confidence Limits

Suppose that we run a series of N tests on a system, and there are no failures. The point failure probability estimate is then $\hat{p} = 0/N = 0$. Clearly, it is difficult to argue that $p = 0$, and that we have a perfect system, for if we were to test $2N$ or $3N$ systems, eventually we would probably get at least one failure. Nevertheless, the test results provide valuable information in bounding the range of values within which p is likely to lie. If the probability of failure is p, then from the binomial distribution the probability β that there are no failures in n trials is seen from Eqs. 2.66 and 2.73 to be

$$\beta = P\{\mathbf{n} = 0|N, p\} = C_0^N p^0(1 - p)^N, \tag{2.94}$$

or simply

$$\beta = (1 - p)^N. \tag{2.95}$$

Thus the larger p is, the less likely it is that no failures will occur. Conversely, we may ask what is the largest value of p, say p_+, for which there is a probability of β or greater of obtaining $n = 0$:

$$p_+ = 1 - \beta^{1/N}. \tag{2.96}$$

The quantity p_+ is called the upper confidence limit on p with probability $1 - \beta$, or simply the upper $1 - \beta$ confidence limit. Stated another way, we are $1 - \beta$ confident that p is less than p_+.

EXAMPLE 2.9

How many systems must be tested without failure to demonstrate that $p < 0.1$ with 90% confidence?

Solution From Eq. 2.96 it follows that $\ln \beta = N \ln(1 - p_+)$ and

$$N = \frac{\ln \beta}{\ln(1 - p_+)}.$$

Since $p_+ = 0.1$ and $(1 - \beta) = 0.9$,

$$N = \frac{\ln(0.1)}{\ln(0.9)} = 21.8.$$

Thus $N = 22$. ◀

A useful graph may be drawn to determine the upper confidence limit on the failure probability from the number of failure-free tests. This is shown in Fig. 2.6. Note the large numbers of tests that must be performed fault-free if a small value of p_+ is to be obtained with a high degree of confidence. The lower confidence limit in these cases is taken to be the same as $\hat{p} = p_- = 0$.

Ordinarily, if we test enough systems, we shall obtain some number $n > 0$, of failures. We may then estimate p with $\hat{p}$ given in Eq. 2.91; $\hat{p}$ is then said to be the point estimator of p. We would also like to obtain some

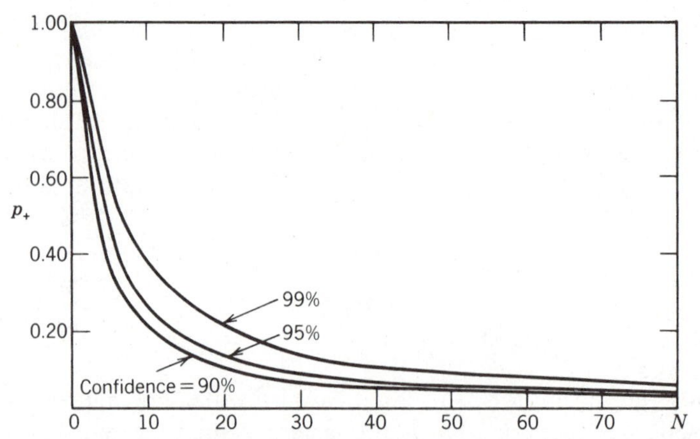

FIGURE 2.6 Upper confidence limit on failure probability p when no failures have occurred.

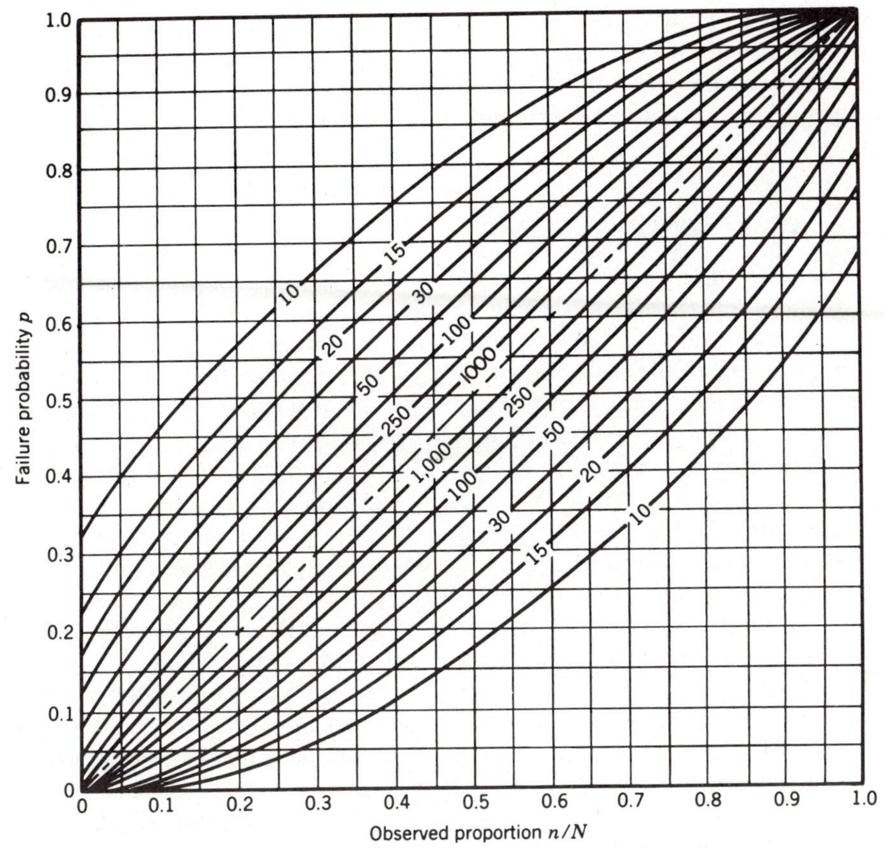

FIGURE 2.7 The 95% confidence intervals for the binomial distribution. [From E. S. Pearson and C. J. Clopper, "The Use of Confidence or Fiducial Limits Illustrated in the Case of the Binomial," *Biometrica*, **26**, 204 (1934). With permission of Biometrica.]

upper and lower confidence limits p_+ and p_-. These limits are determined such that if the true value p does not lie within the interval $p_- \leq p \leq p_+$, obtaining a test result of n failures has a probability of less than α. We are then said to have a two-sided confidence interval of $1 - \alpha$ that p lies in the interval $p_- \leq p \leq p_+$.

The mathematics required to determine p_- and p_+ for a given level of confidence is too lengthy to reproduce here. Fortunately, the results have been tabulated systematically for binomial sampling. The results are shown in Fig. 2.7 for the 95% confidence interval, which corresponds to $\alpha/2 = 0.025$. That is, we are 97.5% confident that $p > p_-$ and 97.5% confident that $p < p_+$. The corresponding graphs for other confidence levels are given in Appendix B.

The results in Fig. 2.7 indicate the limitations of classical sampling methods if highly accurate estimates are needed, particularly when small failure probabilities are involved. Suppose, for example, that 10 items are tested with only one failure; our 95% confidence then is $0.001 \leqslant p \leqslant 0.47$. Thus very large samples are needed to get reasonable error bonds on the parameter p.

2.5 ACCEPTANCE TESTING

Binomial sampling of the type we have discussed has long been associated with the acceptance testing that invariably forms an integral part of reliability and quality control programs. Such sampling is carried out to provide an adequate degree of assurance to the buyer that no more than some specified fraction of a batch of products is defective. Central to the idea of acceptance sampling is that there be a unique pass-fail criterion.

The question naturally arises why all the units are not inspected if it is important that p be small. The most obvious answer is expense. In many cases it may simply be too expensive to inspect every item of large-size batches of mass-produced items. Moreover, for a given budget, much better quality assurance is often achieved if the funds are expended on carrying out thorough inspections, tests, or both on a randomly selected sample instead of carrying out more cursory tests on the entire batch.

When the tests involve reliability-related characteristics, the necessity for performing them on a sample become more apparent, for the tests may be destructive or at least damaging to the sample units. Consider two examples. If safety margins on strength or capacity are to be verified, the tests may involve stress levels far above those anticipated in normal use: large torques may be applied to sample bolts to ensure that failure is by excessive deformation and not fracture; electric insulation may be subjected to a specified but abnormally high voltage to verify the safety factor on the breakdown voltage. If reliability is to be tested directly, each unit of the sample must be operated for a specified time to determine the fraction of failures. This time may be shortened by operating the sample units at higher stress levels, but in either case some sample units will be destroyed, and those that survive the test may exhibit sufficient damage or wear to make them unsuitable for further use.

Binomial Sampling

Typically, an acceptance testing procedure is set up to provide protection for both the producer and the buyer in the following way. Suppose that the buyer's acceptance criteria requires that no more than a fraction p_1 of the total batch fail the test. That is, for the large (theoretically infinite) batch the failure probability must be less than p_1. Since only a finite sample size N is to be tested, there will be some risk that the population will be accepted even though $p > p_1$. Let this risk be denoted by β, the probability of accepting a batch even though $p > p_1$. This is referred to as the buyer's risk; typically, we might take $\beta \approx 10\%$.

The producers of the product may be convinced that their product exceeds the buyer's criteria with a failure fraction of only $p_0 (p_0 < p_1)$. In taking only a finite sample, however, they run the risk that a poor sample will result in the batch being rejected. This is referred to as the producer's risk and it is denoted by α, the probability that a sample will be rejected even though $p < p_0$. Typically, an acceptable risk might be $\alpha \approx 5\%$.

Our object is to construct a binomial sampling scheme in which p_0 and p_1 result in predetermined values of α and β. To do this, we assume that the sample size is much less than the batch size. Let **n** be the random variable denoting the number of defective items, and n_d be the maximum number of defective items allowable in the sample. The buyer's risk β is then the probability that there will be no more than n_d defective items, given a failure probability of p_1:

$$\beta = P\{\mathbf{n} \leq n_d | N, p_1\}. \tag{2.97}$$

Using the binomial distribution, we obtain

$$\beta = \sum_{n=0}^{n_d} C_n^N p_1^n (1 - p_1)^{N-n}. \tag{2.98}$$

Similarly, the producer's risk α is the probability that there will be more than n_d defective items in the batch, even though $p = p_0$:

$$\alpha = P\{\mathbf{n} > n_d | N, p_0\} \tag{2.99}$$

or

$$\alpha = \sum_{n=n_d+1}^{N} C_n^N p_0^n (1 - p_0)^{N-n}. \tag{2.100}$$

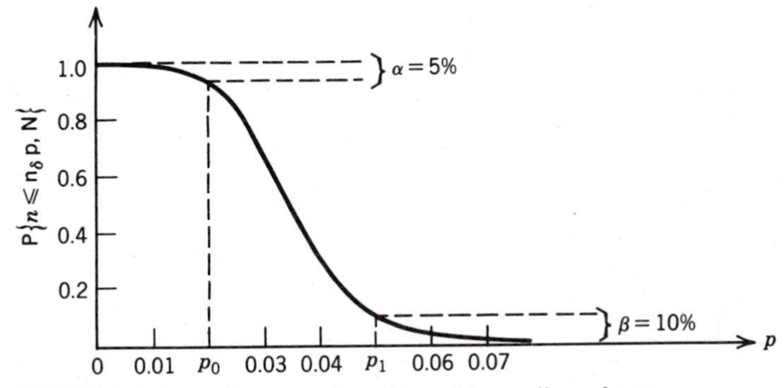

FIGURE 2.8 Operating curve for a binomial sampling scheme.

From Eqs. 2.98 and 2.100 the values of n_d and N for the sampling scheme can be determined. With n_d and N thus determined, the characteristics of the resulting sampling scheme can be presented graphically in the form of an operating curve. The operating curve is just the probability of acceptance versus the value p, the true value of the failure probability:

$$P\{\mathbf{n} \leqslant n_d | N, p\} = \sum_{n=0}^{n_d} C_n^N p^n (1 - p)^{N-n}. \tag{2.101}$$

In Fig. 2.8 is shown a typical operating curve, with β being the probability of acceptance when $p = p_1$, and α the probability of rejection when $p = p_0$.

The Poisson Limit

As in the preceding section, the binomial distribution may be replaced by the Poisson limit when the sample size is very large $N \gg 1$, and the failure probabilities are small $p_0, p_1 \ll 1$. This leads to considerable simplifications in carrying out numerical computations. Defining $m_0 = Np_0$ and $m_1 = Np_1$, we may replace Eqs. 2.98 and 2.100 by the corresponding Poisson distributions:

$$\beta = \sum_{n=0}^{n_d} \frac{m_1^n}{n!} e^{-m_1} \tag{2.102}$$

and

$$\alpha = 1 - \sum_{n=0}^{n_d} \frac{m_0^n}{n!} e^{-m_0}. \tag{2.103}$$

Given α and β, we may solve these equations numerically for m_0 and m_1 with $n_d = 0, 1, 2, \ldots$. The results of such a calculation for $\alpha = 5\%$ and $\beta = 10\%$ are tabulated in Table 2.3. One uses the table by first calculating p_1/p_0; n_d is then read from the first column, and N is determined from $N = (m_0/p_0)$ or $N = (m_1/p_1)$. This is best illustrated by an example.

EXAMPLE 2.10

Construct a sampling scheme for n_d and N, given

$$\alpha = 5\%, \beta = 10\%, p_0 = 0.02, \text{ and } p_1 = 0.05.$$

Solution We have $p_1/p_0 = 0.05/0.02 = 2.5$. Thus from Table 2.3 $n_d = 10$. Now $N = m_0/p_0 = 6.168/0.02 \cong 308$. ◀

Multiple Sampling Methods

We have discussed in detail only situations in which a single sample of size N is used. Acceptance of the items is made, provided that the number of defective items does not exceed n_d, which is referred to as the acceptance

TABLE 2.3[a] Binomial Sampling Chart for $\alpha = 0.05$; $\beta = 0.10$

n_d	m_0	m_1	p_1/p_0	n_d	m_0	m_1	p_1/p_0
0	0.0513	2.303	44.9	13	8.463	18.96	2.24
1	0.3531	3.890	11.0	14	9.246	20.15	2.18
2	0.8167	5.323	6.52	15	10.04	21.32	2.12
3	1.365	6.681	4.89	16	10.83	22.49	2.08
4	1.969	7.994	4.06	17	11.63	23.64	2.03
5	2.613	9.275	3.55	18	12.44	24.78	1.99
6	3.285	10.53	3.21	19	13.25	25.91	1.96
7	3.980	11.77	2.96	20	14.07	27.05	1.92
8	4.695	12.99	2.77	21	14.89	28.20	1.89
9	5.425	14.21	2.62	22	15.68	29.35	1.87
10	6.168	15.45	2.50	23	16.50	30.48	1.85
11	6.924	16.64	2.40	24	17.34	31.61	1.82
12	7.689	17.81	2.32	25	18.19	32.73	1.80

[a] Adapted from E. Schindowski and O. Schürz, *Statistische Qualitätskontrolle,* VEB Verlag Technik, Berlin, 1972.

number. Often more varied and sophisticated sampling schemes may be used to glean additional information without an inordinate increase in sampling effort.* Two such schemes are double sampling and sequential sampling.

In double sampling a sample size N_1 is drawn. The batch, however, need not be rejected or accepted as a result of the first sample if too much uncertainty remains about the quality of the batch. Instead, a second sample N_2 is drawn and a decision made on the combined sample size $N_1 + N_2$. Such schemes often allow costs to be reduced, for a very good batch will be accepted or a very bad batch rejected with the small sample size N_1. The larger sample size $N_1 + N_2$ is reserved for borderline cases.

In sequential sampling the principle of double sampling is further extended. The sample is built up item by item, and a decision is made after each observation to accept, to reject, or to take a larger sample. Such schemes can be expressed as sequential sampling charts, such as the one shown in Fig. 2.9. Sequential sampling has the advantage that very good (or bad) batches can be accepted (or rejected) based on very small sample sizes, with the larger samples being reserved for those situations in which there is more doubt about whether the number of defects will fall within the prescribed limits. Sequential sampling does have a disadvantage. If the test of each item takes a significant length of time, as usually happens in reliability testing, the total test time is likely to take too long. The limited time available then dictates that a single sample be taken and the items tested simultaneously.

* See, for example, A. V. Feigenbaum, *Total Quality Control,* 3rd ed., McGraw-Hill, New York, 1983, Chapter 15.

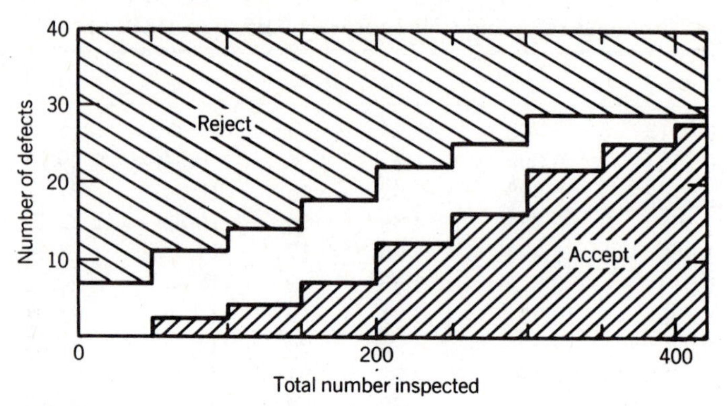

FIGURE 2.9 A sequential sampling chart.

BIBLIOGRAPHY

Feigenbaum, A. V., *Total Quality Control,* 3rd ed., McGraw-Hill, New York, 1983.

Freund, J. E., and R. E. Walpole, *Mathematical Statistics,* 3rd ed., Prentice-Hall, Englewood Cliffs, NJ, 1980.

Ireson, W. G. (ed.), *Reliability Handbook,* McGraw-Hill, New York, 1966.

Lapin, L. L., *Probability and Statistics for Modern Engineering,* Brooks/Cole, Belmont, CA, 1983.

Olkin, I., Z. J. Gleser, and G. Derman, *Probability Models and Applications,* Macmillan, New York, 1980.

Pieruschla, E., *Principles of Reliability,* Prentice-Hall, Englewood Cliffs, NJ, 1963.

Roberts, N. H., *Mathematical Methods in Reliability Engineering,* McGraw-Hill, New York, 1964.

EXERCISES

2.1 Suppose that X and Y are independent events with $P\{X\} = 0.28$ and $P\{Y\} = 0.41$. Find
(*a*) $P\{\tilde{X}\}$, (*b*) $P\{X \cap Y\}$, (*c*) $P\{\tilde{Y}\}$ (*d*) $\{X \cap \tilde{Y}\}$, (*e*) $P\{X \cup Y\}$, (*f*) $P\{\tilde{X} \cap \tilde{Y}\}$.

2.2 Suppose that $P\{X\} = 0.32$, $P\{Y\} = 0.44$, and $P\{XUY\} = 0.58$.
(*a*) Are the events mutually exclusive?
(*b*) Are they independent?
(*c*) Calculate $P\{X|Y\}$.
(*d*) Calculate $P\{Y|X\}$.

2.3 Suppose that $P\{A\} = 1/2$, $P\{B\} = 1/4$, and $P\{A \cap B\} = 1/8$. Determine
(*a*) $P\{A|B\}$, (*b*) $P\{B|A\}$, (*c*) $P\{A \cup B\}$, (*d*) $P\{\tilde{A}|\tilde{B}\}$.

2.4 Two relays with demand failures of $p = 0.15$ are tested.
(*a*) What is the probability that neither will fail?
(*b*) What is the probability that both will fail?

2.5 A particulate monitor has a power supply consisting of two batteries in parallel. Either battery is adequate to operate the monitor. However, since the failure of one battery places an added strain on the other, the conditional probability that the second battery will fail, given the failure of the first, is greater than the probability that the first will fail. On the basis of testing it is known that 7% of the monitors in question will have at least one battery failed by the end of their design life, whereas in 1% of the monitors both batteries will fail during the design life.
(*a*) Calculate the battery failure probability under normal operating conditions.
(*b*) Calculate the conditional probability that the battery will fail, given that the other has failed.

2.6 Two pumps operating in parallel supply secondary cooling water to a condenser. The cooling demand fluctuates, and it is known that each pump is capable of supplying the cooling requirements 80% of the time in case the other fails. The failure probability for each pump is 0.12; the probability of both failing is 0.02. If there is a pump malfunction, what is the probability that the cooling demand can still be met?

2.7 An aircraft landing gear has a probability of 10^{-5} per landing of being damaged from excessive impact. What is the probability that the landing gear will survive a 10,000 landing design life without damage?

2.8 A manufacturer of 16K byte memory boards finds that the reliability of the manufactured boards is 0.98. Assume that the defects are independent.
(*a*) What is the probability of a single byte of memory being defective?
(*b*) If no changes are made in design or manufacture, what reliability may be expected from 128K byte boards?
(*Note:* 16K bytes $= 2^{14}$ bytes, 128K bytes $= 2^{17}$ bytes.)

2.9 For the discrete PMF,

$$f(x_n) = Cx_n^2; \qquad x_n = 1, 2, 3.$$

(*a*) Find C.
(*b*) Find $F(x_n)$.
(*c*) Calculate μ and σ.

2.10 Repeat Exercise 2.9 for

$$f(x_n) = Cx_n(6 - x_n), \qquad x_n = 0, 1, 2, \ldots, 6.$$

2.11 Consider the discrete random variable defined by

x_n	0	1	2	3	4	5
$f(x_n)$	$\frac{11}{36}$	$\frac{9}{36}$	$\frac{7}{36}$	$\frac{5}{36}$	$\frac{3}{36}$	$\frac{1}{36}$

Compute the mean and the variance.

2.12 A discrete random variable **x** takes on the values 0, 1, 2, and 3 with probabilities 0.4, 0.3, 0.2, and 0.1, respectively. Compute the expected values of x, x^2, $2x + 1$, and e^{-x}.

2.13 Evaluate the following:
(*a*) C_3^5, (*b*) C_2^9, (*c*) C_7^{12}, (*d*) C_{19}^{20}.

2.14 A boiler has four identical relief valves. The probability that an individual relief valve will fail to open on demand is 0.06. If the failures are independent:
(*a*) What is the probability that at least one valve will fail to open?
(*b*) What is the probability that at least one valve will open?

2.15 If the four relief valves were to be replaced by two valves in the preceding problem, to what value must the probability of an individual valve's failing be reduced if the probability that no valve will open is not to increase?

2.16 The probability of an engine's failing during a 30-day acceptance test is 0.3 under adverse environmental conditions. Eight engines are included in such a test. What is the probability of the following?
(*a*) None will fail. (*b*) All will fail. (*c*) More than half will fail.

2.17 The probability that a clutch assembly will fail an accelerated reliability test is known to be 0.15. If five such clutches are tested, what is the probability that the error in the resulting estimate will be more than 0.1?

2.18 A manufacturer produces 1000 ball bearings. The failure probability for each ball bearing is 0.002.
(*a*) What is the probability that more than 0.1% of the ball bearings will fail?
(*b*) What is the probability that more than 0.5% of the ball bearings will fail?

2.19 Verify Eqs. 2.88 and 2.89.

2.20 Suppose that the probability of a diode's failing an inspection is 0.006.
(*a*) What is the probability that in a batch of 500, more than 3 will fail?
(*b*) What is the mean number of failures per batch?
(*Note:* Use the Poisson distribution.)

2.21 Diesel engines used for generating emergency power are required to have a high reliability of starting during an emergency. If the failure to start on demand probability of 1% or less is required, how many consecutive successful starts would be necessary to ensure this level of reliability with a 90% confidence?

2.22 An engineer feels confident that the failure probability on a new electromagnetic relay is less than 0.01. The specifications require, however, only that $p < 0.04$. How many units must be tested without failure to prove with 95% confidence that $p < 0.04$?

2.23 A quality control inspector examines a sample of 30 microcircuits from each purchased batch. The shipment is rejected if 4 or more fail. Find the probability of rejecting the batch where the fraction of defective circuits in the entire (large) batch is
(*a*) 0.01, (*b*) 0.05, (*c*) 0.15.

2.24 Suppose that a sample of 20 units passes an acceptance test if no more than 2 units fail. Suppose that the producer guarantees the units for a failure probability of 0.05. The buyer considers 0.15 to be the maximum acceptable failure probability.
(*a*) What is the producers risk?
(*b*) What is the buyer's risk?

2.25 Draw the operating curve for the 2 out of 20 sampling scheme of the preceding problem.
(*a*) What must the failure probability be to obtain a producer's risk of no more than 10%?
(*b*) What must the failure probability be for the buyer to have a risk of no more than 10%?

2.26 Construct a binomial sampling scheme where the producer's risk is 5%, the buyer's risk 10%, and $p_0 = 0.03$ and $p_1 = 0.06$.

2.27 A standard acceptance test is carried out on 20 battery packs. Two fail.
(*a*) What is the 95% confidence interval for the failure probability?
(*b*) Make a rough estimate of how many tests would be required if the 95% confidence interval were to be within ± 0.1 of the true failure probability. Assume the true value is $p = 0.2$.

2.28 A buyer specifies that no more than 10% of large batches of items should be defective. She tests 10 items from each batch and accepts the batch if none of the 10 is defective. What is the probability that she will accept a batch in which more than 10% are defective?

CHAPTER 3

Continuous Random Variables

3.1 INTRODUCTION

In Chapter 2 probabilities of discrete events, most frequently failures, were discussed. The discrete random variables associated with such events are used to estimate the number of events that are likely to take place. In order to proceed further with reliability analysis, however, it is necessary to consider how the probability of failure depends on a variety of other variables that are continuous, the duration of operation time, the strength of the system, the magnitudes of stresses, and so on. If the repeated measurement of such variables is carried out, however, the same value will not be obtained with each test. These values are referred to as continuous random variables, for they cannot be described with certainty, but only with the probability that they will take on values within some range. In Section 3.2 we first introduce the mathematical apparatus that is required to describe random variables. In Section 3.3 some of the most widely used distributions for describing random variables are introduced, and in Section 3.4 we discuss how we may statistically infer the properties of a random variable from limited amounts of test data.

3.2 PROPERTIES OF RANDOM VARIABLES

In this section we examine some of the more important properties of continuous random variables. We first define the quantities that determine the behavior of a single random variable. We then examine how these properties are transformed when the variable is changed. Finally, we consider the case in which two or more random variables must be dealt with simultaneously.

Probability Distribution Functions

We denote a continuous random variable with bold-faced type as $\mathbf{x}$ and the values that $\mathbf{x}$ may take on are specified by x, that is, in normal type. The properties of a random variable are specified in terms of probabilities. For example, $P\{\mathbf{x} < x\}$ is used to designate the probability that $\mathbf{x}$ has a value less than x. Similarly, $P\{a < \mathbf{x} < b\}$ is the probability that $\mathbf{x}$ has a value between a and b. Two particular probabilities are most often used to describe a random variable. The first one,

$$F(x) = P\{\mathbf{x} \leq x\}, \tag{3.1}$$

the probability that $\mathbf{x}$ has a value less than or equal to x, is referred to as the *cumulative distribution function,* or CDF for short. Second, the probability that $\mathbf{x}$ lies between x and $x + \Delta x$ as Δx becomes infinitesimally small is denoted by

$$f(x)\,\Delta x = P\{x \leq \mathbf{x} \leq x + \Delta x\}, \tag{3.2}$$

where $f(x)$ is the *probability density function,* referred to hereafter as the PDF. Since both $f(x)$ and $F(x)$ are probabilities, they must be greater than or equal to zero for all values of x.

These two functions of x are related. Suppose that we allow $\mathbf{x}$ to take on any values $-\infty \leq \mathbf{x} \leq +\infty$. Then the CDF is just the integral of the PDF over all $\mathbf{x} \leq x$:

$$F(x) = \int_{-\infty}^{x} f(x')\,dx'. \tag{3.3}$$

We also may invert this relationship by differentiating to obtain

$$f(x) = \frac{d}{dx}F(x). \tag{3.4}$$

The probability distributions $f(x)$ and $F(x)$ are normalized as follows: We first note that the probability that $\mathbf{x}$ lies between a and b may be obtained by integrating

$$\int_{a}^{b} f(x)\,dx = P\{a \leq \mathbf{x} \leq b\}. \tag{3.5}$$

Now, $\mathbf{x}$ must have some value between $-\infty$ and $+\infty$. Thus

$$P\{-\infty \leq \mathbf{x} \leq \infty\} = 1. \tag{3.6}$$

The combination of this relationship with Eq. 3.5 with $a = -\infty$ and $b = +\infty$ then yields the normalization condition

$$\int_{-\infty}^{\infty} f(x)\,dx = 1. \tag{3.7}$$

Then, setting $\mathbf{x} = \infty$ in Eq. 3.3, we find the corresponding condition on the CDF to be

$$F(\infty) = 1. \tag{3.8}$$

One more function that is often used is the *complementary cumulative distribution function* or CCDF, which is defined as

$$\tilde{F}(x) = P\{\mathbf{x} > x\}, \tag{3.9}$$

where we use the tilde to designate the complementary distribution, since $\mathbf{x} > x$ is the same as $\mathbf{x}$ not $\leqslant x$. The definition of $f(x)$ and Eq. 3.7 allows us to write $\tilde{F}(x)$ as

$$\tilde{F}(x) = \int_x^\infty f(x')\,dx' = 1 - \int_{-\infty}^x f(x')\,dx', \tag{3.10}$$

or combining this expression with Eq. 3.3 yields

$$\tilde{F}(x) = 1 - F(x). \tag{3.11}$$

Thus far we have assumed that $\mathbf{x}$ can take on any value $-\infty \leqslant \mathbf{x} \leqslant +\infty$. In many situations we must deal with variables that are restricted to a smaller domain. For example, time is most often restricted to $0 \leqslant \mathbf{t} \leqslant \infty$. In such cases the foregoing relationships may be modified quite simply. For example, in considering only positive values of time we have

$$F(t) = 0, \qquad t < 0, \tag{3.12}$$

and therefore for time, Eq. 3.3 becomes

$$F(t) = \int_0^t f(t')\,dt'. \tag{3.13}$$

Similarly, the condition of Eq. 3.7 becomes

$$\int_0^\infty f(t)\,dt = 1. \tag{3.14}$$

In Fig. 3.1 the relation between $f(x)$ and $F(x)$ is illustrated for a typical random variable with the restriction that $0 \leqslant \mathbf{x} \leqslant \infty$. In what follows we retain the $\pm\infty$ limits on the random variables, with the understanding that these are to be appropriately reduced in situations in which the domain of the variable is restricted.

EXAMPLE 3.1

The PDF of the lifetime of an appliance is given by

$$f(t) = 0.25te^{-0.5t}, \qquad t \geqslant 0,$$

where t is in years. (*a*) What is the probability of failure during the first year? (*b*) What is the probability of the appliance's lasting at least 5 years? (*c*) If no more than 5% of the appliances are to require warranty services, what is the maximum number of months for which the appliance can be warranted?

Solution First calculate the CDF and CCDF:

$$F(t) = \int_0^t dt\, 0.25te^{-0.5t} = 1 - (1 + 0.5t)e^{-0.5t}.$$

$$\tilde{F}(t) = (1 + 0.5t)e^{-0.5t}.$$

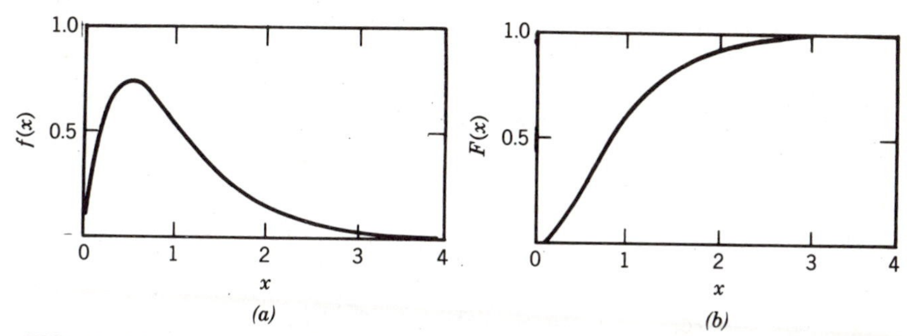

FIGURE 3.1 Continuous probability distribution: (*a*) probability density function (PDF), (*b*) corresponding cumulative distribution function (CDF).

(*a*) $F(1) = 1 - (1 + 0.5 \times 1)e^{-0.5\times 1} = 0.0902$. ◀
(*b*) $\tilde{F}(5) = (1 + 0.5 \times 5)e^{-0.5\times 5} = 0.2873$. ◀
(*c*) We must have $\tilde{F}(t_0) \geq 0.95$, where t_0 is the warranty period in years. From part (*a*) it is clear that the warranty must be less than one year, since $\tilde{F}(1) = 1 - F(1) = 0.91$.

Try 6 months, $t_0 = \frac{6}{12}$; $\tilde{F}(\frac{6}{12}) = 0.973$.
Try 9 months, $t_0 = \frac{9}{12}$; $\tilde{F}(\frac{9}{12}) = 0.945$.
Try 8 months, $t_0 = \frac{8}{12}$; $\tilde{F}(\frac{8}{12}) = 0.955$.

The maximum warranty is 8 months. ◀

Characteristics of a Probability Distribution

Often it is not necessary, or possible, to know the details of the probability density function of a random variable. In many instances it suffices to know certain integral properties. The two most important of these are the mean and the variance.

The mean or expectation value of **x** is defined by

$$\mu = \int_{-\infty}^{\infty} xf(x)\, dx. \tag{3.15}$$

The variance is given by

$$\sigma^2 = \int_{-\infty}^{\infty} (x - \mu)^2 f(x)\, dx. \tag{3.16}$$

The variance is a measure of the dispersion of values about the mean. Note that since the integrand on the right-hand side of Eq. 3.16 is always nonnegative, the variance is always nonnegative. In Fig. 3.2 examples are shown of probability density functions with different mean values and with different values of the variance, respectively.

More general functions of a random variable can be defined. Any func-

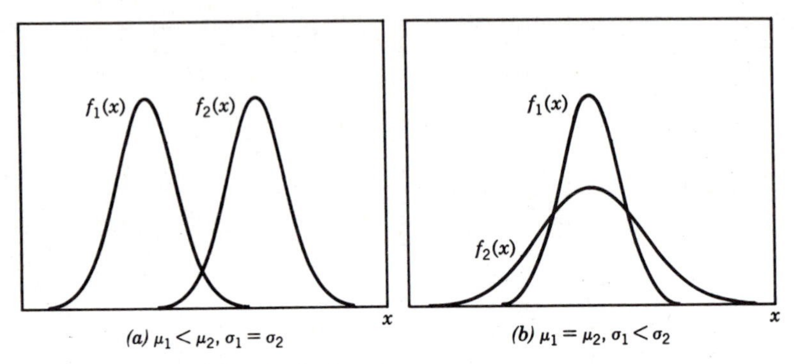

FIGURE 3.2 Probability density functions.

tion, say $g(x)$, that is to be averaged over the values of a random variable we write as

$$E\{g(x)\} \equiv \int_{-\infty}^{\infty} g(x)f(x)\,dx. \tag{3.17}$$

The quantity $E\{g(x)\}$ is referred to as the expected value of $g(x)$. It may be interpreted more precisely as follows. If we sampled an infinite large number of values of $\mathbf{x}$ from $f(x)$ and calculated $g(x)$ for each one of them, the average of these values would be $E\{g\}$. In particular, the nth moment of $f(x)$ is defined to be

$$E\{x^n\} = \int_{-\infty}^{\infty} x^n f(x)\,dx. \tag{3.18}$$

With these definitions we note that $E\{x^0\} = 1$, and the mean is just the first moment:

$$\mu = E\{x\}. \tag{3.19}$$

Similarly, the variance may be expressed in terms of the first and second moments. To do this we first write

$$\sigma^2 = E\{(x - \mu)^2\} = E\{x^2 - 2x\mu + \mu^2\}. \tag{3.20}$$

But since μ is independent of x, it can be brought outside of the integral to yield

$$\sigma^2 = E\{x^2\} - 2E\{x\}\mu + \mu^2. \tag{3.21}$$

Finally, using Eq. 3.19, we have

$$\sigma^2 = E\{x^2\} - E\{x\}^2. \tag{3.22}$$

In addition to the mean and variance, two additional properties are

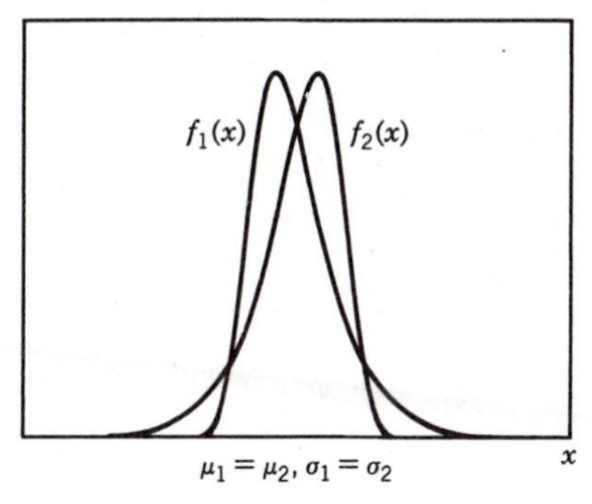

FIGURE 3.3 Probability density functions with skewness of opposite signs.

sometimes used to characterize the PDF of random variable; these are the skewness and the kurtosis. The skewness is defined by

$$\frac{1}{\sigma^3}\int_{-\infty}^{\infty} (x - \mu)^3 f(x)\, dx. \tag{3.23}$$

It is a measure of the asymmetry of a PDF about the mean. In Fig. 3.3 are shown two PDFs with identical values of μ and σ^2, but with values of the skewness that are opposite in sign but of the same magnitude. The kurtosis, like the variance, is a measure of the spread of $f(x)$ about the mean. It is given by

$$\frac{1}{\sigma^4}\int_{-\infty}^{\infty} (x - \mu)^4 f(x)\, dx. \tag{3.24}$$

EXAMPLE 3.2

A lifetime distribution has the form

$$f(t) = \beta t e^{-\alpha t},$$

where t is in years. Find β, μ, and σ in terms of α.

Solution We shall use the fact that (see Appendix A)

$$\int_0^{\infty} d\xi \xi^n e^{-\xi} = n!.$$

From Eq. 3.14,

$$\int_0^{\infty} dt \beta t e^{-\alpha t} = 1.$$

With $\zeta = \alpha t$, we therefore have

$$\frac{\beta}{\alpha^2}\int_0^{\infty} d\zeta \zeta e^{-\zeta} = \frac{\beta}{\alpha^2} \times 1 = 1.$$

Thus $\beta = \alpha^2$ ◀ and we have $f(t) = \alpha^2 t e^{-\alpha t}$.
The mean is determined from Eq. 3.15:

$$\mu \equiv \int_0^\infty dt t f(t) = \alpha^2 \int_0^\infty dt t^2 e^{-\alpha t} = \frac{1}{\alpha}\int_0^\infty d\zeta \zeta^2 e^{-\zeta} = \frac{2!}{\alpha}.$$

Therefore, $\mu = 2/\alpha$. ◀
The variance is found from Eq. 3.22, which reduces to

$$\sigma^2 \equiv \int_0^\infty dt t^2 f(t) - \mu^2,$$

but

$$\int_0^\infty dt t^2 f(t) = \alpha^2 \int_0^\infty dt t^3 e^{-\alpha t} = \frac{1}{\alpha^2}\int_0^\infty d\zeta \zeta^3 e^{-\zeta} = \frac{3!}{\alpha^2} = \frac{6}{\alpha^2}$$

and therefore,

$$\sigma^2 = \frac{6}{\alpha^2} - \left(\frac{2}{\alpha}\right)^2 = \frac{2}{\alpha^2}.$$

Thus $\sigma = \sqrt{2}/\alpha$. ◀

EXAMPLE 3.3

Calculate μ and σ in Example 3.1.

Solution Note that the distribution in Examples 3.1 and 3.2 are identical if $\alpha = 0.5$. Therefore $\mu = 4$ years, ◀ and $\sigma = 2\sqrt{2}$ years. ◀

Transformations of Variables

Frequently, in reliability considerations, the random variable for which data are available is not the one that can be used directly in the reliability estimates. Suppose, for example, that the distribution of speeds of impact $f(v)$ is known for a mechanical snubber. If the wear on the snubber, however, is proportional to the kinetic energy, $e = \frac{1}{2} mv^2$, the energy is also a random variable and it is the distribution of energies $f_e(e)$ that is needed. Such problems are ubiquitous, for much of engineering analysis is concerned with functional relationships that allow us to predict the value of one variable (the dependent variable) in terms of another (the independent variable).

To deal with situations such as the change from speed to energy in the foregoing example, we need a means for transforming one random variable to another. The problem may be stated more generally as follows. Given a distribution $f_{\mathbf{x}}(x)$ or $F_{\mathbf{x}}(x)$ of the random variable $\mathbf{x}$, find the distribution $f_{\mathbf{y}}(y)$ of the random variable $\mathbf{y}$ that is defined by

$$y = y(x). \tag{3.25}$$

We then refer to $f_{\mathbf{y}}(y)$ as the derived distribution. Hereafter, we use subscripts **x** and **y** to distinguish between the distributions whenever there is a possibility of confusion. First, consider the case where the relation between y and x has the characteristics shown in Fig. 3.4; that is, if $x_1 < x_2$, then $y(x_1) < y(x_2)$. Then $y(x)$ is a monotonically increasing function of x; that is, $dy/dx > 0$. To carry out the transformation, we first observe that

$$P\{\mathbf{x} \leqslant x\} = P\{\mathbf{y} \leqslant y(x)\}, \tag{3.26}$$

or simply

$$F_{\mathbf{x}}(x) = F_{\mathbf{y}}(y). \tag{3.27}$$

To obtain the PDF $f_{\mathbf{y}}(y)$ in terms of $f_{\mathbf{x}}(x)$, we first write the preceding equation as

$$\int_{-\infty}^{x} f_{\mathbf{x}}(x')\, dx' = \int_{-\infty}^{y(x)} f_{\mathbf{y}}(y')\, dy'. \tag{3.28}$$

Differentiating with respect to x, we obtain

$$f_{\mathbf{x}}(x) = f_{\mathbf{y}}(y)\frac{dy}{dx} \tag{3.29}$$

or

$$f_{\mathbf{y}}(y) = f_{\mathbf{x}}(x)\left|\frac{dx}{dy}\right|. \tag{3.30}$$

Here we have placed an absolute value about the derivative. With the absolute value, the result can be shown to be valid for either monotonically increasing or monotonically decreasing functions.

The most common transforms are of the linear form

$$y = ax + b, \tag{3.31}$$

and the foregoing equation becomes simply

$$f_{\mathbf{y}}(y) = \frac{1}{|a|} f_{\mathbf{x}}\left(\frac{y - b}{a}\right). \tag{3.32}$$

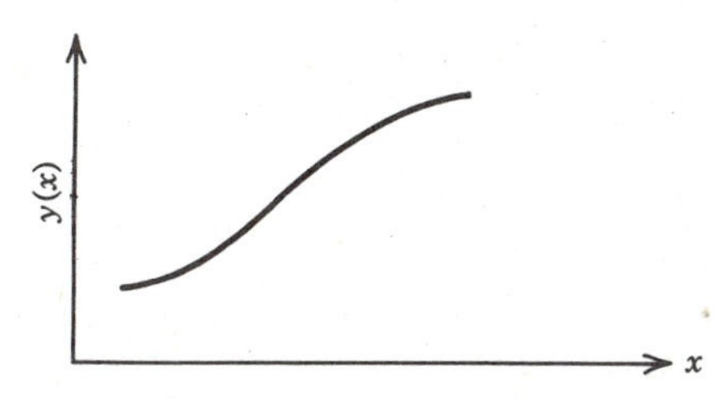

FIGURE 3.4 Function of a random variable x.

Note that once a transformation has been made, new values of the mean and variance must be calculated, since in general

$$\int g(x)f_{\mathbf{x}}(x)\, dx \neq \int g(y)f_{\mathbf{y}}(y)\, dy. \tag{3.33}$$

EXAMPLE 3.4

Consider the distribution $f_{\mathbf{x}}(x) = \alpha e^{-\alpha x}$, $0 \leq x \leq \infty$, $\alpha > 1$.
(*a*) Transform to the distribution $f_{\mathbf{y}}(y)$, where $y = e^x$.
(*b*) Calculate $\mu_{\mathbf{x}}$ and $\mu_{\mathbf{y}}$.

Solution (*a*) $dy/dx = e^x$; therefore, Eq. 3.30 becomes $f_{\mathbf{y}}(y) = e^{-x}f_{\mathbf{x}}(x)$. We also have $x = \ln y$. Therefore,

$$f_{\mathbf{y}}(y) = e^{-\ln y}\alpha e^{-\alpha \ln y} = \frac{\alpha}{y^{\alpha+1}}, \qquad 1 \leq y \leq \infty.$$

(*b*) $$\mu_{\mathbf{x}} = \int_0^\infty x\alpha e^{-\alpha x}\, dx = \frac{1}{\alpha},$$

$$\mu_{\mathbf{y}} = \int_1^\infty y\alpha y^{-(\alpha+1)}\, dy = \frac{\alpha}{\alpha - 1}.$$

Two Independent Variables

Frequently, it is necessary to deal with two, and sometimes more, random variables. In such situations the variables may be continuous, discrete, or combinations of continuous and discrete variables. We consider first the most widely encountered case of two continuous random variables. Then, since it will be employed in later chapters, we also consider the situation of one continuous and one discrete random variable. More general cases may be found in textbooks on probability.

Two Continuous Variables

The starting point for considering two continuous random variables $\mathbf{x}$ and $\mathbf{y}$ is the *joint probability density function* $f(x, y)$. It is defined by

$$f(x, y)\, \Delta x\, \Delta y \equiv P\{(x \leq \mathbf{x} \leq x + \Delta x) \cap (y \leq \mathbf{y} \leq y + \Delta y)\}. \tag{3.34}$$

The corresponding *joint cumulative distribution function* is defined by

$$F(x, y) \equiv P\{(\mathbf{x} \leq x) \cap (\mathbf{y} \leq y)\}. \tag{3.35}$$

It therefore may be expressed in terms of $f(x, y)$ as

$$F(x, y) = \int_{-\infty}^{x} \left[\int_{-\infty}^{y} f(x', y')\, dy' \right] dx'. \tag{3.36}$$

Clearly, from the definition of $F(x, y)$, we have $F(\infty, \infty) = 1$. Moreover, $f(x, y)$ may be expressed in terms of $F(x, y)$ by

$$f(x, y) = \frac{\partial^2}{\partial x\, \partial y} F(x, y). \tag{3.37}$$

The *marginal probability density functions* are defined by integrating over one variable,

$$f_{\mathbf{x}}(x) = \int_{-\infty}^{\infty} f(x, y)\, dy, \tag{3.38}$$

$$f_{\mathbf{y}}(y) = \int_{-\infty}^{\infty} f(x, y)\, dx, \tag{3.39}$$

where we again use subscripts to distinguish between the PDFs. Similarly, the *marginal cumulative distribution functions* are

$$F_{\mathbf{x}}(x) = F(x, \infty), \tag{3.40}$$

$$F_{\mathbf{y}}(y) = F(\infty, y). \tag{3.41}$$

Once we have integrated over one of the random variables to obtain the marginal probability density function, we may define the mean, variance, or other properties in the same way as when only a single random variable is under consideration. For example, from Eq. 3.38, the mean and variance of $\mathbf{x}$ are

$$\mu_{\mathbf{x}} = \int_{-\infty}^{\infty} x f_{\mathbf{x}}(x)\, dx, \tag{3.42}$$

$$\sigma_{\mathbf{x}}^2 = \int_{-\infty}^{\infty} (x - \mu_{\mathbf{x}})^2 f_{\mathbf{x}}(x)\, dx. \tag{3.43}$$

In considering two or more events in Chapter 2, we defined conditional probabilities and independence. These definitions may be generalized to two or more random variables. Corresponding to Eq. 2.4 for the conditional probability, we may now define the conditional PDF $f_{\mathbf{x}}(x|y)$ of x given y by

$$f(x, y) = f_{\mathbf{x}}(x|y) f_{\mathbf{y}}(y), \tag{3.44}$$

or conversely

$$f(x, y) = f_{\mathbf{y}}(y|x) f_{\mathbf{x}}(x), \tag{3.45}$$

subject to the condition that $f_{\mathbf{x}}$ and $f_{\mathbf{y}}$ are not equal to zero.

It follows from these definitions of conditional probability and the definitions of the marginal distributions given by Eqs. 3.38 and 3.39 that

$$\int_{-\infty}^{\infty} f_{\mathbf{x}}(x|y)\, dx = 1, \tag{3.46}$$

and similarly,

$$\int_{-\infty}^{\infty} f_{\mathbf{y}}(y|x)\, dy = 1. \tag{3.47}$$

If the distribution of x is independent of y, then

$$f_{\mathbf{x}}(x|y) \longrightarrow f_{\mathbf{x}}(x), \tag{3.48}$$

and Eq. 3.44 reduces to

$$f(x, y) = f_{\mathbf{x}}(x) f_{\mathbf{y}}(y). \tag{3.49}$$

This, like Eq. 2.8, is the definition of independence.

Analogous to the case of one random variable, the expected value of a function $g(x, y)$ is defined by

$$E\{g(x, y)\} = \int_{-\infty}^{\infty}\int_{-\infty}^{\infty} g(x, y) f(x, y)\, dx\, dy. \tag{3.50}$$

EXAMPLE 3.5

Consider the joint probability density function

$$f(x, y) = ye^{-(x+1)y}, \qquad 0 \leq x \leq \infty, \qquad 0 \leq y \leq \infty.$$

(*a*) Write $f(x, y)$ in the form of Eq. 3.44 and determine $f_{\mathbf{x}}(x|y)$ and $f_{\mathbf{y}}(y)$.
(*b*) Write $f(x, y)$ in the form of Eq. 3.45 and determine $f_{\mathbf{y}}(y|x)$ and $f_{\mathbf{x}}(x)$.
(*c*) Are the variables independent? Why?
(*d*) Determine $P\{(0 \leq \mathbf{x} \leq 1) \cap (1 \leq \mathbf{y} \leq \infty)\}$.

Solution (*a*) To find $f_{\mathbf{y}}(y)$, we use Eq. 3.39:

$$f_{\mathbf{y}}(y) = \int_0^{\infty} f(x, y)\, dx = \int_0^{\infty} ye^{-(x+1)y}\, dx = ye^{-y}\int_0^{\infty} e^{-yx}\, dx$$
$$= e^{-y}.$$

For $f(x|y)$, we use Eq. 3.44. We have

$$f_{\mathbf{x}}(x|y) = \frac{f(x, y)}{f_{\mathbf{y}}(y)} = ye^{-xy}. \blacktriangleleft$$

(*b*) To find $f_{\mathbf{x}}(x)$, we use Eq. 3.38:

$$f_{\mathbf{x}}(x) = \int_0^{\infty} f(x, y)\, dy = \int_0^{\infty} ye^{-(x+1)y}\, dy$$
$$= \frac{1}{(x+1)^2}\int_0^{\infty} \zeta e^{-\zeta}\, d\zeta, \qquad \text{where } \zeta \equiv (x+1)y$$
$$= (x+1)^{-2}. \blacktriangleleft$$

For $f(y|x)$, we use Eq. 3.45. Therefore,

$$f_y(y|x) = \frac{f(x, y)}{f_x(x)} = (x + 1)^2 y e^{-(x+1)y}. \blacktriangleleft$$

(*c*) Since the conditional probability $f_x(x|y)$ depends on $y(x)$, the variables are dependent.

(*d*) $$P\{(0 \leq \mathbf{x} \leq 1) \cap (1 \leq \mathbf{y} \leq \infty)\} = \int_0^1 \left[\int_1^\infty f(x, y)\, dy\right] dx$$

$$= \int_0^1 \left[\int_1^\infty y e^{-(x+1)y}\, dy\right] dx = \int_1^\infty y e^{-y} \left[\int_0^1 e^{-yx}\, dx\right] dy$$

$$= \int_1^\infty y e^{-y} \frac{1}{y} (1 - e^{-y})\, dy = e^{-1} - e^{-2} = 0.2325. \blacktriangleleft$$

If we can express the joint PDF as Eq. 3.49, we are assured that the random variables $\mathbf{x}$ and $\mathbf{y}$ are mutually independent. If they are not, however, it may be desirable to know to what degree they are correlated. Complete correlation signifies that if one of the variables is determined, the other is also specified. Two quantities, the covariance and the correlation coefficient, are used to measure the degree of dependence between variables. The *covariance*, denoted by $\sigma^2_{\mathbf{xy}}$, is the expected value of $(x - \mu_{\mathbf{x}})(y - \mu_{\mathbf{y}})$. It is therefore obtained by setting

$$g(x, y) = (x - \mu_{\mathbf{x}})(y - \mu_{\mathbf{y}}) \tag{3.51}$$

and evaluating Eq. 3.50,

$$\sigma^2_{\mathbf{xy}} \equiv \int_{-\infty}^{\infty} \int_{-\infty}^{\infty} (x - \mu_{\mathbf{x}})(y - \mu_{\mathbf{y}}) f(x, y)\, dx\, dy. \tag{3.52}$$

This may easily be shown to reduce to

$$\sigma^2_{\mathbf{xy}} = \int_{-\infty}^{\infty} \int_{-\infty}^{\infty} xy f(x, y)\, dx\, dy - \mu_{\mathbf{x}}\mu_{\mathbf{y}}, \tag{3.53}$$

where $\mu_{\mathbf{x}}$ is defined by Eq. 3.42, and similarly for $\mu_{\mathbf{y}}$. For independent variables, substitution of Eq. 3.49 into this expression immediately leads to $\sigma^2_{\mathbf{xy}} = 0$.

The *correlation coefficient* is a normalized form of the covariance. It is defined by

$$\rho_{\mathbf{xy}} = \frac{\sigma^2_{\mathbf{xy}}}{\sigma_{\mathbf{x}}\sigma_{\mathbf{y}}}. \tag{3.54}$$

It ranges between minus one and one. It has, of course, a value of zero for no correlation, that is, for complete independencies. It is positive if an increase in one variable causes an increase in the other, and negative if an increase in one variable leads to a decrease in the other.

One Continuous and One Discrete Variable

In Chapter 5 we encounter a situation in which **x** is a continuous random variable, but **y** is a discrete random variable that can take on only values y_n, $n = 0, 1, 2, \ldots$. Formulas analogous to those for two continuous random variables are applicable, the primary difference being that integrals over y are replaced by sums over n. Instead of Eq. 3.34, our starting point is

$$f(x, y_n)\,\Delta x = P\{(x \leqslant \mathbf{x} \leqslant x + \Delta x) \cap (\mathbf{y} = y_n)\}. \tag{3.55}$$

The joint cumulative probability distribution,

$$F(x, y_n) = P\{(\mathbf{x} \leqslant x) \cap (\mathbf{y} \leqslant y_n)\}, \tag{3.56}$$

is then given by

$$F(x, y_n) = \sum_{n'=0}^{n} \int_{-\infty}^{x} f(x', y_{n'})\,dx', \tag{3.57}$$

with the normalization condition

$$\sum_{n=0}^{\infty} \int_{-\infty}^{\infty} f(x, y_n)\,dx = 1. \tag{3.58}$$

The marginal probability density function in **x** is obtained by summing over the discrete variable y_n,

$$f_{\mathbf{x}}(x) = \sum_{n} f(x, y_n), \tag{3.59}$$

whereas the marginal probability mass function for **y** results from integrating over all x,

$$f_{\mathbf{y}}(y_n) = \int_{-\infty}^{\infty} f(x, y_n)\,dx. \tag{3.60}$$

As for two continuous variables, we may define conditional probabilities and independence. Analogous to Eqs. 3.44 and 3.45, we have the conditional probabilities $f_{\mathbf{x}}(x|y_n)$ and $f_{\mathbf{y}}(y_n|x)$ defined by

$$f(x, y) = f_{\mathbf{x}}(x|y_n) f_{\mathbf{y}}(y_n) \tag{3.61}$$

or

$$f(x, y) = f_{\mathbf{y}}(y_n|x) f_{\mathbf{x}}(x), \tag{3.62}$$

subject to the condition that $f_{\mathbf{x}}$ and $f_{\mathbf{y}}$ are not equal to zero. Independence is then defined in the same way as in Eqs. 3.48 and 3.49.

The normalization conditions are found by combining Eqs. 3.61 and 3.62,

$$\int_{-\infty}^{\infty} f_x(x|y_n)\, dx = 1, \tag{3.63}$$

and Eq. 3.62 and 3.59,

$$\sum_n f_y(y_n|x) = 1. \tag{3.64}$$

Finally, expected value for any function $g(x, y_n)$ is found from

$$E\{g(x, y_n)\} = \sum_n \int_{-\infty}^{\infty} g(x, y_n) f(x, y_n)\, dx. \tag{3.65}$$

where the mean, variance, covariance, and related quantities can be determined by taking the expected values of appropriate functions of **x** and **y**.

3.3 NORMAL AND RELATED DISTRIBUTIONS

Continuous random variables find extensive use in reliability analysis for the description of survival times, system loads and capacities, repair rates, and a variety of other phenomena. Moreover, a substantial number of standardized probability distributions are employed to model the behavior of these variables. For the most part we shall introduce these distributions as they are needed for model reliability phenomena in the following chapters. We introduce here the normal distribution and the related lognormal and Dirac delta distributions, for they appear in a variety of different contexts throughout the book. Moreover, they provide convenient vehicles for applying the concepts of the foregoing discussion.

The Normal Distribution

Unquestionably, the normal distribution is the most widely applied in statistics. It is frequently referred to as the Gaussian distribution. To introduce the normal distribution, we first consider the following function of the random variable **x**,

$$f(x) = \frac{1}{\sqrt{2\pi}\, b} \exp\left[-\frac{1}{2} \left(\frac{x - a}{b} \right)^2 \right], \qquad -\infty \leqslant x \leqslant \infty, \tag{3.66}$$

where a and b are parameters that we have yet to specify. It may be shown that $f(x)$ meets the conditions for a probability density function. First, it is clear that $f(x) \geqslant 0$ for all x. Second, by performing the integral

$$\int_{-\infty}^{\infty} \frac{1}{\sqrt{2\pi}\, b} \exp\left[-\frac{1}{2} \left(\frac{x - a}{b} \right)^2 \right] dx = 1. \tag{3.67}$$

It may be shown that the condition on the PDF given by Eq. 3.7 is met. The evaluation of Eq. 3.67 cannot be carried out by rudimentary means; rather, the method of residues from the theory of complex variables must be employed. For convenience, some of the more common integrals involving the normal distribution are included in Appendix A.

A unique feature of the normal distribution is that the mean and variance appear explicitly as the two parameters a and b. To demonstrate this fact, we insert Eq. 3.66 into the definitions of the mean and variance, Eqs. 3.15 and 3.16. Using the evaluated integrals in Appendix A, we find

$$\mu \equiv \int_{-\infty}^{\infty} dx \frac{x}{\sqrt{2\pi}\, b} \exp\left[-\frac{1}{2}\left(\frac{x-a}{b}\right)^2\right] = a, \tag{3.68}$$

$$\sigma^2 \equiv \int_{-\infty}^{\infty} dx (x-\mu)^2 \frac{1}{\sqrt{2\pi}\, b} \exp\left[-\frac{1}{2}\left(\frac{x-a}{b}\right)^2\right] = b^2. \tag{3.69}$$

Consequently, we may write the normal PDF directly in terms of the mean and variance as

$$f(x) = \frac{1}{\sqrt{2\pi}\,\sigma} \exp\left[-\frac{1}{2}\left(\frac{x-\mu}{\sigma}\right)^2\right], \qquad -\infty \leqslant x \leqslant \infty. \tag{3.70}$$

Similarly, the CDF corresponding to Eq. 3.66 is

$$F(x) = \int_{-\infty}^{x} \frac{1}{\sqrt{2\pi}\,\sigma} \exp\left[-\frac{1}{2}\left(\frac{x'-\mu}{\sigma}\right)^2\right] dx'. \tag{3.71}$$

When we use the normal distribution, it is often beneficial to make a change of variables first in order to express $F(x)$ in a standardized form. To this end, we define the random variable **u** in terms of **x** by

$$u \equiv (x - \mu)/\sigma. \tag{3.72}$$

Recalling that PDFs transform according to Eq. 3.30, we have

$$f_{\mathbf{u}}(u) = f(x)\left|\frac{dx}{du}\right| = \frac{1}{\sqrt{2\pi}\,\sigma} \exp\left[-\frac{1}{2}\left(\frac{x-\mu}{\sigma}\right)^2\right] |\sigma|, \tag{3.73}$$

which for $x = \mu + \sigma u$

$$f_{\mathbf{u}}(u) = \frac{1}{\sqrt{2\pi}} \exp(-\tfrac{1}{2}u^2). \tag{3.74}$$

This implies that for the reduced variate u, $\mu_{\mathbf{u}} = 0$ and $\sigma_{\mathbf{u}}^2 = 1$.

The PDF is plotted in Fig. 3.5. Its appearance causes it to be referred to frequently as the bell-shaped curve. The standardized form of the CDF may also be found by applying Eq. 3.72 to $F(x)$,

$$F(x) \equiv \Phi[(x - \mu)/\sigma], \tag{3.75}$$

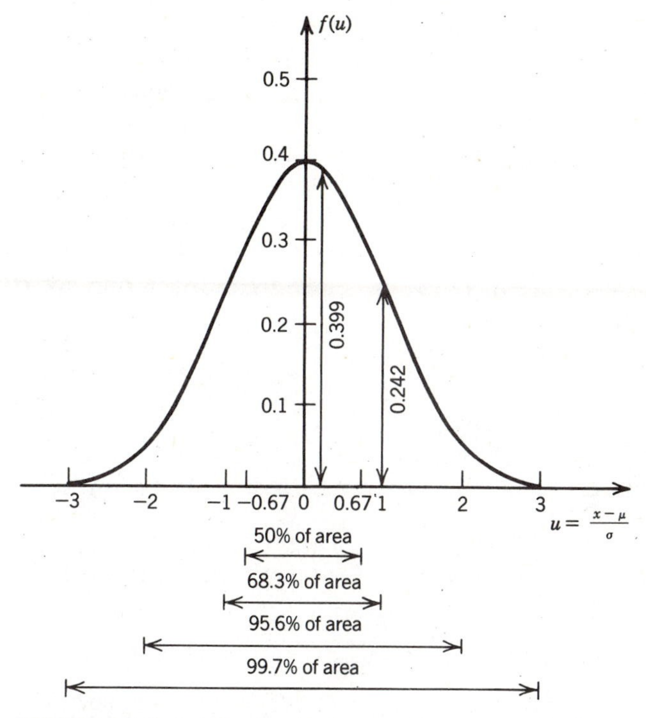

FIGURE 3.5 Probability density function for a standardized normal distribution.

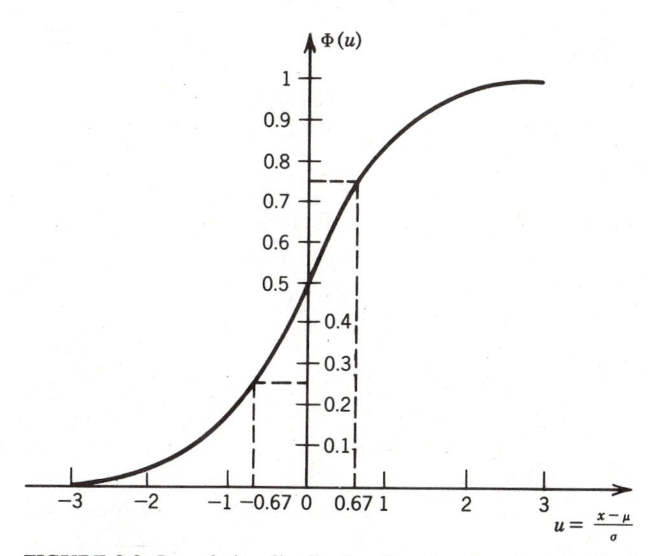

FIGURE 3.6 Cumulative distribution function for a standardized normal distribution.

where the standardized error function on the right is defined as

$$\Phi(u) = \frac{1}{\sqrt{2\pi}} \int_{-\infty}^{u} \exp(-\tfrac{1}{2}\zeta^2)\, d\zeta. \tag{3.76}$$

The integrand of this expression is just the standardized normal PDF. A graph of $\Phi(u)$ is given in Fig. 3.6; note that each unit on the horizontal axis corresponds to one standard deviation σ, and that the mean value is now at the origin. A tabulation of $\Phi(u)$ is included in Appendix C. Although values for $u < 0$ are included in Appendix C, this is only for convenience, since for the normal distribution we may use the property $f(-u) = f(u)$ to obtain $\Phi(-u)$ from

$$\Phi(-u) = 1 - \Phi(u). \tag{3.77}$$

EXAMPLE 3.6

The time to wear out of a cutting tool edge is distributed normally with $\mu = 2.8$ hr and $\sigma = 0.6$ hr.
(*a*) What is the probability that the tool will wear out in less than 1.5 hr?
(*b*) How often should the cutting edges be replaced to keep the failure rate less than 10% of the tools?

Solution (*a*) $P\{\mathbf{t} < 1.5\} = F(1.5) = \Phi(u)$, where

$$u = (t - \mu)/\sigma, \qquad u = (1.5\text{–}2.8)/0.6 = -2.1667 \blacktriangleleft$$

From Appendix C: $\Phi(-2.1667) = 0.0151$. ◀
(*b*) $P\{\mathbf{t} < t\} = 0.10$; $\Phi(u) = 0.10$. Then from Appendix C, $u \approx -1.28$. Therefore, we have

$$-t + \mu = 1.28\sigma, \qquad t = \mu - 1.28\sigma = 2.8 - 1.28 \times 0.6 = 2.03 \text{ hr.} \blacktriangleleft$$

The normal distribution arises in many contexts. It may be expected to occur whenever the random variable $\mathbf{x}$ arises from the sum of a number of random effects, no one of which dominates the total. It is widely used to represent measurement errors, dimensional variability in manufactured goods, material properties, and a host of other phenomena.

A specific illustration might be as follows. Suppose that an elevator cable consists of strands of wire. The strength of the cable is then

$$x = x_1 + x_2 + \cdots x_i + \cdots x_N, \tag{3.78}$$

where x_i is the strength of the ith strand. Even though the PDF of the individual strands x_i is not a normal distribution, the strength of the cable will be given by a normal distribution, provided that N, the number of strands, is sufficiently large.

The normal distribution also has the following property. If $\mathbf{x}$ and $\mathbf{y}$ are random variables that are normally distributed, then

$$z = ax + by, \tag{3.79}$$

where a and b are constants, is also distributed normally. Moreover, it may be shown that the mean and variance of z are related to those of x and y by

$$\mu_{\mathbf{z}} = a\mu_{\mathbf{x}} + b\mu_{\mathbf{y}} \tag{3.80}$$

and

$$\sigma_{\mathbf{z}}^2 = a^2\sigma_{\mathbf{x}}^2 + b^2\sigma_{\mathbf{y}}^2. \tag{3.81}$$

The same relationships may be extended to linear combinations of three or more random variables.

Often the normal distribution is adopted as a convenient approximation, even though there may be no sound physical basis for assuming that the previously stated conditions are met. In some situations this may be justified on the basis that it is the limiting form of several other distributions, the binomial and the Poisson, to name two. More important, if one is concerned only with very general characteristics and not the details of the shape, the normal distribution may sometimes serve as a widely tabulated, if rough, approximation to empirical data. One must take care, however, not to pursue too far the idea that the normal distribution is generally a reasonable representation for empirical data. If the data exhibit a significant skewness, the normal distribution is not likely to be a good choice. Moreover, if one is interested in the "tails" of the distribution, where $|(x - \mu)/\sigma| \gg 1$, improper use of the normal distribution is likely to lead to large errors. Extreme values of distribution must often be considered when determining safety factors and related phenomena. Distributions appropriate to such extreme-value problems are taken up in Chapter 6.

The Dirac Delta Distribution

If the normal distribution is used to describe a random variable $\mathbf{x}$, the mean μ is the measure of the average value of x and the standard deviation σ is a measure of the dispersion of x about μ. Suppose that we consider a series of measurements of a quantity μ with increasing precision. The PDF for the measurements might look similar to Fig. 3.7. As the precision is increased—

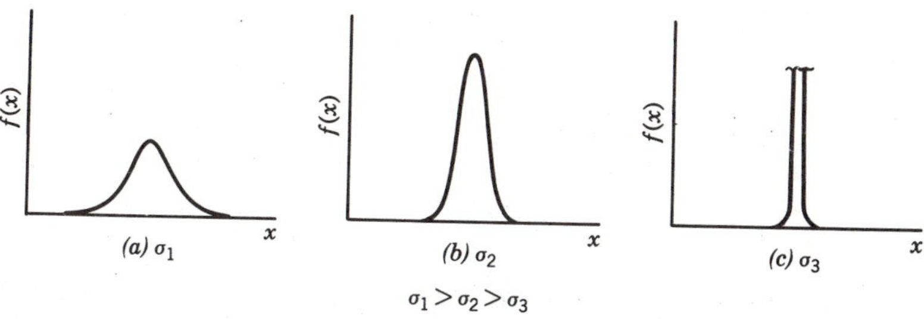

FIGURE 3.7 Normal distributions with different values of the variance.

decreasing the uncertainty—the value of σ decreases. In the limit where there is no uncertainty $\sigma \rightarrow 0$, $\mathbf{x}$ is no longer a random variable, for we know that $\mathbf{x} = \mu$.

The Dirac delta function is used to treat this situation. It may be defined as

$$\delta(x - \mu) = \lim_{\sigma \to 0} \frac{1}{\sqrt{2\pi}\,\sigma} \exp\left[-\frac{1}{2\sigma^2} (x - \mu)^2 \right]. \tag{3.82}$$

Two most important properties immediately follow from this definition:

$$\delta(x - \mu) = \begin{cases} \infty, & x = \mu, \\ 0, & x \neq \mu, \end{cases} \tag{3.83}$$

and

$$\int_{\mu-\epsilon}^{\mu+\epsilon} \delta(x - \mu)\, dx = 1, \qquad \epsilon > 0. \tag{3.84}$$

Specifically, even though $\delta(0)$ is infinite, the area under the curve is equal to one.

The primary use of the Dirac delta function in this book is to simplify integrals in which one of the variables has a fixed value. This appears, for example, in the treatment of expected values.

Suppose that we want to calculate the expected value of $g(x)$, as given by Eq. 3.17 when $f(x) = \delta(x - x_0)$; then

$$E\{g(x)\} = \int_{-\infty}^{\infty} g(x)\delta(x - x_0)\, dx \tag{3.85}$$

may be written as

$$E\{g(x)\} = \int_{x_0-\epsilon}^{x_0+\epsilon} g(x)\delta(x - x_0)\, dx, \qquad \epsilon > 0, \tag{3.86}$$

since $\delta(x - x_0) = 0$ away from $x = x_0$. If $g(x)$ is continuous, we may pull it outside the integral for very small ϵ to yield

$$E\{g(x)\} \simeq g(x_0) \int_{x_0-\epsilon}^{x_0+\epsilon} \delta(x - x_0)\, dx. \tag{3.87}$$

Therefore, for arbitrarily small ϵ, we obtain

$$E\{g(x)\} \equiv \int_{x_0-\epsilon}^{x_0+\epsilon} g(x)\delta(x - x_0)\, dx = g(x_0). \tag{3.88}$$

A more rigorous proof may be provided by using Eq. 3.82 in Eq. 3.85 and expanding $g(x)$ in a power series about x_0.

The Lognormal Distribution

As indicated earlier, if a random variable $\mathbf{x}$ can be expressed as a sum of the random variables, x_i, $i = 1, 2, \ldots, N$ where no one of them is dominant, then $\mathbf{x}$ can be described as a normal distribution, even though the $\mathbf{x}_i$ are described by nonnormal distributions that may not even be the same for different values of i. A second frequently arising situation consists of a random variable $\mathbf{y}$ that is a product of the random variables $\mathbf{y}_i$:

$$y = y_1 y_2 \cdots y_N. \tag{3.89}$$

For example, the wear on a system may be proportional to the product of the magnitudes of the demands that have been made on it. Suppose that we take the natural logarithm of Eq. 3.89:

$$\ln y = \ln y_1 + \ln y_2 + \cdots + \ln y_N. \tag{3.90}$$

The analogy to the normal distribution is clear. If no one of the terms on the right-hand side has a dominant effect, then $\ln y$ should be distributed normally. Thus, if we define

$$x \equiv \ln y, \tag{3.91}$$

then x is distributed normally and y is said to be distributed lognormally.

To obtain the lognormal distribution for y, we first write the normal distribution for x,

$$f_{\mathbf{x}}(x) = \frac{1}{\sqrt{2\pi}\,\sigma_{\mathbf{x}}} \exp\left[-\frac{1}{2\sigma_{\mathbf{x}}^2}(x - \mu_{\mathbf{x}})^2\right], \tag{3.92}$$

where $\mu_{\mathbf{x}}$ is the mean value of $\mathbf{x}$, and $\sigma_{\mathbf{x}}^2$ is the variance of the distribution in $\mathbf{x}$. Now suppose that we let x be the natural logarithm of the variable y. In order to find the PDF in y, we must transform the distribution according to Eq. 3.30:

$$f_{\mathbf{y}}(y) = f_{\mathbf{x}}(x) \left|\frac{dx}{dy}\right|. \tag{3.93}$$

Noting that

$$\frac{dx}{dy} = \frac{d}{dy}\ln y = \frac{1}{y}, \tag{3.94}$$

and using $x = \ln y$ to eliminate x from Eqs. 3.92 and 3.93, we obtain

$$f_{\mathbf{y}}(y) = \frac{1}{\sqrt{2\pi}\,sy} \exp\left\{-\frac{1}{2s^2}\left[\ln\left(\frac{y}{y_0}\right)\right]^2\right\}, \tag{3.95}$$

where we have made the replacements

$$\mu_{\mathbf{x}} \equiv \ln y_0; \qquad \sigma_{\mathbf{x}} = s. \tag{3.96}$$

The corresponding CDF is obtained by integrating over y with a lower limit of $y = 0$. The results can be expressed in terms of the standardized normal integral as

$$F_y(y) = \Phi\left[\frac{1}{s}\ln\left(\frac{y}{y_0}\right)\right]. \tag{3.97}$$

The PDF and the CPD for the lognormal distribution are plotted as a function of y in Fig. 3.8. Note that for small values of s, the lognormal and normal distributions have very similar appearances.

The mean of the lognormal distribution may be obtained by applying Eq. 3.15 to Eq. 3.95:

$$\mu_y = y_0 \exp(s^2/2). \tag{3.98}$$

Note that it is not equal to the parameter y_0 for which the distribution is a maximum. On the contrary, y_0 may be shown to be the median value of $\mathbf{y}$. Similarly, the variance in $\mathbf{y}$ is not equal to s but rather is

$$\sigma_y^2 = y_0^2\, exp(s^2)[\exp(s^2) - 1]. \tag{3.99}$$

Lognormal distributions are widely applied in reliability engineering to describe failure caused by fatigue, uncertainties in failure rates, and a variety of other phenomena. It has the property that if a variable x and y have lognormal distributions, the product random variable $z = xy$ is also lognormally distributed.

The lognormal distribution also finds frequent use in the following manner. Suppose that the best estimate of a variable is y_0 and there is a 90% certainty that y_0 is known within a factor of n. That is, there is a probability of 0.9 that it lies between y_0/n and $y_0 n$, where $n > 1$. We then have

$$0.05 = \int_0^{y_0/n} \frac{1}{\sqrt{2\pi}\, sy} \exp\left\{-\frac{1}{2s^2}\left[\ln\left(\frac{y}{y_0}\right)\right]^2\right\} dy. \tag{3.100}$$

With the change of variables $\zeta = (1/s)\ln(y/y_0)$ Eq. 3.100 may be written as

$$0.05 = \int_{-\infty}^{-(1/s)\ln n} \frac{1}{\sqrt{2\pi}} \exp(-\tfrac{1}{2}\zeta^2)\, d\zeta. \tag{3.101}$$

This integral is the CDF for the standardized normal distribution, given by Eq. 3.76. Thus we have

$$0.05 = \Phi\left(-\frac{1}{s}\ln n\right), \tag{3.102}$$

where Φ is the standardized normal CDF. Similarly, it may be shown that

$$0.95 = \Phi\left(+\frac{1}{s}\ln n\right). \tag{3.103}$$

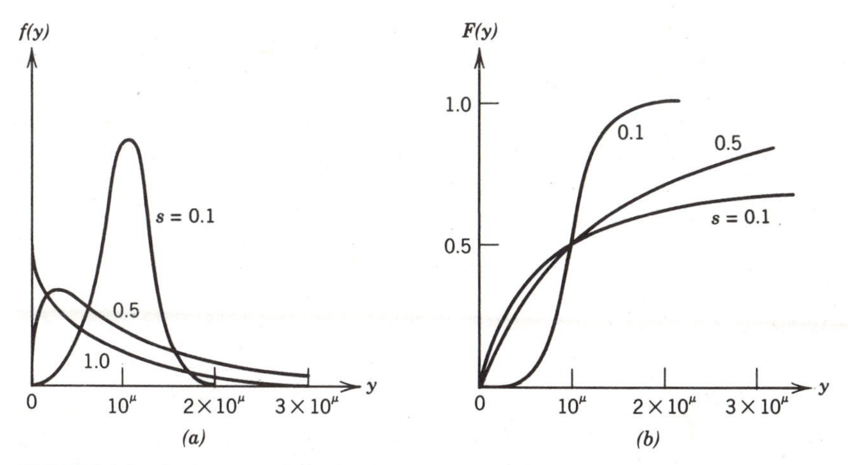

FIGURE 3.8 The lognormal distribution: (*a*) probability density function (PDF), (*b*) cumulative distribution function (CDF).

From the table in Appendix C it is seen that the argument for which Φ = 0.05 or 0.95 is ∓ 1.645. Thus we have

$$\frac{1}{s} \ln n = 1.645. \tag{3.104}$$

Therefore, the parameter s is given by

$$s = \frac{1}{1.645} \ln n. \tag{3.105}$$

With y_0 and s determined, the μ_y can be determined from Eq. 3.98.

EXAMPLE 3.7

Fatigue life data for an industrial rocker arm is fit to a lognormal distribution. The following parameters are obtained: $y_0 = 2 \times 10^7$ cycles, $s = 2.3$. (*a*) To what value should the design life be set if the probability of failure is not to exceed 1.0%? (*b*) If the design life is set to 1.0×10^6 cycles, what will the failure probability be?

Solution (*a*) Let y be the number of cycles for which the failure probability is 1%. Then, from Eq. 3.97, we have

$$0.01 = F_y(y) = \Phi\left[\frac{1}{2.3} \ln\left(\frac{y}{2 \times 10^7}\right)\right].$$

From Appendix C we find

$$\Phi(-2.32) \approx 0.01.$$

Thus

$$-2.32 = \frac{1}{2.3} \ln\left(\frac{y}{2 \times 10^7}\right)$$

and

$$y = 2 \times 10^7 \exp(-2.32 \times 2.3)$$
$$= 9.63 \times 10^4 \text{ cycles. } \blacktriangleleft$$

(*b*) In Eq. 3.97 we have

$$u \equiv \frac{1}{s} \ln\left(\frac{y}{y_0}\right) = \frac{1}{2.3} \ln\left(\frac{10^6}{2.0 \times 10^7}\right)$$
$$= -1.302.$$

From Appendix C, $\Phi(-1.302) \approx 0.096$ so that

$$F_y(y) = 0.096 \blacktriangleleft \text{ probability of failure.}$$

3.4 DATA AND DISTRIBUTIONS

In the preceding sections the laws of probability are examined as they apply to continuous random variables, and the characteristics of some of the more common distribution functions are illustrated. Generally, however, we have assumed that the distribution function is known. We now take up the questions of statistics: Given a set of data, as for example the stopping distances in Table 3.1, how do we infer the properties of the underlying distribution from which the data have been drawn? In approaching such problems, we must deal with three classes of uncertainty: intrinsic, model, and parameter.

Clearly, there is intrinsic uncertainty in the data of Table 3.1, for the stopping distances will vary between vehicles and between drivers. Even if the data were the result of repeated trials with the same vehicle and driver, some scatter in the data would still exist because of the variability in the driver's reflex from trial to trial. Thus the variability is intrinsic; it is not due simply to our inability to measure the stopping distances exactly. In contrast, if several measurements were to be made of a basic physical property, say Young's modulus for a pure material, the results would also show some variation from measurement to measurement. But these would be due not to intrinsic variability in the quantity but rather to measurement imprecision, the inability to completely control the laboratory environment, or both. Thus the variability in the Young's modulus measurement is not intrinsic.

The choice of a probability model to describe the data is the second source of uncertainty. We may be uncertain about our choice whether or

TABLE 3.1 Raw Data: 70 Measurements, in Feet of Stopping Distances

39	54	21	42	66	50	56
62	59	40	41	75	63	58
32	43	51	60	65	48	61
27	46	60	73	36	38	54
60	36	35	76	54	55	45
71	54	46	47	42	52	47
62	55	49	39	40	69	58
52	78	56	55	62	32	57
45	84	36	58	64	67	62
51	36	73	37	42	53	49

Source: Erich Pieruschka, *Principles of Reliability*, © 1963, p. 5, with permission from Prentice-Hall, Englewood Cliffs, NJ.

not the phenomena under study possess intrinsic variability. In the present example, we must decide what probability density function should be used to describe the stopping distance, which is a random variable: Is a normal, a lognormal, or some other distribution more appropriate? Conversely, even though Young's modulus of a pure material is not a random variable, the measured value is. Therefore, we again face uncertainty in choosing a probability distribution to use in describing the variability of the measured values.

Suppose that we are able to make a rational selection of a particular probability density function for describing a set of data. We must still determine the values of the parameters—the mean and variance, for example—to use in the distribution. It is here that the third type of uncertainty is encountered, for as we shall see, if there is only a finite amount of data, we cannot determine the parameters exactly. We can only make a best estimate and also estimate the uncertainty with a confidence interval, with the interval decreasing in width as the amount of available data increases.

As a rule, the data that we deal with in reliability engineering describe phenomena that are intrinsically variable. In what follows, we concentrate our attention first on the selection of a probability model (i.e., on the choice of a PDF) and second on the estimate of the model parameters. The aforementioned stopping distance data serve as an illustration.

Model Selection

In analyzing random data, we must select a probability distribution. In this a number of factors may guide our choices. Frequently, a great deal of experience has already been gained in fitting distributions to data from very similar tests. If this is the case, the distribution that has proved effective in the past may be chosen again. If, for example, stopping distance data from

TABLE 3.2 Frequency Table

Class interval, ft	*Tally*	*Frequency*
20–29	//	2
30–39	///// ///// /	11
40–49	///// ///// ///// /	16
50–59	///// ///// ///// /////	20
60–69	///// ///// ////	14
70–79	///// /	6
80–89	/	1

Source: Erich Pieruschka, *Principles of Reliability*, © 1963, p. 5, with permission from Prentice-Hall, Englewood Cliffs, NJ.

other types or models of vehicles had been successfully described by lognormal distributions, a reasonable assumption might be also to apply the lognormal distribution to the data in Table 3.1. As an alternative, we may be able to choose between distributions on the basis of the nature of the phenomena. If the phenomenon may be viewed as the sum of many small effects, for instance, the normal distribution may be suitable. If it is the product of many factors, the lognormal distribution may be more appropriate. Similar arguments can be made for the exponential, extreme-value, and other distribution functions that will be introduced later in the book.

If neither of the foregoing techniques is adequate, histograms may serve as a vehicle for choosing a distribution. The histogram may be constructed as follows. We first find the range of the data (i.e., the maximum minus the minimum value). Knowing the range, we choose an interval width such that the data can be divided into some number N of groups. If the interval for the data in Table 3.1 is chosen to be 10 ft, for example, a table can be made

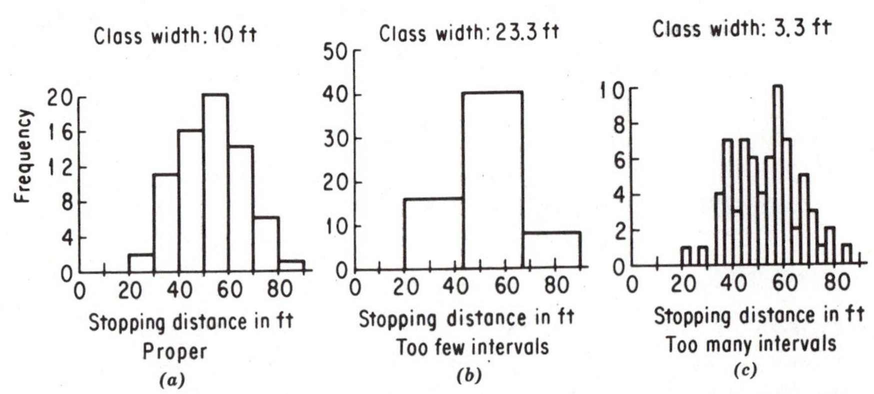

FIGURE 3.9 Effect of the choice of the number of class intervals. (From Eric Pieruschka, *Principles of Reliability*, © 1963, p. 6, with permission from Prentice-Hall, Englewood Cliffs, NJ.)

up according to how many data points fall in each interval. This is carried out in Table 3.2, with the data falling into seven intervals. A histogram, referred to as a frequency diagram, may then be drawn as indicated in Fig. 3.9*a*.

In order to glean as much information as possible about the nature of the distribution, the number of groups into which the data are divided must be reasonable. If too few intervals are used, as indicated in Fig. 3.9*b*, the nature of the distribution is obscured by the lack of resolution. If the number is too large, as in Fig. 3.9*c*, the large fluctuations in frequency hide the nature of the distribution. In general, the more data points that are available, the larger is the number of intervals that can be used effectively, and the better the representation of the distribution. Although there is no precise rule for determining the optimum number of the intervals, the following rule of thumb may be used.* If N is the number of data points and r is the range of the data, a reasonable interval width Δ is

$$\Delta \approx r[1 + 3.3 \log_{10}(N)]^{-1}. \tag{3.106}$$

One method for estimating which distribution should be used to describe the data consists of plotting the analytical form of the distribution over the histogram. Before this can be accomplished, however, the frequency diagram must be normalized to approximate $f(x)$. This is done by requiring that the histogram satisfy the normalization condition Eq. 3.7. Suppose that $n_1, n_2, \ldots$ are the frequencies with which the data appear in the various intervals, and $\Sigma_i\, n_i = N$.

If we want to approximate $f(x)$ by f_i in the ith interval, f_i must be proportional to n_i:

$$f_i = an_i, \tag{3.107}$$

where a is the proportionality constant to be determined. For the histogram to satisfy Eq. 3.7 we have

$$\sum_i f_i\, \Delta = 1. \tag{3.108}$$

Combining the two preceding equations, we have

$$1 = \sum_i an_i\, \Delta = a\, \Delta \sum_i n_i = a\, \Delta\, N. \tag{3.109}$$

Hence $a = 1/(N\, \Delta)$, and

$$f_i = \frac{1}{\Delta}\frac{n_i}{N}. \tag{3.110}$$

In Fig. 3.10 the histogram that approximates $f(x)$ is plotted. To deter-

* H. A. Sturges, "The Choice of a Class Interval," *J. Am. Stat. Assoc.*, **21**, 65–66 (1926); see also E. Pieruschka, *Principles of Reliability*, Prentice-Hall, Englewood Cliffs, NJ, 1963.

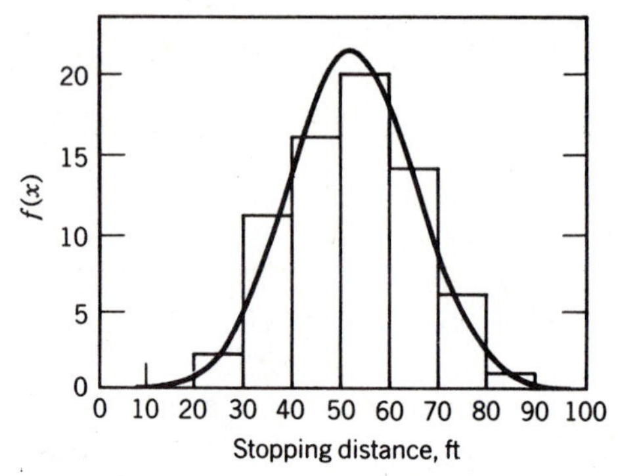

FIGURE 3.10 Normal distribution and histogram for the data in Table 3.1.

mine whether the normal distribution reasonably describes the data, we have plotted the PDF for a normal distribution; the estimates of the mean and variance are obtained using the methods described in the following subsection.

When the number of data points is relatively small, say fewer than a dozen or so, constructing a histogram gives us little idea of what distribution to use. Unfortunately, such situations occur frequently in reliability engineering, particularly when an expensive piece of equipment must be tested to failure for each data point. More powerful graphical techniques utilizing specialized graph paper may be employed to assist in determining which distribution most closely describes the data. We defer treatment of such techniques to Chapter 5, where the problems of analyzing data from reliability testing are discussed in some detail.

Parameter Estimation

After we have decided which distribution to use in describing data, we must estimate the parameters appearing in this distribution. There are two types of parameter estimates.* In point estimates we attempt to find the "best" estimate, in some sense, of the parameter to be made from the available data. In interval estimates we attempt to put an upper and lower bound on the uncertainty of the parameter value, consistent with the data at hand. Clearly, the more data that are available, the closer the parameter estimate is likely to be to the true value, and the tighter the bounds on uncertainty will be. In what follows we concentrate on determining the mean and the variance of the distribution, for this is all that is necessary to specify the

* For further discussion, see L. L. Lapin, *Reliability and Statistics for Modern Engineering*, Brooks/Cole, Monterey, CA, 1983.

two-parameter distributions that appear most frequently in reliability estimates.

Point Estimates

Consider first the point estimate of the mean, μ. Suppose that a sample of N values, $x_1, x_2, \cdots, x_N$, has been obtained from a distribution $f(x)$. We estimate μ, the mean of this distribution by the *sample mean* defined as

$$\hat{\mu} = \frac{1}{N}\sum_{n=1}^{N} x_n. \tag{3.111}$$

As in Chapter 2 we use the caret to signify estimators. There are at least two reasons for choosing Eq. 3.111 over the median value of the x_n or some other estimator. The first is that it has the correct expected value:

$$E\{\hat{\mu}\} = \mu. \tag{3.112}$$

Specifically, this indicates that if many samples of size N are taken, they will average to the true value μ. An estimator with this property is said to be unbiased. In addition to being unbiased, the sample mean has a second desirable property. It is efficient in the sense that the likelihood of $\hat{\mu}$ being close to μ is great, even though the sample size N is not excessively large.

To examine the efficiency of the estimator $\hat{\mu}$ more closely, we need to return to the idea of the sampling distribution introduced in Chapter 2. Suppose that we take many samples, each of size N. Each of the resulting values of $\hat{\mu}$ is also randomly distributed, and therefore $\hat{\mu}$ is a random variable. In principle, there exists a PDF $f_{\hat{\mu}}(\hat{\mu})$ for $\hat{\mu}$. This is called the sampling distribution for the estimator $\hat{\mu}$. If $\hat{\mu}$ is an unbiased estimator of μ, the mean value of $f_{\hat{\mu}}(\hat{\mu})$ must be equal to μ. Thus

$$\mu = \int_{-\infty}^{\infty} \hat{\mu} f_{\hat{\mu}}(\hat{\mu})\, d\hat{\mu}. \tag{3.113}$$

Two very powerful theorems enable us to say a good deal more about how good an estimate Eq. 3.111 provides. We simply state them here, since their formal proofs would require the introduction of extensive additional background material.* The first theorem states that regardless of the type of distribution that $f(x)$ is, if σ^2, the variance of $f(x)$, exists, the variance

$$\sigma_{\hat{\mu}}^2 \equiv \int_{-\infty}^{\infty} (\hat{\mu} - \mu)^2 f_{\hat{\mu}}(\hat{\mu})\, d\hat{\mu} \tag{3.114}$$

is given by

$$\sigma_{\hat{\mu}}^2 = \sigma^2/N. \tag{3.115}$$

* L. L. Lapin, op. cit.

Thus we are assured that as we increase sample size N, the probability that our result will have a large error becomes smaller with the standard error $\sigma_{\hat{\mu}}$ dropping off as a $1/\sqrt{N}$.

A second powerful theorem may be employed to estimate the uncertainty of μ. The central limit theorem states that regardless of the form of $f(x)$, for sufficiently large N the sampling distribution becomes a normal distribution. Thus, with $\sigma_{\hat{\mu}}$ of this distribution given by Eq. 3.115, we may write

$$f_{\hat{\mu}}(\hat{\mu}) = \frac{\sqrt{N}}{\sqrt{2\pi}\,\sigma} \exp\left[-\frac{N}{2\sigma^2}(\hat{\mu} - \mu)^2\right]. \tag{3.116}$$

As a rule of thumb, N is usually large enough for this result to apply if $N \geqslant 30$.

This result applies regardless of the form of the distribution $f(x)$. Suppose, for example, that we consider the frequently used exponential distribution

$$f(x) = \frac{1}{\gamma} e^{-x/\gamma}, \tag{3.117}$$

for which it may be shown that $\mu = \gamma$, and $\sigma^2 = \gamma^2$. The sampling distribution for $N > 30$ will then be approximated by Eq. 3.116:

$$f_{\hat{\mu}}(\hat{\mu}) = \frac{\sqrt{N}}{\sqrt{2\pi}\,\gamma} \exp\left[-\frac{N}{2\gamma^2}(\hat{\mu} - \mu)^2\right]. \tag{3.118}$$

This is true even though $f(x)$ is far from a normal distribution.

Equation 3.116 is applicable to discrete as well as continuous random variables. For example, in Chapter 2, it is shown that the sampling distribution for Bernoulli trials is given by Eq. 2.92, where $\hat{p}$ is the sample mean for N samples, each with a value of $x_n = 0$ or $x_n = 1$. If the sample size N is large, Eq. 2.92 may be shown to reduce to

$$f_{\hat{p}}(\hat{p}) \simeq \frac{\sqrt{N}}{\sqrt{2\pi\, p(1-p)}} \exp\left[-\frac{N}{2p(1-p)}(\hat{p} - p)^2\right], \hat{p} = \frac{1}{N}, \frac{2}{N}, \cdots \tag{3.119}$$

where the mean and variance of $f_{\hat{p}}(\hat{p})$ are

$$\mu_{\hat{p}} = p \tag{3.120}$$

and

$$\sigma_{\hat{p}}^2 = \frac{1}{N} p(1 - p), \tag{3.121}$$

as derived in Chapter 2. This normal behavior is apparent in Fig. 2.5*d*, for sample size $N = 40$.

Estimates are often required for parameters other than the mean. Probably the most common of these is the variance, since for any two parameter distributions, such as the normal or lognormal, only the mean and variance need to be determined.

The standard deviation may be estimated from

$$\hat{\sigma}^2 = \frac{1}{N} \sum_{n=1}^{N} (x_n - \mu)^2. \tag{3.122}$$

Provided that the mean μ is known, this is an unbiased estimator; that is,

$$E\{\hat{\sigma}^2\} = \sigma^2. \tag{3.123}$$

Invariably, if we do not know the value of σ^2, neither do we know the true value of μ. Therefore, it must be estimated by $\hat{\mu}$. If μ is replaced by $\hat{\mu}$ in Eq. 3.122, however, the result is no longer an unbiased estimate; Eq. 3.123 does not hold. Although the proof is rather lengthy, it may be shown that if we use as the definition of the *sample variance*

$$\hat{\sigma}^2 = \frac{1}{N - 1} \sum_{n=1}^{N} (x_n - \hat{\mu})^2, \tag{3.124}$$

then Eq. 3.123 holds, and the estimator therefore is unbiased. Of course, when N is reasonably large, this change makes little difference in the numerical value of $\hat{\sigma}^2$.

EXAMPLE 3.8

Find the sample mean and variance for the stopping data given in Table 3.1.

Solution First calculate the sums

$$\sum_{n=1}^{70} x_n = 3611, \qquad \sum_{n=1}^{70} x_n^2 = 203{,}095.$$

Therefore from Eq 3.111, we have

$$\hat{\mu} = \tfrac{3611}{70} = 51.6. \blacktriangleleft$$

From Eq. 3.124,

$$\begin{aligned}
\hat{\sigma}^2 &= \frac{1}{N-1}\left(\sum_{n=1}^{N} x_n^2 - 2\hat{\mu}\sum_{n=1}^{N} x_n + N\hat{\mu}^2\right) \\
&= \frac{1}{N-1}\left(\sum_{n=1}^{N} x_n^2 - N\hat{\mu}^2\right) \\
&= \frac{N}{N-1}\left(\frac{1}{N}\sum_{n=1}^{N} x_n^2 - \hat{\mu}^2\right)
\end{aligned}$$

$$\hat{\sigma}^2 = \frac{70}{69}\left[\frac{1}{70}\,203{,}095 - (51.6)^2\right] = 242.3. \blacktriangleleft$$

Interval Estimates

The sample mean and sample variance, in Eqs. 3.111 and 3.124, are point estimates of μ and σ^2, respectively. Often it is also necessary to place some upper and lower bounds on the values that the parameter is likely to take. The process of doing this leads to the notion of confidence limits that are discussed in Chapter 2 in conjunction with binomial distributions. Here we discuss confidence limits for the mean, and then more generally for other parameters that we may wish to estimate.

Suppose that we consider only the instance in which sample size N is sufficiently large (e.g., $N \geqslant 30$) that the sampling distribution of the mean is given by the normal distribution, Eq. 3.116. From the definition of a PDF and CDF we then have

$$P\{\hat{\mu} \leqslant \mu_+\} = \int_{-\infty}^{\mu_+} \frac{1}{\sqrt{2\pi}\,\sigma_{\hat{\mu}}} \exp\left[\frac{-1}{2\sigma_{\hat{\mu}}^2}(\hat{\mu} - \mu)^2\right] d\hat{\mu}. \qquad (3.125)$$

The change of variables to $\zeta = (\hat{\mu} - \mu)/\sigma_{\hat{\mu}}$ then yields

$$P\{\hat{\mu} \leqslant \mu_+\} = \frac{1}{\sqrt{2\pi}} \int_{-\infty}^{(\mu_+ - \mu)/\sigma_{\hat{\mu}}} \exp(-\tfrac{1}{2}\zeta^2)\, d\zeta = \Phi\left(\frac{\mu_+ - \mu}{\sigma_{\hat{\mu}}}\right). \qquad (3.126)$$

Letting $u = (\mu_+ - \mu)/\sigma_{\hat{\mu}}$, we may rewrite this expression as

$$P\{\hat{\mu} \leqslant \sigma_{\hat{\mu}} u + \mu\} = \frac{1}{\sqrt{2\pi}} \int_{-\infty}^{u} \exp(-\tfrac{1}{2}u')^2\, du' \equiv \Phi(u). \qquad (3.127)$$

Rearranging the inequality, we have equivalently,

$$P\{\mu \geqslant \hat{\mu} - \sigma_{\hat{\mu}} u\} = \Phi(u). \qquad (3.128)$$

By similar argument, we may show that

$$P\{\mu \leqslant \hat{\mu} + \sigma_{\hat{\mu}} u\} = \Phi(u). \qquad (3.129)$$

Suppose that we choose u such that

$$\Phi(u) = 1 - \alpha/2. \qquad (3.130)$$

We then say that we are $1 - \alpha/2$ confident that $\mu \leqslant \hat{\mu} + \sigma_{\hat{\mu}} u$. We are also $1 - \alpha/2$ confident that $\mu > \hat{\mu} - \sigma_{\hat{\mu}} u$. Or simply that $\mu \pm \sigma_{\hat{\mu}} u$ are the upper and lower $1 - \alpha/2$ confidence limits on $\hat{\mu}$. Finally, we may say that $\hat{\mu} \pm \sigma_{\hat{\mu}} u$ has the $1 - \alpha$ confidence interval for $\hat{\mu}$.

$$P\{\hat{\mu} - \sigma_{\hat{\mu}} u \leqslant \mu \leqslant \hat{\mu} + \sigma_{\hat{\mu}} u\} = 1 - \alpha. \qquad (3.131)$$

To summarize, if we choose α for a large sample size, we can determine u from the standardized normal distribution, where $\hat{\mu}$ is known. One problem remains: We have assumed that $\sigma_{\hat{\mu}}$ is known, when in fact it is not. For large N (>30) we may replace $\sigma_{\hat{\mu}}$ by its unbiased estimator,

$$N\sigma_{\hat{\mu}}^2 \approx \hat{\sigma}^2 = \frac{1}{N-1}\sum_{n=1}^{N}(x_n - \hat{\mu})^2 \tag{3.132}$$

to obtain

$$P\left\{\hat{\mu} - \frac{u\hat{\sigma}}{\sqrt{N}} \leq \mu \leq \hat{\mu} + \frac{u\hat{\sigma}}{\sqrt{N}}\right\} = 1 - \alpha. \tag{3.133}$$

Thus far we have applied the idea of confidence only to the mean, and only when a fairly large amount of data is available. Indeed, it is often necessary to determine confidence intervals on the mean when relatively few data are available, and for the standard deviations or for other parameters as well. Binomial sampling was discussed in Chapter 2. For other distributions more sophisticated techniques are required.* In general, the $1 - \alpha$ confidence interval on any parameter θ is presented in the form

$$P\{L_{\alpha/2,N} < \theta < U_{\alpha/2,N}\} = 1 - \alpha, \tag{3.134}$$

where $L_{\alpha/2,N}$ and $U_{\alpha/2,N}$ are the lower and upper confidence limits on θ for sample size N.

We must be specific about the preceding probability statements, for

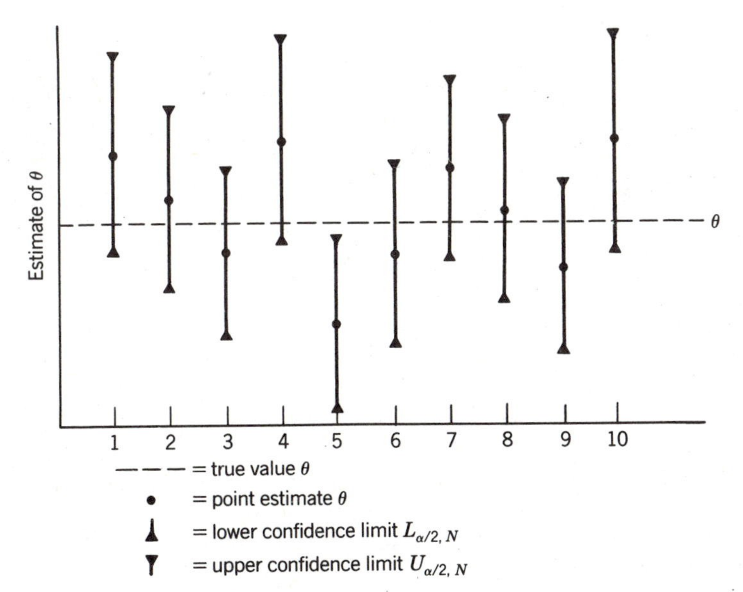

FIGURE 3.11 Confidence limits for repeated estimates of a parameter.

* See, for example, K. C. Kapur and L. R. Lamberson, *Reliability in Engineering Design*, Wiley, New York, 1977.

they define the meaning of confidence intervals. Equation 3.134, in particular, may be understood with the aid of Fig. 3.11 as follows. Suppose that a large number of samples each of size N are taken, and that $U_{\alpha/2,N}$ and $L_{\alpha/2,N}$ are calculated. In Fig. 3.11 we have plotted these quantities for 10 such samples along with a point estimator $\hat{\theta}$. If $U_{\alpha/2,N}$ and $L_{\alpha/2,N}$ define the 90% confidence interval (i.e., $\alpha = 0.1$), for 90% of the samples of size N the true value of θ will lie within the intervals indicated by vertical lines.

Bibliography

Ang, A. H. S., and W. H. Tang, *Probability Concepts in Engineering Planning and Design*, Vol. 1, Wiley, New York, 1975.

Benjamin, J. R., and C. A. Cornell, *Probability, Statistics, and Decisions for Civil Engineers*, McGraw-Hill, New York, 1970.

Freund, J. E., and R. E. Walpole, *Mathematical Statistics*, 3rd ed., Prentice-Hall, Englewood Cliffs, NJ, 1980.

Lapin, L. L., *Probability and Statistics for Modern Engineering*, Brooks/Cole, Belmont, CA, 1983.

Olkin, I., Z. J. Gleser, and G. Derman, *Probability Models and Applications*, Macmillan, New York, 1980.

Pieruschla, E., *Principles of Reliability*, Prentice-Hall, Englewood Cliffs, NJ, 1963.

Sandler, G. H., *System Reliability Engineering*, Prentice-Hall, Englewood Cliffs, NJ, 1963.

EXERCISES

3.1 For the PDF

$$f(x) = \begin{cases} bx(1 - x), & 0 \leq x \leq 1, \\ 0, & \text{otherwise}, \end{cases}$$

determine b, μ, and σ.

3.2 A motor is known to have an operating life (in hours) that fits the distribution

$$f(t) = \frac{a}{(t + b)^3}, \qquad t \geq 0.$$

The mean life of the motor has been estimated to be 3000 hr.

(*a*) Find a and b.

(*b*) What is the probability that the motor will fail in less than 2000 hr?

(*c*) If the manufacturer wants no more than 5% of the motors returned for warranty service, how long should the warranty be?

3.3 Suppose that

$$F(x) = 1 - e^{-0.2x} - 0.2xe^{-0.2x}, \qquad 0 \leq x \leq \infty.$$

(*a*) Find $f(x)$.
(*b*) Determine μ and σ^2.
(*c*) Find the expected value of e^{-x}.

3.4 Suppose that the maximum flaw size in steel bars is given by

$$f(x) = 4xe^{-2x}, \qquad 0 \leq x \leq \infty,$$

where x is in microns.
(*a*) What is the mean value of the maximum flaw size?
(*b*) If flaws of lengths greater than 1.5 microns are detected and the bars rejected, what fraction of the bars will be accepted?
(*c*) What is the mean value of the maximum flaw size for the bars that are accepted?

3.5 The following PDF has been proposed for the distribution of pit depths in a tailpipe of thickness x_0:

$$f(x) = A \sinh[\alpha(x_0 - x)], \qquad 0 \leq x \leq x_0.$$

(*a*) Determine A in terms of α.
(*b*) Determine $F(x)$, the CDF.
(*c*) Determine the mean pit depth. What is the probability that there will be a pit of more than twice the mean depth?

3.6 The PDF for the maximum depths of undetected cracks in steel piping is

$$f(x) = \frac{1}{\gamma} \frac{e^{-x/\gamma}}{(1 - e^{-\tau/\gamma})},$$

where τ is the pipe thickness and $\gamma = 6.25$ mm.
(*a*) What is the CDF?
(*b*) For a 20-mm-thick pipe, what is the probability that a crack will penetrate more than half of the pipe thickness?

3.7 Under design pressure the minimum unflawed thickness of a pipe required to prevent failure is τ_0.
(*a*) Using the maximum crack depth PDF from Exercise 3.6, show that if the probability of failure is to be less than ϵ, the total pipe thickness must be at least

$$\tau = \gamma \ln\left[1 + \frac{1}{\epsilon}(e^{\tau_0/\gamma} - 1)\right].$$

(*b*) For $\gamma = 6.25$ mm and a minimum unflawed thickness of $\tau_0 = 4$ cm, what must the total thickness be if the probability of failure is 0.1%?

(*c*) Repeat part *b* for a probability of failure of 0.01%.
(*d*) Show that for $\tau_0 \gg \gamma$ and $\epsilon \ll 1$, τ is approximately $\tau_0 + \gamma \ln(1/\epsilon)$.

3.8 Express the skewness in terms of the moments $E\{x^n\}$.

3.9 The beta distribution is defined by

$$f(x) = \frac{1}{B} x^{r-1}(1 - x)^{t-r-1}, \qquad 0 \leq x \leq 1.$$

Show
(*a*) that if *t* and *r* are integers,

$$B = \frac{(r - 1)!(t - r - 1)!}{(t - 1)!}.$$

(*b*) That $\mu = r/t$.
(*c*) That

$$\sigma^2 = \frac{\mu(1 - \mu)}{t + 1} = \frac{r(t - r)}{t^2(t + 1)}.$$

(*d*) That if *t* and *r* are integers, $f(x)$ may be written in terms of the binomial distribution:

$$f(x) = (t - 1)C_{r-1}^{t-2} x^{r-1}(1 - x)^{t-r-1}.$$

3.10 Transform the beta distribution given in the Exercise 3.9 by $y = a + (b - a)x$ so that $a \leq y \leq b$.
(*a*) Find $f_y(y)$. (*b*) Find μ_y.

3.11 A PDF of impact velocities is given by $\alpha\, e^{-\alpha v}$. Find the PDF for impact kinetic energies E, where $E = \frac{1}{2} mv^2$.

3.12 An elastic bar is subjected to a force l. The resulting strain energy is given by

$$\epsilon = cl^2,$$

where c is $d/2AE$, with d the length of the bar, A the area, and E the modulus of elasticity. Suppose that the PDF of the force can be represented by standardized normal from $f(l)$, given by Eq. 3.70. Find the PDF $f_\epsilon(\epsilon)$ for the strain energy.

3.13 A joint probability distribution function is given by

$$F(x, y) = (1 - e^{-x^2})(1 - e^{-y}), \qquad \begin{cases} 0 \leq x \leq \infty, \\ 0 \leq y \leq \infty. \end{cases}$$

(*a*) Determine the joint PDF.
(*b*) Determine $P\{(\mathbf{x} < 1) \cap (\mathbf{y} > 2)\}$.
(*c*) Are **x** and **y** independent?

3.14 Given the joint PDF

$$f(x, y) = \tfrac{2}{3}(2x + y), \qquad \begin{cases} 0 \leq x \leq 1, \\ 0 \leq y \leq 1. \end{cases}$$

(*a*) Calculate the marginal PDF in $\mathbf{x}$ and $\mathbf{y}$.
(*b*) Determine $f_{\mathbf{x}}(x|y)$ and $f_{\mathbf{y}}(y|x)$.
(*c*) Are the variables independent?

3.15 Calculate the correlation coefficient for Exercise 3.14.

3.16 Verify the reduction of Eq. 3.52 to Eq. 3.53.

3.17 Consider the joint PDF

$$f(x, y) = \tfrac{3}{8}(y - x)^2, \qquad \begin{cases} -1 \leq x \leq 1 \\ -1 \leq y \leq 1 \end{cases}$$

(*a*) Determine $\mu_{\mathbf{x}}$ and $\mu_{\mathbf{y}}$.
(*b*) Determine $\sigma_{\mathbf{x}}^2$ and $\sigma_{\mathbf{y}}^2$.
(*c*) Determine $\sigma_{\mathbf{xy}}^2$ and $\rho_{\mathbf{xy}}$.

3.18 Verify Eq. 3.77.

3.19 Suppose that a batch of ball bearings is produced for which the diameters are distributed normally. The acceptance testing procedures remove all those for which the diameter is more than 1.5 standard deviations from the mean value. Therefore, the truncated distribution of the diameters of the delivered ball bearings is

$$f(x) = \begin{cases} \dfrac{A}{\sqrt{2\pi}\, c} \exp\left[-\dfrac{1}{2c^2}(x - \mu)^2 \right], & |x - \mu| < 1.5c, \\ 0, & |x - \mu| > 1.5c. \end{cases}$$

(*a*) What fraction of the ball bearings is accepted?
(*b*) What is the value of A?
(*c*) What fraction of the accepted ball bearings will have diameters between $\mu - c$ and $\mu + c$?
(*d*) What is the variance of $f(x)$, the PDF of delivered ball bearings?

3.20 The total load on a building may often be represented as the sum of three contributions: the dead load $\mathbf{d}$, from the weight of the structure; the live load $\mathbf{l}$, from human beings, furniture, and other movable weights; and the wind load $\mathbf{w}$. Suppose that the loads from each of the sources on a support column are represented as normal distributions with the following properties:

$$\mu_d = 6.0 \text{ kips} \qquad \sigma_d = 0.4 \text{ kips},$$

$$\mu_l = 9.2 \text{ kips} \qquad \sigma_l = 1.2 \text{ kips},$$

$$\mu_w = 4.6 \text{ kips} \qquad \sigma_w = 1.1 \text{ kips}.$$

Determine the mean and standard deviation of the total load.

3.21 Verify that μ and σ^2 appearing in Eq. 3.70 are indeed the mean and variance of $f(x)$; that is, verify Eqs. 3.68 and 3.69.

3.22 If the strength of a structural member is known with 90% confidence to a factor of 3, to what factor is it known with (*a*) 99% confidence, (*b*) with 50% confidence? Assume a lognormal distribution.

3.23 Verify Eqs. 3.97 through 3.99.

3.24 Consider the following response time data measured in seconds.*

1.48	1.46	1.49	1.42	1.35
1.34	1.42	1.70	1.56	1.58
1.59	1.59	1.61	1.25	1.31
1.66	1.58	1.43	1.80	1.32
1.55	1.60	1.29	1.51	1.48
1.61	1.67	1.36	1.50	1.47
1.52	1.37	1.66	1.44	1.29
1.80	1.55	1.46	1.62	1.48
1.64	1.55	1.65	1.54	1.53
1.46	1.57	1.65	1.59	1.47
1.38	1.66	1.59	1.46	1.61
1.56	1.38	1.57	1.48	1.39
1.62	1.49	1.26	1.53	1.43
1.30	1.58	1.43	1.33	1.39
1.56	1.48	1.53	1.59	1.40
1.27	1.30	1.72	1.48	1.66
1.37	1.68	1.77	1.62	1.33

(*a*) Compute the mean and the variance.
(*b*) Use the Sturges formula to make a histogram approximating $f(x)$.

3.25 Fifty measurements of the ultimate tensile strength of wire are given in the accompanying table.
(*a*) Group the data and make an appropriate histogram to approximate the PDF.
(*b*) Calculate $\hat{\mu}$ and $\hat{\sigma}^2$ for the distribution from the ungrouped data.

* Data from A. E. Green and A. J. Bourne, *Reliability Technology*, Wiley, New York, 1972.

(*c*) Using $\hat{\mu}$ and $\hat{\sigma}$ from part *b*, draw a normal distribution through the histogram.

Ultimate Tensile Strength

103,779	102,325	102,325	103,799
102,906	104,651	105,377	100,145
104,796	105,087	104,796	103,799
103,197	106,395	106,831	103,488
100,872	100,872	105,087	102,906
97,383	104,360	103,633	101,017
101,162	101,453	107,848	104,651
98,110	103,779	99,563	103,197
104,651	101,162	105,813	105,337
102,906	102,470	108,430	101,744
103,633	105,232	106,540	106,104
102,616	106,831	101,744	100,726
103,924		101,598	

Source: Data from E. B. Haugen, *Probabilistic Mechanical Design*, Wiley, New York, 1980.

CHAPTER 4

Reliability and Rates of Failure

4.1 INTRODUCTION

Generally, reliability is defined as the probability that a system will perform properly for a specified period of time under a given set of operating conditions. Implied in this definition is a clear-cut criterion for failure, from which we may judge at what point the system is no longer functioning properly. Similarly, the treatment of operating conditions requires an understanding both of the loading to which the system is subjected and of the environment within which it must operate. Perhaps the most important variable to which we must relate reliability, however, is time. For it is in terms of the rates of failure that most reliability phenomena are understood.

In this chapter we examine reliability as a function of time, and this leads to the definition of the failure rate. Examining the time dependence of failure rates allows us to gain additional insight into the nature of failures—whether they be early failures, failures random in time, or failures brought on by aging. Similarly, the time dependence of failures can be viewed in terms of failure modes to differentiate between failures caused by different mechanisms, and those caused by different components of a system. Then, once the time dependence of the failures has been designated for a system, expressions can be derived for the probable number of failures taking place for systems that are repaired or replaced after each failure. Finally, with the understanding gained, we examine how reliability considerations relate to the problems of engineering design and provide a prologue for the chapters that follow.

4.2 RELIABILITY CHARACTERIZATIONS

We begin by defining reliability and failure rates in terms of the PDF for the time to failure of a system. We then examine the properties of failure rates and other reliability parameters from the standpoint of the mechanism causing failure.

Basic Definitions

For a given set of operating conditions, the reliability as defined in Chapter 1 may be paraphrased as the probability that a system survives for some specified period of time. This may be expressed in terms of the random variable $\mathbf{t}$, the time to system failure. The PDF, $f(t)$, then has the physical meaning

$$f(t)\,\Delta t = P\{t \le \mathbf{t} \le t + \Delta t\} = \begin{array}{l}\text{probability that failure}\\ \text{takes place at a time between}\\ t \text{ and } t + \Delta t.\end{array} \tag{4.1}$$

From Eq. 3.1 we see that the CDF now has the meaning

$$F(t) = P\{\mathbf{t} \le t\} = \begin{array}{l}\text{probability that failure}\\ \text{takes place at a time less}\\ \text{than or equal to } t.\end{array} \tag{4.2}$$

We define the *reliability* as

$$R(t) = P\{\mathbf{t} > t\} = \begin{array}{l}\text{probability that a system}\\ \text{operates without failure for}\\ \text{a length of time } t.\end{array} \tag{4.3}$$

Since a system that does not fail for $\mathbf{t} \le t$ must fail at some $\mathbf{t} > t$, we have

$$R(t) = 1 - F(t), \tag{4.4}$$

or equivalently either

$$R(t) = 1 - \int_0^t f(t')\,dt' \tag{4.5}$$

or

$$R(t) = \int_t^\infty f(t')\,dt'. \tag{4.6}$$

From the properties of the PDF, it is clear that

$$R(0) = 1 \tag{4.7}$$

and

$$R(\infty) = 0. \tag{4.8}$$

We see that the reliability is the CCDF of $\mathbf{t}$; that is, $R(t) = \tilde{F}(t)$. Similarly, since $F(t)$ is the probability that the system will fail before $\mathbf{t} = t$, it is often referred to as the unreliability; at times we may denote the unreliability or failure probability as

$$\tilde{R}(t) \equiv 1 - R(t) = F(t). \tag{4.9}$$

Equation 4.5 may be inverted by differentiation to give the PDF of failure times in terms of the reliability:

$$f(t) = -\frac{d}{dt} R(t). \tag{4.10}$$

Insight is normally gained into failure mechanisms by examining the behavior of the system failure rate. This *failure rate,* $\lambda(t)$, may be defined in terms of the reliability or the PDF of the time to failure as follows. Let $\lambda(t)$ Δt be the probability that the system will fail at some time $\mathbf{t} < t + \Delta t$, *given* that it has not yet failed at $\mathbf{t} = t$. Thus it is the conditional probability

$$\lambda(t)\, \Delta t = P\{\mathbf{t} < t + \Delta t | \mathbf{t} > t\}. \tag{4.11}$$

Using Eq. 2.5, the definition of a conditional probability, we have

$$P\{t < \mathbf{t} + \Delta t | \mathbf{t} > t\} = \frac{P\{(\mathbf{t} > t) \cap (\mathbf{t} < t + \Delta t)\}}{P\{\mathbf{t} > t\}}. \tag{4.12}$$

The numerator on the right-hand side is just an alternative way of writing the PDF; that is,

$$P\{(\mathbf{t} > t) \cap (\mathbf{t} < t + \Delta t)\} \equiv P\{t < \mathbf{t} < t + \Delta t\} = f(t)\, \Delta t. \tag{4.13}$$

The denominator of Eq. 4.12 is just $R(t)$, as may be seen by examining Eq. 4.3. Therefore, combining equations, we obtain

$$\lambda(t) = \frac{f(t)}{R(t)}. \tag{4.14}$$

This quantity, the *failure rate,* is also referred to as the hazard function or instantaneous failure rate.

The most useful way to express the reliability and the failure PDF is in terms of the failure rate. To do this, we first eliminate $f(t)$ from Eq. 4.14 by inserting Eq. 4.10 to obtain the failure rate in terms of the reliability,

$$\lambda(t) = -\frac{1}{R(t)} \frac{d}{dt} R(t). \tag{4.15}$$

Then multiplying by dt, we obtain

$$\lambda(t)\, dt = -\frac{dR(t)}{R(t)}. \tag{4.16}$$

Integrating between zero and t yields

$$\int_0^t \lambda(t')\, dt' = -\ln[R(t)], \tag{4.17}$$

since $R(0) = 1$. Finally, exponentiating results in the desired expression for the reliability

$$R(t) = \exp\left[-\int_0^t \lambda(t')\, dt'\right]. \tag{4.18}$$

To obtain the probability density function for failures, we simply insert Eq. 4.18 into Eq. 4.14 and solve for $f(t)$:

$$f(t) = \lambda(t) \exp\left[-\int_0^t \lambda(t')\, dt'\right]. \tag{4.19}$$

Probably the single most-used parameter to characterize reliability is the *mean time to failure* (or MTTF). It is just the expected or mean value $E\{t\}$ of the failure time t. Hence

$$\text{MTTF} = \int_0^\infty t f(t)\, dt. \tag{4.20}$$

The MTTF may be written directly in terms of the reliability by substituting Eq. 4.10 into Eq. 4.20 and integrating by parts:

$$\text{MTTF} = -\int_0^\infty t \frac{dR}{dt}\, dt = -tR(t)\Big|_0^\infty + \int_0^\infty R(t)\, dt. \tag{4.21}$$

Clearly, the $tR(t)$ term vanishes at $t = 0$. Similarly, from Eq. 4.18, we see that $R(t)$ will decay exponentially or faster, since the failure rate $\lambda(t)$ must be greater than zero. Thus $tR(t) \rightarrow 0$ as $t \rightarrow \infty$. Therefore, we have

$$\text{MTTF} = \int_0^\infty R(t)\, dt. \tag{4.22}$$

EXAMPLE 4.1

An engineer approximates the reliability of a cutting assembly by

$$R(t) = \begin{cases} (1 - t/t_0)^2, & 0 \le t < t_0, \\ 0 & t \ge t_0. \end{cases}$$

(*a*) Determine the failure rate. (*b*) Does the failure rate increase or decrease with time? (*c*) Determine the MTTF.

Solution (*a*) From Eq. 4.10,

$$f(t) = -\frac{d}{dt}(1 - t/t_0)^2 = \frac{2}{t_0}(1 - t/t_0), \qquad 0 \le t < t_0.$$

and from Eq. 4.14,

$$\lambda(t) = \frac{f(t)}{R(t)} = \frac{2}{t_0(1 - t/t_0)}, \blacktriangleleft \qquad 0 \le t < t_0.$$

(*b*) The failure rate increases from $2/t_0$ at $t = 0$ to infinity at $t = t_0$. ◀
(c) From Eq. 4.22

$$\text{MTTF} = \int_0^{t_0} dt(1 - t/t_0)^2 = t_0/3. \blacktriangleleft$$

The Bathtub Curve

The behavior of $\lambda(t)$ with time is quite revealing with respect to the causes of failure. Unless a system has redundant components, such as those discussed in Chapter 7, it will invariably have the general characteristics of a "bathtub" curve such as shown in Fig. 4.1. The bathtub curve is an ubiquitous characteristic both of inanimate, complex engineering devices and of living creatures. Much of the terminology, moreover, comes from human mortality distributions, to which the foregoing formalism is applicable, provided that failure is defined as death.

The short period of time on the left-hand side of Fig. 4.1 is a region of high but decreasing failure rates. This is referred to as the period of *early failures* or in human populations infant mortality. These infant deaths are caused primarily by congenital defects or weaknesses. The death rate decreases with time as the weaker infants die and are lost from the population or their defects are detected and repaired. Similarly, defective pieces of equipment, prone to failure because they were not manufactured or constructed properly, cause the high initial failure rates of engineering devices.

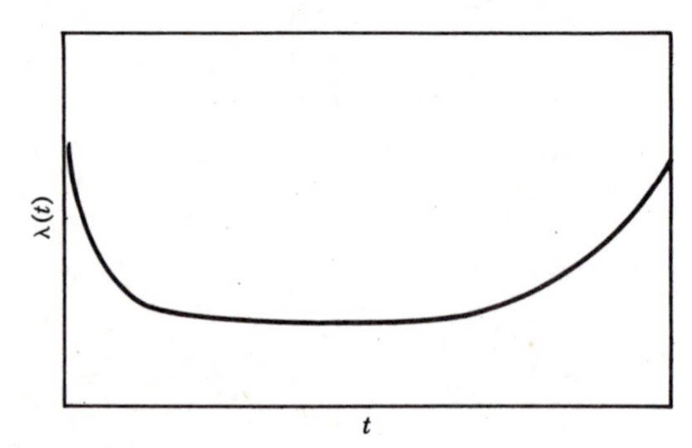

FIGURE 4.1 A "bathtub" curve representing a time-dependent failure rate.

Missing parts, substandard material batches, components which are out of tolerance, and damage in shipping are a few of the quality control shortcomings that may cause excessive failure rates near the beginning of design life. These problems in engineering devices are usually dealt with by specifying a period of time during which the device undergoes wearin. During this time loading and use are controlled in such a way that weaknesses are likely to be detected and repaired without failure, or so that failures attributable to defective manufacture or construction will not cause inordinate harm or financial loss.

The middle section of the bathtub curve contains the smallest and most nearly constant failure rates; this is referred to as the useful life. Failures during this period of time are frequently referred to as *random failures.* They are likely to stem from unavoidable loads coming from without, rather than from any inherent defect in the device or system under consideration. In human populations, deaths during this part of the bathtub curve are likely to be due to accidents or to infectious disease. In engineering devices, external loading on the system or its components is usually the cause of failure. These may take a wide variety of forms, depending on the type of system under consideration: earthquakes, power surges, vibration, mechanical impact, temperature fluctuations, and moisture variation are a few of the possible causes.

On the right of the bathtub curve is a region of increasing failure rates. During this period of time *aging failures* are said to take place. Again, with obvious analogy to human populations, the failures tend to be dominated by cumulative effects such as corrosion, embrittlement, fatigue cracking, and diffusion of materials. The onset of rapidly increasing failure rates normally forms the basis for determining when parts should be replaced and for specifying the system's design life.

Although Fig. 4.1 displays the general features present in failure rate curves for many types of devices, one of the three mechanisms may predominant for a particular class of system. Examples of three such curves are given in Fig. 4.2. The curve in Fig. 4.2*a* is representative of much computer and other electronic hardware. In particular, after a rather inconspicuous wearin period, there is a long span of time over which the failure rate is essentially constant. For systems of this type, the primary

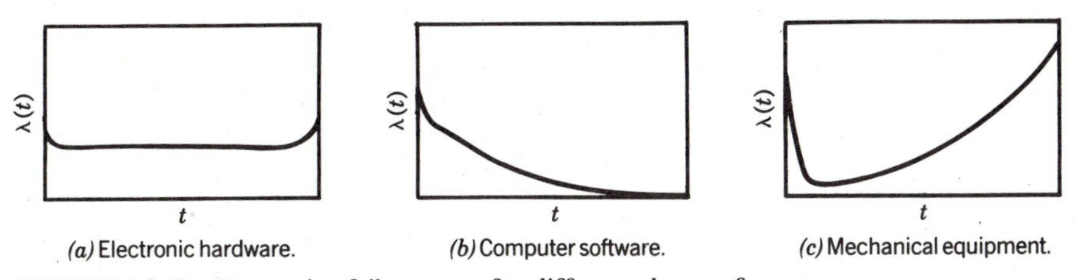

FIGURE 4.2 Representative failure rates for different classes of systems:

concerns are with random failures, and with methods for controlling the environment and external loading to minimize their occurrence.

The curve shown in Fig. 4.2*b* is more likely to be appropriate for computer software. Programs or other software do not wear out. Neither are external loads likely to be substantial contributors to the failure rate. Rather, most failures are due to "bugs" or defects built into the software from the beginning. Thus a major task is in the "debugging" or wearin of the software in order to eliminate the defects.

The failure rate curve in Fig. 4.2*c* is typical of valves, pumps, engines, and other pieces of equipment that are primarily mechanical in nature. Their initial wearin period is followed by a long span of time with a monotonically increasing failure rate. In these systems, for which the primary failure mechanisms are fatigue, corrosion, and other cumulative effects, the primary concern is in estimating safe and economical operating lives, and in determining prudent schedules for preventive maintenance and for replacing parts.

In the following sections models for representing failure rates with one or at most a few parameters are discussed. These are particularly useful when most of the failures are caused by early failures, by random events, or by aging effects. Even when more than one mechanism contributes substantially to the failure rate curve, however, the models have been shown to be very useful. For then, as we shall see, the failure rate curve can be viewed as a superposition of curves for different failure modes. Each failure mode rate curve may then be represented by an analytical expression.

4.3 RANDOM FAILURES

Random failures constitute the most widely used model for describing reliability phenomena. They are defined by the assumption that the rate of failure of a system is independent of its age and other characteristics of its operating history. For continuously operating systems this implies a constant failure rate, whereas for demand failures it requires that the failure probability per demand be independent of the number of demands.

The constant failure rate approximation is often quite adequate even though a system or some of its components may exhibit moderate early-failures or aging effects. The magnitude of early-failure effects is limited by strict quality control in manufacture and installation and may be further reduced by a wearin period before actual operations are begun. Similarly, in many systems aging effects can be sharply limited by careful preventive maintenance, with timely replacement of the parts or components in which the wear effects are concentrated. Conversely, if components are replaced as they fail, the overall failure rate of a many-component system will appear nearly constant, for the failure of the components will be randomly distributed in time as will the ages of the replacement parts. Finally, even though

of the motor to start when required is best described by a failure on demand. In Chapter 2 an expression is derived for the reliability of a system that is subjected to n repeated demands:

$$R_n = e^{-np}. \tag{4.28}$$

In order for this expression to be valid, the probability of failure on demand p must be small, and the demands must be independent of one another. Specifically, the probability that the system will fail on a particular demand must be independent of its past history. This is precisely the same condition that must hold for a constant failure rate λ to be obtained for a continuously operating system. It is therefore not surprising that we often can combine constant failure rates and demand models. What is needed is a measure of how frequently the demands are made.

A particularly simple result is obtained when we know, on the average, how many times a demand is made on the system per unit time. Suppose that there are m demands per unit time. Then the number of demands in time t is

$$n = mt, \tag{4.29}$$

and we may rewrite Eq. 4.28 as

$$R(t) = e^{-mpt}. \tag{4.30}$$

This allows us to write the reliability in the same form as Eq. 4.25, provided that we define the demand failure rate as

$$\lambda_d = mp. \tag{4.31}$$

Composite Models

In order to express the reliability as a function of time, it is often necessary to combine failure rates and demand failures. For example, with a motor it may be necessary to take into account both the probability of failure to start on demand and the failure rate per hour of operation. In performing such calculations, however, we must clearly state what measure of time is being used. If the reliability is to be expressed, in calendar time rather than operating time the duty cycle or capacity factor c, defined as the fraction of time that the engine is running, must also enter the calculations.

Consider as an example a refrigerator motor that runs some fraction c of the time; the failure rate is λ_0 per unit operating time. The contribution to the total failure rate from failures while the refrigerator is operating will then be $c\lambda_0$ per unit calendar time. If the demand failure is also to be taken into account, we must know how many times the motor is turned on. Suppose that the average length of time that the motor runs when it comes on is $\bar{t}$. Then the average number of times that the motor is turned on per unit operating time is $1/\bar{t}_0$. The average number of times that it is turned on per unit calendar time is $m = c/\bar{t}_0$. To obtain the total failure rate, we add the demand and operating failure rates. Consequently, the composite failure rate to be used in Eqs. 4.23 through 4.25 is

$$\lambda = \frac{c}{\bar{t}_0} p + c\lambda_0. \tag{4.32}$$

In the foregoing development we have neglected the possibility that the motor may fail while it is not operating, that is, while it is in a standby mode. Often such failure rates are small enough to be neglected. However, for systems that are operated only a small fraction of the time, such as an emergency generator, failure in the standby mode may be quite significant. To take this into account, we define λ_s as the failure rate in the standby mode. Since the system in our example is in the standby mode for a fraction $1 - c$ of the time, we add a contribution of $(1 - c)\lambda_s$ to the composite failure rate in Eq. 4.32:

$$\lambda = \frac{c}{\bar{t}_0} p + c\lambda_0 + (1 - c)\lambda_s. \tag{4.33}$$

EXAMPLE 4.3

A pump on a volume control system at a chemical process plant operates intermittently. The pump has an operating failure rate of 0.0004/hr and a standby failure rate of 0.00001/hr. The probability of failure on demand is 0.0005. The times at which the pump is turned on t_u and turned off t_d over a 24-hr period are listed in the following table.

t_u	0.78	1.69	2.89	3.92	4.71	5.97	6.84	7.76
t_d	1.02	2.11	3.07	4.21	5.08	6.31	7.23	8.12
t_u	8.91	9.81	10.81	11.87	12.98	13.81	14.87	15.97
t_d	9.14	10.08	11.02	12.14	13.18	14.06	15.19	16.09
t_u	16.69	17.71	18.61	19.61	20.56	21.49	22.58	23.61
t_d	16.98	18.04	19.01	19.97	20.91	21.86	22.79	23.89

Assuming that these data are representative, (*a*) Calculate a composite failure rate for the pump under these operating conditions. (*b*) What is the probability of the pump's failing during any 1-month (30-day) period?

Solution (*a*) From the data given we first calculate

$$\sum_{i=1}^{M} t_{di} = 301.50 \quad \text{and} \quad \sum_{i=1}^{M} t_{ui} = 294.36,$$

where $M = 24$ is the number of operations.

The average operating time $\bar{t}_0$ of the pump is estimated for the data to be

$$\bar{t}_0 = \frac{1}{M}\sum_{i=1}^{M}(t_{di} - t_{ui}) = \frac{1}{M}\left(\sum_{i=1}^{M} t_{di} - \sum_{i=1}^{M} t_{ui}\right)$$

$$= \frac{1}{24}(301.50 - 294.36) = 0.2975 \text{ hr.}$$

Then the capacity factor is

$$c = \frac{M\bar{t}_0}{24} = \frac{24 \times 0.2975}{24} = 0.2975.$$

Thus the failure rate from Eq. 4.33 is

$$\lambda = \frac{0.2975}{0.2975} \times 0.0005 + 0.2975 \times 0.0004 + (1 - 0.2975) \times 0.00001$$

$$= 6.26 \times 10^{-4} \text{ hr}^{-1}. \blacktriangleleft$$

(*b*) The reliability is

$$R = \exp(-\lambda \times 24 \times 30) = \exp(-0.4507) = 0.637,$$

yielding a 30-day failure probability of

$$1 - R = 0.363. \blacktriangleleft$$

4.4 TIME-DEPENDENT FAILURE RATES

We move now to the variety of situations in which the explicit treatment of early failures or aging effects, or both, requires the use of time-dependent failure rate models. When these phenomena are present, the reliability of a device or system becomes a strong function of its age. This may be illustrated by considering the effect of the accumulated operating time T_0 on the probability that a device can survive an additional time t.

Suppose that we define $R(t|T_0)$ as the reliability of a device that has previously been operated for a time T_0. We may therefore write

$$R(t|T_0) = P\{\mathbf{t}' > T_0 + t|\mathbf{t}' > T_0\}, \tag{4.34}$$

where $t' = T_0 + t$ is the time elapsed at failure since the device was new. From the definition given in Eq. 2.5, we may write the conditional probability as

$$P\{\mathbf{t}' > T_0 + t|\mathbf{t}' > T_0\} = \frac{P\{(\mathbf{t}' > T_0 + t) \cap (\mathbf{t}' > T_0)\}}{P\{\mathbf{t}' > T_0\}}. \tag{4.35}$$

However, since $(\mathbf{t}' > T_0 + t) \cap (\mathbf{t}' > T_0) = \mathbf{t}' > T_0 + t$, we may combine equations to obtain

$$R(t|T_0) = \frac{P\{\mathbf{t}' > T_0 + t\}}{P\{\mathbf{t}' > T_0\}}. \tag{4.36}$$

Since the reliability of a new device is just

$$R(t) \equiv R(t|T_0 = 0) = P\{\mathbf{t}' > t\}, \tag{4.37}$$

we obtain

$$R(t|T_0) = \frac{R(t + T_0)}{R(T_0)}. \tag{4.38}$$

Finally, using Eq. 4.18, we obtain

$$R(t|T_0) = \exp\left[-\int_{T_0}^{t+T_0} \lambda(t')\, dt'\right]. \tag{4.39}$$

The significance of this result may be interpreted as follows. Suppose that we view T_0 as a wearin time undergone by a device before being put into service, and t as the service time. Now we ask whether the wearin time decreases or increases the service life reliability of the device. To determine this, we take the derivative of $R(t|T_0)$ with respect to the wearin period and obtain

$$\frac{\partial}{\partial T_0} R(t|T_0) = [\lambda(T_0) - \lambda(T_0 + t)]R(t|T_0). \tag{4.40}$$

Thus increasing the wearin period improves the reliability of the device only if the failure rate is decreasing [i.e., $\lambda(T_0) > \lambda(T_0 + t)$]. If the failure rate increases with time, wearin only adds to the deterioration of the device with age and the service life reliability decreases.

To model early failures or wear effects more explicitly, we must turn to specific distributions of the time to failure. In contrast to the exponential distribution used for random failures, these distributions must have at least two parameters. Although the normal and lognormal distributions are frequently used to model aging effects, the Weibull distribution is probably the most universally employed. With it we may model early failures and random failures as well as aging effects.

The Normal Distribution

To describe the time dependence of reliability problems, we write the PDF for the normal distribution given by Eq. 3.70 with **t** as the random variable,

$$f(t) = \frac{1}{\sqrt{2\pi}\,\sigma} \exp\left[-\frac{(t - \mu)^2}{2\sigma^2}\right], \tag{4.41}$$

where μ is now the MTTF. The corresponding CDF is

$$F(t) = \int_{-\infty}^{t} \frac{1}{\sqrt{2\pi}\,\sigma} \exp\left[-\frac{(t - \mu)^2}{2\sigma^2}\right] dt, \tag{4.42}$$

or in the standardized normal form given in Appendix C,

$$F(t) = \Phi\left(\frac{t - \mu}{\sigma}\right). \tag{4.43}$$

From Eq. 4.4 the reliability for the normal distribution is found to be

$$R(t) = 1 - \Phi\left(\frac{t - \mu}{\sigma}\right), \tag{4.44}$$

and the associated failure rate is obtained by substituting this expression into Eq. 4.14:

$$\lambda(t) = \frac{1}{\sqrt{2\pi}\sigma} \exp\left[-\frac{1}{2}\frac{(t-\mu)^2}{\sigma^2}\right]\left[1 - \Phi\left(\frac{t-\mu}{\sigma}\right)\right]^{-1}. \tag{4.45}$$

The failure rate along with the reliability and the PDF for times to failure are plotted in Fig. 4.4. As indicated by the behavior of the failure rate, normal distributions are used to describe the reliability of equipment that is quite different from that to which constant failure rates are applicable. It is useful in describing the reliability in situations in which there is a reasonably well-defined wearout time, μ. This may be the case, for example, in describing the life of a tread on a tire or the cutting edge on a machine tool. In these situations the life may be given as a mean value and an uncertainty. When normal distribution is used, the uncertainty in the life is measured in terms of intervals in time. For instance, if we say that there is a 90% probability that the life will fail between $\mu - \Delta t$ and $\mu + \Delta t$, then

$$P\{\mu - \Delta t \le \mathbf{t} \le \mu + \Delta t\} = 0.9. \tag{4.46}$$

If the times to failures are normally distributed, it is equally probable that the failure will take place before $\mu - \Delta t$ or after $\mu + \Delta t$. Moreover, we can determine the failure distribution time from the standardized curve. Equation 4.46 implies that

$$\Delta t = 1.645\sigma. \tag{4.47}$$

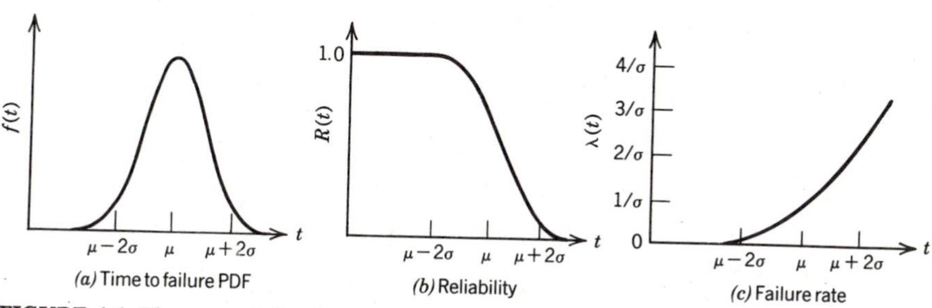

FIGURE 4.4 The normal distribution.

TABLE 4.1 Confidence Intervals for a Normal Distribution

Standard deviations	Confidence interval, %
$\pm 0.5\sigma$	0.3830
$\pm 1.0\sigma$	0.6826
$\pm 1.5\sigma$	0.8664
$\pm 2.0\sigma$	0.9544
$\pm 2.5\sigma$	0.9876
$\pm 3.0\sigma$	0.9974

Therefore, σ can be determined. The corresponding values for several other probabilities are given in Table 4.1. Once μ and σ are known, the reliability can be determined as a function of time from Eq. 4.44.

EXAMPLE 4.4

A tire manufacturer estimates that there is a 90% probability that his tires will wear out between 25,000 and 35,000 miles. Assuming a normal distribution, find μ and σ.

Solution Assume that 5% of failures are at fewer than 25×10^3 miles and 5% at more than 35×10^3 miles:

$$\Phi(u_1) = 0.05, \quad u_1 = \frac{25 - \mu}{\sigma}, \quad \Phi(u_2) = 0.95, \quad u_2 = \frac{35 - \mu}{\sigma}.$$

From Appendix C, $u_1 = -1.65$, $u_2 = +1.65$. Hence

$$-1.65\sigma = 25 - \mu, \qquad +1.65\sigma = 35 - \mu,$$

and the solutions are $\mu = 30$ thousand miles, ◀ $\sigma = 3.03$ thousand miles. ◀

The Lognormal Distribution

As we have indicated, the normal distribution is particularly useful for describing aging when we can specify a time to failure along with an uncertainty, Δt. The lognormal is a related distribution that has been found to be useful in describing failure distributions for a variety of situations. It is particularly appropriate under the following set of circumstances. If the time to failure has associated with it a large uncertainty, so that, for example, the variance of the distribution is a large fraction of the MTTF, the use of the normal distribution is problematical. However, it still may be possible to state a failure time and to estimate with it the probability that the time to failure lies within some factor, say n, of this value. For example, if it is known that 90% of the failures are within a factor of n of some time t_0,

$$P\left\{\frac{t_0}{n} \leq \mathbf{t} \leq nt_0\right\} = 0.9. \tag{4.48}$$

As indicated in Chapter 3, the lognormal distribution describes such situations. The PDF for the time to failure is then

$$f(t) = \frac{1}{\sqrt{2\pi}\, st} \exp\left\{-\frac{1}{2s^2}\left[\ln\left(\frac{t}{t_0}\right)\right]^2\right\} \tag{4.49}$$

and the corresponding CDF by

$$F(t) = \Phi\left[\frac{1}{s}\ln\left(\frac{t}{t_0}\right)\right] \tag{4.50}$$

Now, however, t_0 is not the MTTF; rather, they are related as indicated in Chapter 3, by

$$\text{MTTF} \equiv \mu = t_0 \exp(s^2/2). \tag{4.51}$$

Similarly, the variance of $f(t)$ is not equal to s^2, but rather to

$$\sigma^2 = t_0^2 \exp(s^2)[\exp(s^2) - 1]. \tag{4.52}$$

When the time to failure is known to within a factor of n, t_0 and s are determined by the same procedure discussed in Chapter 3. If it is assumed that 90% of the failures occur between $t_- = t_0/n$ and $t_+ = t_0 n$, then t_0 is the geometric mean,

$$t_0 = [t_- \times t_+]^{1/2}, \tag{4.53}$$

and

$$s = \frac{1}{1.645} \ln n. \tag{4.54}$$

The PDF for time to failure, the reliability, and the failure rate $\lambda(t)$ for the lognormal distribution are plotted in Fig. 4.5. Note that the failure rate can

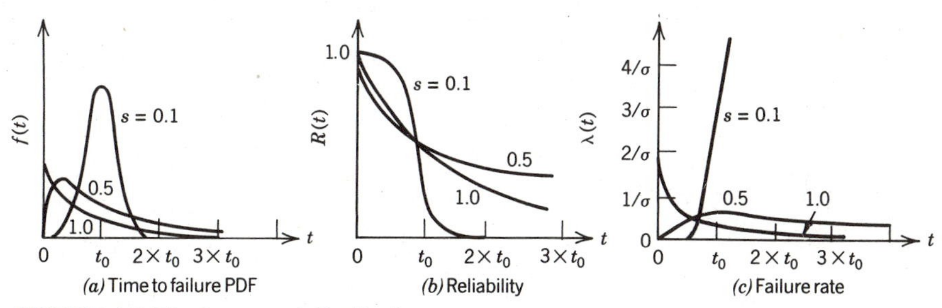

FIGURE 4.5 The lognormal distribution.

be increasing or decreasing depending on the value of s. The lognormal distribution is frequently used to describe fatigue and other phenomena that are caused by aging or wear and that result in failure rates that increase with time.

EXAMPLE 4.5

It is known that 90% of the truck axles of a particular type will suffer fatigue failure between 120,000 and 180,000 miles. Assuming that the failures may be fit to a lognormal distribution,
(*a*) To what factor n is the fatigue life known with 90 percent confidence?
(*b*) What are the parameters t_0 and s of the lognormal distribution?
(*c*) What is the MTTF?

Solution (*a*) For 90% certainty, $t_0 n = 180$ and $t_0/n = 120$. Taking the quotients of these equations yields

$$n^2 = \frac{180}{120}$$

$$n = 1.2247. \blacktriangleleft$$

(*b*) Taking the products of $t_0 n$ and t_0/n, we have

$$t_0^2 = 180 \times 120$$

$$t_0 = 146.97 \times 10^3 \text{ miles.} \blacktriangleleft$$

For 90% confidence Eq. 4.54 gives

$$s = \frac{1}{1.645} \ln n = \frac{\ln(1.2247)}{1.645} = 0.1232. \blacktriangleleft$$

(*c*) From Eq. 4.51,

$$\text{MTTF} = 146.97 \times \exp(\tfrac{1}{2} \times 0.1232^2) = 148.09 \times 10^3 \text{ miles.} \blacktriangleleft$$

The Weibull Distribution

The Weibull distribution is one of the most widely used distributions in reliability calculations, for through the appropriate choice of parameters a variety of failure rate behaviors can be modeled. These include, as a special case, the constant failure rate, in addition to failure rates modeling both wearin and wearout phenomena. The Weibull distribution may be formulated in either a two- or a three-parameter form. We treat the two-parameter form first.

The two-parameter Weibull distribution assumes that the failure rate is in the form of a power law:

$$\lambda(t) = \frac{m}{\theta}\left(\frac{t}{\theta}\right)^{m-1}. \tag{4.55}$$

From this failure rate we may use Eq. 4.19 to obtain the PDF:

$$f(t) = \frac{m}{\theta}\left(\frac{t}{\theta}\right)^{m-1} \exp\left[-\left(\frac{t}{\theta}\right)^{m}\right]. \tag{4.56}$$

Then, integrating over the time variable from zero to t, we obtain the CDF to be

$$F(t) = 1 - \exp(-(t/\theta)^m] \tag{4.57}$$

and a reliability of

$$R(t) = \exp[-(t/\theta)^m]. \tag{4.58}$$

The mean and the variance of the Weibull distribution may be shown to be

$$\mu = \theta\Gamma(1 + 1/m) \tag{4.59}$$

and

$$\sigma^2 = \theta^2\left\{\Gamma\left(1 + \frac{2}{m}\right) - \left[\Gamma\left(1 + \frac{1}{m}\right)\right]^2\right\}. \tag{4.60}$$

In these expressions the complete gamma function $\Gamma(v)$ is given by

$$\Gamma(v) = \int_0^\infty \zeta^{v-1}e^{-\zeta}\, d\zeta. \tag{4.61}$$

In Fig. 4.6 is shown the dependence of $\Gamma(v)$ on the value of v, for $0 \leq v < 1$. For $v > 1$, $\Gamma(v)$ can be obtained from the identity:

$$\Gamma(v) = (v - 1)\Gamma(v - 1). \tag{4.62}$$

Figure 4.7 shows the properties of $f(t)$ and $R(t)$ for a number of values of m, as well as the associated time-dependent failure rates. From these figures and the foregoing equations it is clear that the Weibull distribution provides a good deal of flexibility in fitting failure rate data. When $m = 1$, the exponential distribution corresponding to a constant failure rate is obtained. For values of $m < 1$ failure rates are typical of wearin phenomena and decrease, and for $m > 1$ failure rates are typical of aging effects and

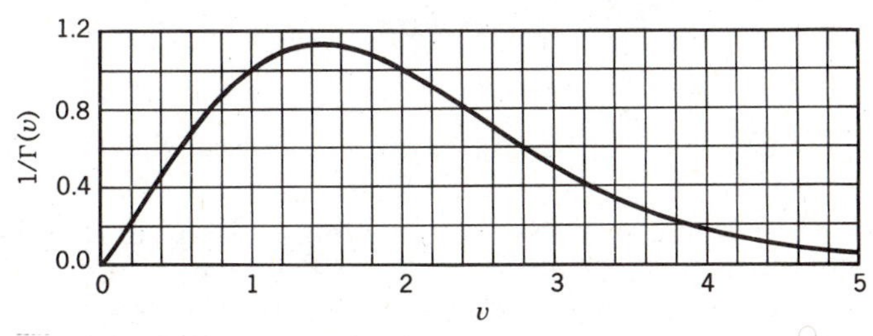

FIGURE 4.6 The gamma function.

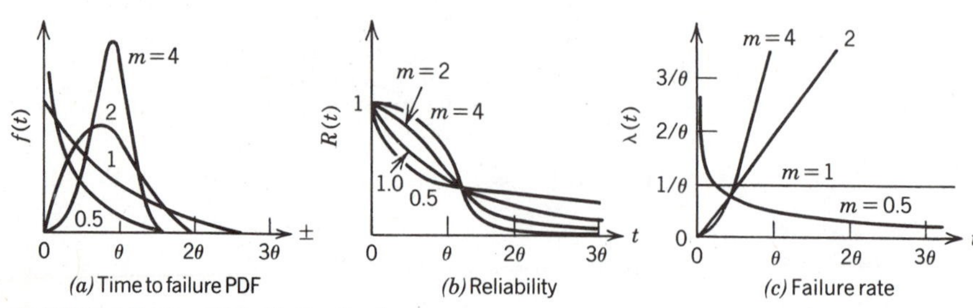

FIGURE 4.7 The Weibull distribution.

increase. Finally, as m becomes large, say $m > 4$, a normal PDF is approximated. In the following chapter we discuss the use of experimental data to estimate the parameters of the Weibull distribution.

EXAMPLE 4.6

A device has a decreasing failure rate characterized by a two-parameter Weibull distribution with $\theta = 180$ years and $m = \frac{1}{2}$. The device is required to have a design-life reliability of 0.90. (*a*) What is the design life if there is no wearin period? (*b*) What is the design life if the device is first subject to a wearin period of one month?

Solution (*a*) $R(T) = \exp[-(T/\theta)^m]$. Therefore, $T = \theta\{\ln[1/R(T)]\}^{1/m}$. Then

$$T = 180[\ln(1/0.9)]^2 = 2.00 \text{ years.} \blacktriangleleft$$

(*b*) The reliability with wearin time T_0 is given by Eq. 4.38. With the Weibull distribution it becomes

$$R(t|T_0) = \frac{\exp\left[-\left(\dfrac{t+T_0}{\theta}\right)^m\right]}{\exp\left[-\left(\dfrac{T_0}{\theta}\right)^m\right]}.$$

Setting $t = T$, the design life, we solve for T,

$$\begin{aligned} T &= \theta\left\{\ln\left[\frac{1}{R(T)}\right] + \left(\frac{T_0}{\theta}\right)^m\right\}^{1/m} - T_0 \\ &= 180\left[\ln\left(\frac{1}{0.9}\right) + \left(\frac{1}{12 \times 180}\right)^{1/2}\right]^2 - \frac{1}{12} \\ &= 2.81 \text{ years.} \blacktriangleleft \end{aligned}$$

Thus a wearin period of 1 month adds nearly 10 months to the design life.

The three-parameter Weibull distribution is useful in describing phenomena for which some time must elapse before there can be failures. In practice, a threshold amount of wear may have to take place before failures begin to occur. To obtain this distribution, we simply translate the origin to the right by an amount t_0 on the time axis. Thus we have

$$\lambda(t) = \begin{cases} 0, & t < t_0, \\ \dfrac{m}{\theta}\left(\dfrac{t - t_0}{\theta}\right)^{m-1}, & t \geq t_0, \end{cases} \tag{4.63}$$

$$f(t) = \begin{cases} 0, & t < t_0, \\ \dfrac{m}{\theta}\left(\dfrac{t - t_0}{\theta}\right)^{m-1} \exp\left[-\left(\dfrac{t - t_0}{\theta}\right)^{m}\right], & t \geq t_0, \end{cases} \tag{4.64}$$

and

$$F(t) = \begin{cases} 0, & t < t_0, \\ 1 - \exp\left[-\left(\dfrac{t - t_0}{\theta}\right)^{m}\right], & t \geq t_0. \end{cases} \tag{4.65}$$

The variance is the same as for the two-parameter distribution given in Eq. 4.60, and the mean is obtained simply by adding t_0 to the right-hand side of Eq. 4.59.

4.5 FAILURE MODES

In Sections 4.3 and 4.4 the quantitative behavior of reliability is modeled for situations with constant and time-dependent failure rates, respectively. In real systems, however, a system may fail through a number of different mechanisms, causing the curve to take a bathtub shape too complex to be described by any single one of the distributions discussed thus far. The mechanisms may be physical phenomena within a single monolithic structure, such as the tread wear, puncture, and defective sidewalls in an automobile tire. Or physically distinct components of a system, such as the processor unit, disk drives, and memory, of a computer, may fail. In either case it is usually possible to separate the failures according to the mechanism or the components that caused them. It is then possible, provided that the failures are independent, to generalize and treat the system reliability in terms of mechanisms or component failures. We refer to these collectively as failure modes.

Failure Mode Rates

In Section 2.2 the reliability of a system that may fail owing to M independent events was introduced. There, the independent events are associated with failures of different components, but the idea may be generalized to treat independent "modes" whether they correspond to failures of different components or to failures of different mechanisms in the same components. The reliability of a system with M different failure modes is

$$R(t) = P\{X_1 \cap X_2 \cap \cdots \cap X_M\}, \tag{4.66}$$

where X_i is the event that the ith failure mode does *not* occur before time t. Hence $\tilde{X}_i$ is the event that failure is caused by mode i before time t.

For independent modes

$$R(t) = P\{X_1\}P\{X_2\} \cdots P\{X_M\}. \tag{4.67}$$

The probability of surviving mode i may be written as the mode reliability,

$$R_i(t) = P\{X_i\}, \tag{4.68}$$

yielding

$$R(t) = \prod_i R_i(t). \tag{4.69}$$

Naturally, if mode i is the failure of component i, then $R_i(t)$ is just the component reliability.

For each mode—mechanism or component—we may define a mode PDF for time to failure, $f_i(t)$, and an associated failure rate, $\lambda_i(t)$. The derivation is exactly the same as in Section 4.2, yielding

$$R_i(t) = 1 - \int_0^t f_i(t')\, dt', \tag{4.70}$$

$$\lambda_i(t) = \frac{f_i(t)}{R_i(t)}, \tag{4.71}$$

$$R_i(t) = \exp\left[-\int_0^t \lambda_i(t')\, dt'\right] \tag{4.72}$$

and

$$f_i(t) = \lambda_i(t) \exp\left[-\int_0^t \lambda_i(t')\, dt'\right]. \tag{4.73}$$

The fundamental relationship that we seek results from combining Eq.

4.72 with Eq. 4.68. We may then write

$$R(t) = \exp\left[-\int_0^t \lambda(t')\,dt'\right], \tag{4.74}$$

where

$$\lambda(t) = \sum_i \lambda_i(t). \tag{4.75}$$

Specifically, to obtain the system reliability we simply add the mode failure rates.

When we consider the failure rate of a system to consist of a superposition of failure mode terms, it is easy to understand the "bathtub" behavior illustrated in Fig. 4.1. For this purpose Weibull distributions are particularly useful. If we write

$$\int_0^t \lambda(t')\,dt' = \left(\frac{t}{\theta_a}\right)^{m_a} + \left(\frac{t}{\theta_b}\right)^{m_b} + \left(\frac{t}{\theta_c}\right)^{m_c} \tag{4.76}$$

where $0 < m_a < 1$, $m_b = 1$, and $m_c > 1$, the three terms correspond, respectively, to contributions to the failure rate $\lambda(t)$ that decrease, remain flat, and increase with time. These correspond to early failures, random failures, and wear failures, respectively. We shall consider each of these types of failures in more detail in Chapter 6, where they are related to the loading and capacity of the system.

When the situation is such that the mode failure rates are constant, $\lambda_i(t) \rightarrow \lambda_i$, the reliability is given by Eq. 4.25 with

$$\lambda = \sum_i \lambda_i, \tag{4.77}$$

and Eq. 4.26 may be used to determine the system's mean time to failure. Note that if we define the mode mean time to failure as

$$\text{MTTF}_i = \frac{1}{\lambda_i}, \tag{4.78}$$

then the system mean time to failure is related by

$$\frac{1}{\text{MTTF}} = \sum_i \frac{1}{\text{MTTF}_i}. \tag{4.79}$$

Component Counts

The ability to add failure rates is most widely applied in situations in which each failure mode corresponds to a component or part failure. Often, failure rate data may be available at a component level but not for an entire system.

This is true, in part, because several professional organizations collect and publish failure rate estimates for frequently used items, whether they be diodes, switches, and other electrical components; pumps, valves, and similar mechanical devices; or a number of other types of components. At the same time the design of a new system may involve new configurations and numbers of such standard items. Thus the foregoing equations allow reliability estimates to be made before the new design is built and tested. In this chapter we consider only systems without redundance in which failure of any component implies system failure. In systems with redundant components, the idea of failure mode is still applicable in a more general sense. We reserve the treatment of systems with redundant components to Chapter 7.

When component failure rates are available, the most straightforward, but crudest, estimate of reliability comes from the parts count method. We simply count the number n_j of parts of type j in the system. The system's failure rate is then

$$\lambda = \sum_j n_j \lambda_j, \tag{4.80}$$

where the sum is over the types of parts in the system.

TABLE 4.2 Components and Failure Rates for Computer Circuit Card[a]

Component type	Quantity	Failure rate/ 10^6 hr	Total failure rate/10^6 hr
Capacitor tantalum	1	0.0027	0.0027
Capacitor ceramic	19	0.0025	0.0475
Resistor	5	0.0002	0.0010
J–K, M–S flip flop	9	0.4667	4.2003
Triple Nand gate	5	0.2456	1.2286
Diff line receiver	3	0.2738	0.8214
Diff line driver	1	0.3196	0.3196
Dual Nand gate	2	0.2107	0.4214
Quad Nand gate	7	0.2738	1.9166
Hex invertor	5	0.3196	1.5980
8-bit shift register	4	0.8847	3.5388
Quad Nand buffer	1	0.2738	0.2738
4-bit shirt register	1	0.8035	0.8035
And-or-inverter	1	0.3196	0.3196
PCB connector	1	4.3490	4.3490
Printed wiring board	1	1.5870	1.5870
Soldering connections	1	0.2328	0.2328
Total	67		21.6720 ◄

[a]"Reprinted from 'Mathematical Modelling' by A. H. K. Ling, *Reliability and Maintainability of Electronic Systems*, edited by Arsenault and Roberts with the permission of the publisher Computer Science Press, Inc., 1803 Research Boulevard, Rockville, Maryland 20850, USA."

EXAMPLE 4.7

A computer-interface circuit card assembly for airborne application is made up of interconnected components in the quantities listed in the first column of Table 4.2. If the assembly must operate in a 50°C environment, the component failure rates are given in column 2 of Table 4.2. Calculate (*a*) the assembly failure rate, (*b*) the reliability for a 12-hr mission, and (*c*) the MTTF.

Solution (*a*) We have calculated the total failure rate $n_j\lambda_j$ for each component type with Eq. 4.80 and listed them in the third column of Table 4.2. For a nonredundant system the assembly failure rate is just the sum of these numbers, or, as indicated, $\lambda = 21.6720 \times 10^{-6}$/hr. ◀

(*b*) The 12-hour reliability is calculated from $R = e^{-\lambda t}$ to be

$$R(12) = \exp(-21.672 \times 12 \times 10^{-6}) = 0.9997. ◀$$

(*c*) For constant failure rates the MTTF is

$$\text{MTTF} = \frac{1}{\lambda} = \frac{10^6}{21.672} = 46{,}142 \text{ hr.} ◀$$

The parts count method, of course, is no better than the available failure rate data. Moreover, the failure rates must be appropriate to the particular conditions under which the components are to be employed. For electronic equipment, extensive computerized data bases have been developed that allow the designer to take into account the various factors of stress and environment, as well as the quality of manufacture.* For military procurement such procedures have been formalized as the parts stress analysis method.†

In parts stress analysis each component failure rate λ_i is expressed as a base failure rate λ_b and as a series of multiplicative correction factors:

$$\lambda_i = \lambda_b \Pi_E \Pi_Q \cdots \Pi_N. \tag{4.81}$$

The base failure rate, λ_b, takes into account the temperature at which the component operates as well as the primary electrical stresses (i.e., voltage, current, or both) to which it is subjected. Figure 4.8 shows qualitatively the effects these variables might have on a particular component type.

The correction factors, indicated by the Π's in Eq. 4.81, take into account environmental, quality, and other variables that are designated as having a significant impact on the failure rate. For example, the environmental factor Π_E accounts for environmental stresses other than temperature; it is related

* *Reliability Prediction of Electronic Equipment*, MIL-HDBK-217D, U.S. Department of Defense, 1982.

† R. T. Anderson, *Reliability Design Handbook*, U.S. Department of Defense Reliability Analysis Center, 1976; see also J. E. Arsenault and J. A. Roberts, "Allocation and Estimation," *Reliability and Maintainability of Electronic Systems*, J. E. Arsenault and J. A. Roberts (eds.), Computer Science Press, Potomac, MD, 1980.

to the vibration, humidity, and other conditions encountered in operation. For purposes of military procurement, there are 11 environmental categories, as listed in Table 4.3. For each component type there is a wide range of values of Π_E; for example, for microelectronic devices Π_E ranges from 0.2 for "Ground, benign" to 10.0 for "Missile, launch."

Similarly, the quality multiplier Π_Q takes into account the level of specification, and therefore the level of quality control under which the component has been produced and tested. Typically, $\Pi_Q = 1$ for the highest levels of specification and may increase to 100 or more for commercial parts procured under minimal specifications. Other multiplicative corrections also

TABLE 4-3. Environmental Symbol Identification and Description

Environment	π_E symbol	Nominal environmental conditions	π_E value[a]
Ground, benign	G_B	Nearly zero environmental stress with optimum engineering operation and maintenance.	0.2
Space, flight	S_F	Earth orbital. Approaches G_B conditions without access for maintenance. Vehicle neither under powered flight nor in atmospheric reentry.	0.2
Ground, fixed	G_F	Conditions less than ideal: installation in permanent racks with adequate cooling air, maintenance by military personnel, and possible installation in unheated buildings.	1.0
Ground, mobile (and portable)	G_M	Conditions less favorable than those for G_F, mostly through vibration and shock. The cooling air supply may be more limited and maintenance less uniform.	4.0
Naval, sheltered	N_S	Surface ship conditions similar to G_F but subject to occasional high levels of shock and vibration.	4.0
Naval, unsheltered	N_U	Nominal surface shipborne conditions but with repetitive high levels of shock and vibration.	5.0
Airborne, inhabited	A_I	Typical cockpit conditions without environmental extremes of pressure, temperature, shock and vibration.	4.0
Airborne, uninhabited	A_U	Bomb-bay, tail, or wing installations, where extreme pressure, temperature, and vibration cycling may be aggravated by contamination from oil, hydraulic fluid, and engine exhaust.	6.0
Missile, launch	M_L	Severe noise, vibration, and other stresses related to missile launch, boosting space vehicles into orbit, vehicle reentry, and landing by parachute. Conditions may also apply to installation near main rocket engines during launch operations.	10.0

[a] Values for monolithic microelectronic devices.
Source: From R. T. Anderson, *Reliability Design Handbook* RDH-376, Rome Air Development Center, Griffiss Air Force Base, NY, 1976.

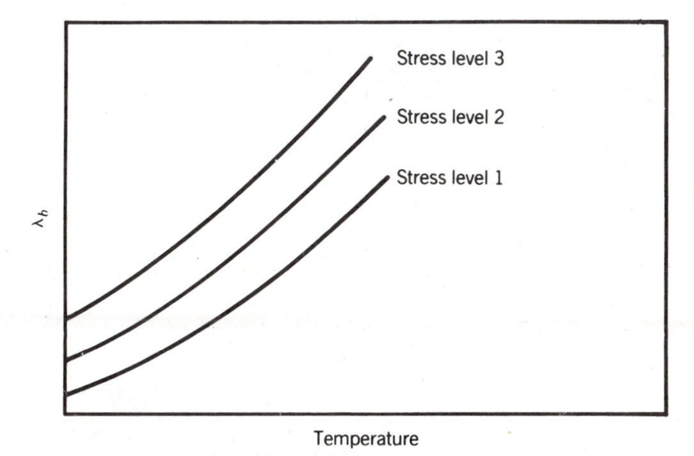

FIGURE 4.8 Failure rate versus temperature for different levels of applied stress (power, voltage, etc.).

are used. These include Π_A, the application factor to take into account stresses found in particular applications, and factors to take into account cyclic loading, system complexity, and a variety of other relevant variables.

4.6 REPLACEMENTS

Thus far we have considered the distribution of the failure times given that the system is new at $t = 0$. In many situations, however, failure does not constitute the end of life. Rather, the system is immediately replaced or repaired and operation continues. In such situations a number of new pieces of information became important. We may want to know the expected number of failures over some specified period of time in order to estimate the costs of replacement parts. More important, it may be necessary to estimate the probability that more than a specific number of failures N will occur over a period of time. Such information allows us to maintain an adequate inventory of repair parts.

In modeling these situations, we restrict our attention to the constant failure rate approximation. In this the failure rate is often given in terms of the *mean time between failures* (MTBF), as opposed to the mean time to failure or MTTF. In fact, they are both the same number if, when a system fails at t_n, it is assumed to be repaired immediately to an as-good-as-new condition. The time between the nth and the $(n + 1)$th failure is calculated from the probability that the $(n + 1)$th failure will occur between t and $t + dt$, given that the nth failure occurred at t_n:

$$f^{(n)}(t) = \begin{cases} 0, & t < t_n, \\ \lambda \exp[-\lambda(t - t_n)], & t \geq t_n. \end{cases} \tag{4.82}$$

The mean time between the n and the $n + 1$ is

$$\text{MTBF} = \int_{t_n}^{\infty} (t - t_n) f^{(n)}(t)\, dt, \tag{4.83}$$

or combining equations,

$$\text{MTBF} = \frac{1}{\lambda}. \tag{4.84}$$

In what follows we use the foregoing constant failure rate model to derive $p_n(t)$, the probability of there being n failures during a time interval of length t. We refer, however, to "events" instead of failures and replace λ by γ, the constant event rate. We do this because the derivation leads again to the ubiquitous Poisson distribution, and many phenomena other than failures of interest in reliability, such as the frequency of earthquakes, are described by the same distribution. The phenomena are collectively referred to as Poisson processes.

The Poisson Distribution

We first consider the times at which the events (e.g., failures) take place, and therefore the number of events that take place within any given span of time. Suppose that we let $\mathbf{n}$ be a discrete random variable representing the number of events that take place between $t = 0$ and a time t. Let

$$p_n(t) = P\{\mathbf{n} = n \mid t\} \tag{4.85}$$

be the probability that exactly n events have taken place before time t. Clearly, if we start counting events at time zero, we must have

$$p_0(0) = 1, \tag{4.86}$$

$$p_n(0) = 0, \qquad n = 1, 2, 3, \ldots, \infty. \tag{4.87}$$

In addition, at any time

$$\sum_{n=0}^{\infty} p_n(t) = 1. \tag{4.88}$$

For small Δt, let $\gamma\, \Delta t$ be the probability that the $(n + 1)$th event will take place during the time increment between t and $t + \Delta t$, given that exactly n events have taken place before time t. Then the probability that no event will occur during Δt is $1 - \gamma\, \Delta t$. From this we see that the probability that no events have occurred before $t + \Delta t$ may be written as

$$p_0(t + \Delta t) = (1 - \gamma\, \Delta t) p_0(t). \tag{4.89}$$

Then noting that

$$\frac{d}{dt} p_n(t) = \lim_{\Delta t \to 0} \frac{p_n(t + \Delta t) - p_n(t)}{\Delta t}, \tag{4.90}$$

we obtain the simple differential equation

$$\frac{d}{dt} p_0(t) = -\gamma p_0(t). \tag{4.91}$$

Using the initial condition, Eq. 4.86, we find

$$p_0(t) = e^{-\gamma t}. \tag{4.92}$$

With $p_0(t)$ determined, we may now solve successively for $p_n(t)$, $n = 1, 2, 3, \ldots$. in the following manner. We first observe that if n events have taken place before time t, the probability that the $(n + 1)$th event will take place between t and $t + \Delta t$ is $\gamma \Delta t$. Therefore, since this transition probability is independent of the number of previous events, we may write

$$p_n(t + \Delta t) = \gamma \Delta t \, p_{n-1}(t) + (1 - \gamma \Delta t) p_n(t). \tag{4.93}$$

The last term accounts for the probability that no event takes place during Δt. For sufficiently small Δt we can ignore the possibility of two or more events taking place.

Using the definition of the derivative once again, we may reduce Eq. 4.93 to the differential equation

$$\frac{d}{dt} p_n(t) = -\gamma p_n(t) + \gamma p_{n-1}(t). \tag{4.94}$$

This equation allows us to solve for $p_n(t)$ in terms of $p_{n-1}(t)$. To do this we multiply both sides by the integrating factor $e^{\gamma t}$. Then noting that

$$\frac{d}{dt} [e^{\gamma t} p_n(t)] = e^{\gamma t} \left[\frac{d}{dt} p_n(t) + \gamma p_n(t) \right], \tag{4.95}$$

we have

$$\frac{d}{dt} [e^{\gamma t} p_n(t)] = \gamma p_{n-1}(t) e^{\gamma t}. \tag{4.96}$$

Multiplying both sides by dt and integrating between 0 and t, we obtain

$$e^{\gamma t} p_n(t) - p_n(0) = \gamma \int_0^t p_{n-1}(t') e^{\gamma t'} \, dt'. \tag{4.97}$$

But, since from Eq. 4.87 $p_n(0) = 0$, we have

$$p_n(t) = \gamma e^{-\gamma t} \int_0^t p_{n-1}(t') e^{\gamma t'} \, dt'. \tag{4.98}$$

This recursive relationship allows us to calculate the p_n successively. For p_1, insert Eq. 4.92 on the right-hand side and carry out the integral to obtain

$$p_1(t) = \gamma t e^{-\gamma t}. \tag{4.99}$$

Repeating this procedure for $n = 2$ yields

$$p_2(t) = \frac{(\gamma t)^2}{2} e^{-\gamma t}, \tag{4.100}$$

and so on. It is easily shown that Eq. 4.98 is satisfied for all $n > 0$ by

$$p_n(t) = \frac{(\gamma t)^n}{n!} e^{-\gamma t}, \tag{4.101}$$

and these quantities in turn satisfy the initial conditions given by Eqs. 4.86 and 4.87.

The probabilities $p_n(t)$ are the same as the Poisson distribution $f(n)$ derived in Chapter 2, provided that we set $\mu = \gamma t$. We may therefore use Eqs. 2.56 through 2.58 to determine the mean and the variance of the number n of events occurring over a time span t. Thus the expected number of events during time t is

$$\mu_{\mathbf{n}} \equiv E\{n\} = \gamma t, \tag{4.102}$$

and the variance of n is

$$\sigma_{\mathbf{n}}^2 = \gamma t. \tag{4.103}$$

Of course, since $p_n(t)$ are the probability functions of a discrete variable $\mathbf{n}$, we must have, according to Eq. 2.51,

$$\sum_{n=0}^{\infty} p_n(t) = 1. \tag{4.104}$$

Number of Failures

By setting $\lambda = \gamma$ in the Poisson distribution, we may calculate the expected number of failures or related quantities from the $p_n(t)$. For example, we may want to know the expected number of failures or the probability that there will be more than N failures during some time span t.

The expected number of failures during t is obtained by setting $\gamma = \lambda$ in Eq. 4.102:

$$\mu_{\mathbf{n}} = \lambda t. \tag{4.105}$$

With the definition of Eq. 4.84 for the mean time between failures,

$$\mu_{\mathbf{n}} = \frac{t}{\text{MTBF}}. \tag{4.106}$$

We have derived the expression relating $\mu_{\mathbf{n}}$ and the MTBF assuming a constant failure rate. It has, however, much more general validity.* Al-

* See, for example, R. E. Barlow and F. Proschan, *Mathematical Theory of Reliability*, Wiley, New York, 1965.

though the proof is beyond the scope of this book, it may be shown that Eq. 4.106 is also valid for time-dependent failure rates in the limiting case that $t >> \text{MTBF}$. Thus, in general, the MTBF may be determined from

$$\text{MTBF} = \frac{t}{n}, \tag{4.107}$$

where n, the number of failures, is large.

We may also require the probability that more than N failures have occurred. It is

$$P\{\mathbf{n} > N\} = \sum_{n=N+1}^{\infty} \frac{(\lambda t)^n}{n!} e^{-\lambda t}. \tag{4.108}$$

Instead of writing this infinite series, however, we may use Eq. 4.104 to write

$$P\{\mathbf{n} > N\} = 1 - \sum_{n=0}^{N} \frac{(\lambda t)^n}{n!} e^{-\lambda t}. \tag{4.109}$$

EXAMPLE 4.8

In an industrial plant there is a dc power supply in continuous use. It is known to have a failure rate of $\lambda = 0.40/\text{year}$. If replacement supplies are delivered at 6-month intervals, and if the probability of running out of replacement power supplies is to be limited to 0.01, how many replacement power supplies should the operations engineer have on hand at the beginning of the 6-month interval.

Solution First calculate the probability that the supply will have more than n failures with $t = 0.5$ year,

$$\lambda t = 0.4 \times 0.5 = 0.2; \qquad e^{-0.2} = 0.819.$$

Now use Eq. 4.105:

$$P\{\mathbf{n} > 0\} = 1 - e^{-\lambda t} = 0.181,$$

$$P\{\mathbf{n} > 1\} = 1 - e^{-\lambda t}(1 + \lambda t) = 0.018,$$

$$P\{\mathbf{n} > 2\} = 1 - e^{-\lambda t}[1 + \lambda t + \tfrac{1}{2}(\lambda t)^2] = 0.001.$$

There is less than a 1% probability of more than two power supplies failing. Therefore, two spares ◀ should be kept on hand.

4.7 RELIABILITY IN ENGINEERING DESIGN

In the preceding sections we have examined the rudiments of how probabilistic methods may be applied to the analysis of the reliability of a system, component, or part. In terms of failure rates we are able to visualize the effects of the three general categories of failure: early failures, random failures, and aging effects. And with not too severe constraints, reliabilities can very often be treated using a constant failure rate model. Indeed, within the framework of the failure rate formalism, the dependencies of reliability on failure modes and on numbers of components become clear, and we may treat not only single but also repeated failures assuming the Poisson process model.

Having dwelt on the methods of analysis, we must now relate reliability back to the problems of engineering design. As discussed in Chapter 1, design is most concerned with two quantities, performance capability and cost. Performance capability may be defined roughly as the adequacy of a design to perform required functions when it is operating properly. The system design must meet performance requirements at minimum cost. How then does reliability enter into the design process to make a three-way trade-off between performance capability, cost, and reliability? In attempting to answer this question, we must examine how reliability requirements are set, how the designer may attempt to build reliability into a design, and how we may verify that the requirements have been achieved. Concomitant with these topics, it is necessary to examine not only the probability of failure, but also the implications and varied consequences of different modes of failure.

Reliability Requirements

Reliability requirements vary greatly depending on the type of equipment or system under consideration. They may sometimes be set by the designer or more broadly by the firm responsible for the design. In other situations the requirements may be imposed by the buyer of the product; and in some instances third outside parties, such as insurance underwriters or government agencies, may play a large role.

For any given product there are likely to be trade-offs between capital costs and maintenance costs. Generally, the costs will behave qualitatively as shown in Fig. 4.9. The more effort put into making the product reliable, the higher will be its initial or capital cost. At the same time, the more reliable a product is, the lower the costs for maintenance and repair are likely to be. We might be tempted to say that the solution is optimal when the total cost is minimized. In practice, however, the considerations that go into making such a trade-off vary greatly, depending on the nature of the product. The varying requirements may be understood more concretely by considering the factors that may come into play in the following three diverse examples: a consumer appliance, an industrial robot, and a jet engine.

The reliability requirements for an appliance may depend to a great extent on consumer psychology. To what extent will the buyer be willing to pay a higher price in order to save later repair or replacement costs? The price is obvious, but how will the reliability be impressed on the general public: by information and advertising about the appliance's features, or by the willingness to offer a longer warranty than the competition? Moreover, what will the sales consequences be if the cost is too high because excessive

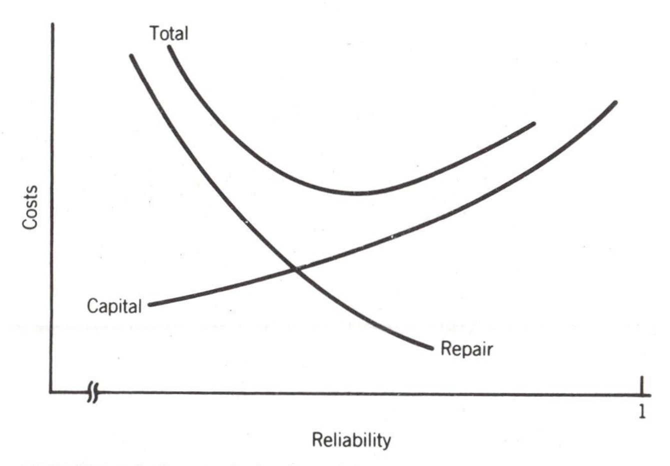

FIGURE 4.9 Cost optimization with respect to reliability.

reliability is built into the appliance? On the other hand, if the public buys the item at a lower price and then comes to resent the inconvenience and cost of breakdowns, will the firm's other products acquire a lingering reputation for poor design or shoddy workmanship? This, in turn, may depend significantly on the extent and efficiency of the service organization responsible for maintenance. These are complex issues that must be decided either implicitly or explicitly. The decisions are most likely made by the manufacturer with the help of market surveys or other input to ascertain the public's preferences with regard to price and reliability.

The situation is quite different for an industrial robot or other equipment that is designed primarily for sale to large organizations, whether they be manufacturing firms, public utilities, or government agencies. For, if the equipment is expensive, the buyer is likely to have a staff of engineers or outside consultants who are able to assess the cost trade-offs and determine their impact on the buyer's operations. With the robot, for example, a trade-off must be made between capital cost and production lost through robot breakdown. The direct cost of repair may be small in comparison to the costs of lost production. In effect, the speed at which the buyer's maintenance organization is able to accomplish repairs will be a central consideration. Thus, for these types of products, design may center not only on reliability, but also on maintainability. Are they able to be repaired in a short period of time? Increasing the reliability at the expense of greatly increasing the repair time may in some situations cause greater losses in production.

For the aircraft engine the consequences of failure are generally so severe that a much higher standard of reliability is required, even though the cost of the engine is made much higher than it otherwise would be. Similarly, preventive maintenance to enhance reliability will be the primary consideration, rather than the time required to repair a failed engine. Avoiding failure is likely to be so important that it may not be left entirely in the hands of the design or manufacturing organization, or to the buyer, to set reliability specifications. Insurance underwriters and government agencies are likely to be much more closely involved. In the design more emphasis also will be placed on early warning of incipient failures so that the engine can be taken out of service before an accident occurs.

Thus far we have discussed only the reliability requirements on a system as a whole. In setting reliability requirements and in designing to ensure that they are met, we should allocate reliability among major subsystems, and then on down to lower levels of components and parts. For example, in designing an aircraft we might decide that the reliabilities of the propulsion, structural, control, and guidance systems should be approximately comparable. Then if the required reliability for the total system is to be equal to R, we would have $R = R_0^4$, where R_0 is the reliability requirement for each of the subsystems. Each subsystem reliability requirement would then be $R_0 = R^{1/4}$. Such reliability allocations are important particularly when the subsystems or components are to be designed or manufactured, or both, by different groups or by different corporations.

After the requirements are set, we need means to ensure that they are being met. The parts count technique discussed earlier and related techniques can be very valuable in predicting the reliability of a preliminary design. Even though the component data may not be very accurate, and the absolute numbers subject to a great deal of uncertainty, they can be usefully compared for alternative configurations.

After prototype models are built, there is usually a text-fix-test-fix process in which prototypes are tested to failure, the failure mode examined, the design modified, and the test repeated. In this way reliability is said to grow, for in the design refinement the more common failure modes should be eliminated. After the design has become final and again when the item is being manufactured in mass, it is generally necessary to test the actual product. Such testing is valuable, but it may also be problematical, particularly if production deadlines are short or the individual units of the system are very expensive. We discuss reliability testing and its limitations in more detail in Chapter 5. Consider, however, that by the time the design process is completed, the reliability weaknesses discovered through testing programs are likely to be expensive to correct, particularly if they require major redesign. Preferably, the reliability is built in from the start. We now discuss some of the techniques pursuant to that goal.

Reliability Enhancement

The most important ingredient, of course, for the creation of a reliable product is the knowledge of the designers. In addition to having a well-thought-out set of specifications for the performance requirements of the product, the designers must have a thorough understanding of the types and magnitudes of the loading that are likely to occur, and of the range of environmental conditions under which the product will operate. Moreover, they must understand the physics of the potential modes of failure to prevent unanticipated mechanisms from lessening reliability. Within this framework there are three broad classes of techniques that the knowledgeable designer may apply to enhance reliability. Within each class, trade-offs must again be considered between increased reliability, increased cost, and a lessening of the system's capacities.

Design Margins

The reliability of a system can clearly be increased by increasing the ratio of the capacity of the components relative to the loads placed upon them. By such action we can decrease the chance that the system will fail through uncertainties in either quantity. The probabilistic approach to design margins and its relation to reliability is discussed in some detail in Chapter 6. The net outcome, however, might be visualized as in Fig. 4.10, where for an assumed operating environment, the failure rate decreases as the component load is reduced. This procedure of deliberately reducing the loading is referred to as derating. The terminology stems from the deliberate reduction of voltages of electrical systems, but it is also applicable to mechanical, thermal, or other classes of loads as well. Conversely, the chance of component failure is decreased if the capacity or strength of the component is increased.

Redundancy

Reliability can be increased by adding redundance to a system at the system level or at the component level; components are added in a parallel configuration so that one or more components can fail without causing system

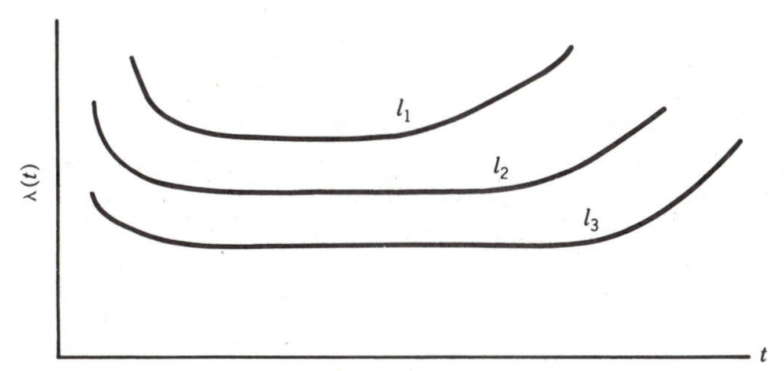

FIGURE 4.10 Time-dependent failure rates at different levels of loading: $l_1 > l_2 > l_3$.

failure. Redundancy is discussed in detail in Chapter 7. For now it is sufficient to observe that redundant components can increase reliability dramatically, but only if adequate precautions can be taken to ensure that the redundant components have very little chance of failing simultaneously from a common cause. Moreover, the effects of the additional components on system costs and on other performance parameters, such as weight, must also be considered.

Maintenance

As discussed earlier, preventive maintenance can significantly reduce failure rates, and in the event of failure, prompt repair can often limit the consequences. These matters are taken up in Chapter 8. Preventive maintenance, testing, and repair are considered in conjunction with redundant component configurations to achieve highly reliable systems. In employing these three techniques, the designer must be cognizant of the additional price of each to ensure that the gain in reliability is commensurate. Higher-capacity components, redundant components, and increased maintenance schedules may all increase costs.

Safety and Reliability

Thus far our discussion has been success-oriented in the sense that all failures are treated alike. We have been interested only in the reliability, in the probability that the system will function properly. In reality not all failures can be treated alike, for there may be a wide range of consequences, depending on the mode of failure. In most system failures we may hope that the result is simply inconvenience and some economic loss. However, the designer must also consider carefully the possibility that a particular failure mode may endanger individuals or possibly the health or safety of a larger human population. The emphasis then becomes failure-oriented, for the primary goal is to prevent the system from causing an accident by the manner in which it fails.

The study of the prevention and mitigation of the accidents that may result from system failure is taken up in detail in Chapter 10, titled "System Safety Analysis." At present we return to our diverse examples of an appliance, an industrial robot, and a jet engine to contrast safety analysis to more conventional reliability considerations. In most failures of an appliance its operation simply ceases. However, an electric short in an appliance may cause overheating and fire or even electrocute the user. The possibility of such accidents must be eliminated or reduced to a very small probability through careful design and safety analysis. More than customer dissatisfaction is involved. If a faulty product causes such accidents, it may have to be withdrawn from the market, and the company may be brought to court in costly liability lawsuits.

The failure of the industrial robot is somewhat analogous, except that accidents are likely to be occupational risks to the workers involved, rather than risks to members of the general public. In contrast, when an aircraft engine fails there is little distinction between reliability and safety analysis; for if reliability is defined in terms of engine function, any failure has major safety implications because it will increase significantly the probability of a crash.

The safety analysis techniques discussed in Chapter 10 have been applied most extensively to large centralized facilities, where the consequences of accidents may be very great: airliners, guided missiles, and other aerospace systems, large chemical processing plants, nuclear reactors, and so on. For such systems not only design and manufacturing must be emphasized, but also errors in maintenance and operation, and the complex relations between hardware failures and human error.

The relation between safety and reliability can be complex. As an example, consider a large process plant that manufactures a toxic chemical. Reliability is more closely related to the frequency with which the plant must be shut down because of equipment failure and to the length of time required to make repairs. Most such failures are likely to cause economic loss through lost production. But some may develop into accidents in which toxic chemicals are released to the atmosphere, creating a health risk to the public. Clearly, the interests of reliability and safety coincide in preventing equipment failures, for when there are no failures no production time is lost and no public risk is caused.

In the design of protective equipment and operating procedures, however, the two desires, for reliability and for safety, may sometimes come into conflict. From a safety standpoint, design and procedures may be tilted toward shutting down the plant and securing it at the first signs of trouble so that risk to the public is minimized. From a reliability standpoint, however, management may be inclined to keep the plant on line and wait until actual failure before shutting down, or to make repairs whenever possible while the plant is on line. Criteria must be carefully worked out for risk management so that the public is protected, and so that the plant can be operated without frequent and unwarranted shutdowns.

Bibliography

Anderson, R. T., *Reliability Design Handbook,* U.S. Department of Defense Reliability Analysis Center, 1976.

Arsenault, J. E., and J. A. Roberts (eds.), *Reliability and Maintainability of Electronic Systems,* Computer Science Press, Potomac, MD 1980.

Barlow, R. E., and F. Proschan, *Mathematical Theory of Reliability,* Wiley, New York, 1965.

Bazovsky, I., *Reliability Theory and Practice,* Prentice-Hall, Englewood Cliffs, NJ, 1961.

Billinton, R., and R. N. Allan, *Reliability Evaluation of Engineering Systems,* Plenum Press, New York, 1983.

Dillon, B. S., and C. Singh, *Engineering Reliability,* Wiley, New York, 1981.

Green, A. E., and A. J. Bourne, *Reliability Technology,* Wiley, New York, 1972.

Henley, E. J., and H. Kumanoto, *Reliability Engineering and Risk Assessment,* Prentice-Hall, Englewood Cliffs, NJ, 1981.

Reliability Prediction of Electronic Equipment MIL-HDBK-217D, U.S. Department of Defense, 1982.

Shooman, M. L., *Probabilistic Reliability: An Engineering Approach,* McGraw-Hill, New York, 1968.

EXERCISES

4.1 The PDF for the time to failure of an appliance is

$$f(t) = \frac{32}{(t+4)^3}, \qquad t > 0,$$

where t is in years
(*a*) Find the reliability of $R(t)$.
(*b*) Find the failure rate $\lambda(t)$.
(*c*) Find the MTTF.

4.2 The failure rate for a high-speed fan is given by

$$\lambda(t) = (2 \times 10^{-4} + 3 \times 10^{-6} t)/\text{hr},$$

where t is in hours of operation. The required design-life reliability is 0.95.
(*a*) How many hours of operation should the design life be?
(*b*) If, by preventive maintenance, the wear contribution to the failure rate can be eliminated, to how many hours can the design life be extended?
(*c*) By placing the fan in a controlled environment, we can reduce the contribution of random failures to $\lambda(t)$ by a factor of two. Then, without preventive maintenance, to how many hours may the design life be extended?
(*d*) What is the extended design life with both preventive maintenance and a controlled environment?

4.3 Repeat Exercise 4.2, but fix the design life at 100 hr and calculate the design-life reliability for conditions *a*, *b*, *c*, and *d*.

4.4 A logic circuit is known to have a decreasing failure rate of the form

$$\lambda(t) = \tfrac{1}{20}\, t^{-1/2},/\text{year},$$

where t is in years.

(*a*) If the design life is one year, what is the reliability?

(*b*) If the component undergoes wearin for one month before being put into operation, what will the reliability be for a one-year design life?

4.5 A device has a constant failure rate of 0.7/year.

(*a*) What is the probability that the device will fail during the *second* year of operation?

(*b*) If upon failure the device is immediately replaced, what is the probability that there will be more than one failure in 3 years of operation?

4.6 The failure rate for a hydraulic component is given empirically by

$$\lambda(t) = 0.001(1 + 2e^{-2t} + e^{t/40})/\text{year}$$

where t is in years. If the system is installed at $t = 0$, calculate the probability that it will have failed by time t. Plot your results for 40 years.

4.7 The failure rate on a new brake drum design is estimated to be

$$\lambda(t) = 1.2 \times 10^{-6} \exp(10^{-4}t)$$

per set, where t is in kilometers of normal driving. Forty vehicles are each test-driven for 15,000 km.

(*a*) How many failures are expected, assuming that the vehicles with failed drives are removed from the test?

(*b*) What is the probability that more than two vehicles will fail?

4.8 A home computer manufacturer determines that his machine has a constant failure rate of $\lambda = 0.4$ year in normal use. For how long should the warranty be set if no more than 5% of the computers are to be returned to the manufacturer for repair?

4.9 A one-year guarantee is given based on the assumption that no more than 10% of the items will be returned. Assuming an exponential distribution, what is the maximum failure rate that can be tolerated?

4.10 There is a contractual requirement to demonstrate with 90% confidence that a vehicle can achieve a 100-km mission with a reliability of 99%. The acceptance test is performed by running 10 vehicles over a 50,000-km test track.

(*a*) What is the contractual MTTF?

(*b*) What is the maximum number of failures that can be experienced on the demonstration test without violating the contractual requirement? (*Note:* Assume an exponential distribution.)

4.11 Suppose that amplifiers have a constant failure rate of $\lambda = 0.08/\text{month}$. Suppose that four such amplifiers are tested for 6 months. What is the probability that more than one of them will fail? Assume that when they fail, they are not replaced.

4.12 The reliability for the Rayleigh distribution is

$$R(t) = e^{-(t/\theta)^2}.$$

Find the MTTF in terms of θ.

4.13 A manufacturer determines that the average television set is used 1.8 hr/day. A one-year warranty is offered on the picture tube having a MTTF of 2000 hr. If the distribution is exponential, what fraction of the tubes will fail during the warranty period?

4.14 Night watchmen carry an industrial flashlight 8 hr per night, 7 nights per week. It is estimated that on the average the flashlight is turned on about 20 min per 8-hr shirt. The flashlight is assumed to have a constant failure rate of 0.08/hr while it is turned on and of 0.005/hr when it is turned off but being carried.
(*a*) In working hours, estimate the MTTF of the light.
(*b*) What is the probability of the light's failing during one 8-hr shift?
(*c*) What is the probability of its failing during one month (30 days) of 8-hr shifts?

4.15 A motor-operated valve has a failure rate λ_0 while it is open and λ_c while it is closed. It also has a failure probability p_0 to open on demand and a failure probability p_c to close on demand. Develop an expression for the composite failure rate similar to Eq. 4.33 for the valve.

4.16 A failure PDF for an appliance is assumed to be a normal distribution with $\mu = 5$ years and $\sigma = 0.8$ years. Set the design life for (*a*) a reliability of 90%, (*b*) a reliability of 99%.

4.17 A designer is 90% confident that a new piece of machinery will fail at some time between 2 years and 10 years.
(*a*) Fit a lognormal distribution to this belief.
(*b*) What is the MTTF?

4.18 The life of a rocker arm is assumed to be 4 million cycles. This is known to a factor of two with 90% confidence. If the reliability is to be 0.95, how many cycles should the design life be?

4.19 Two components have the same MTTF; the first has a constant failure rate λ_0 and the second follows a Rayleigh distribution, for which

$$\int_0^t \lambda(t')\, dt' = \left(\frac{t}{\theta}\right)^2.$$

(*a*) Find θ in terms of λ_0.
(*b*) If for each component the design-life reliability must be 0.9, how much longer (in percentage) is the design life of the second (Rayleigh) component?

4.20 Consider the two components in Exercise 4.19.
(*a*) For what design life reliability are the design lives of the two components equal?
(*b*) On the same graph plot reliability versus time for the two components.

4.21 The two-parameter Weibull distribution with $m = 2$ is known as the Rayleigh distribution. For a nonredundant system made of N components, each described by the same Rayleigh distribution, find the system MTTF in terms of N and the component θ.

4.22 The one-month reliability on an indicator lamp is 0.95 with the failure rate specified as constant. What is the probability that more than two spare bulbs will be needed during the first year of operation? (Ignore replacement time.)

4.23 A servomechanism has an MTBF of 2000 hr.
(*a*) What is the reliability for a 125-hr mission?
(*b*) Neglecting repair time, what is the probability that more than one failure will occur during a 125-hr mission?
(*c*) That more than two failures will occur during a 125-hr mission?

4.24 A relay circuit has an MTBF of 0.8 yr. Assuming random failures,
(*a*) Calculate the probability that the circuit will survive one year without failure.
(*b*) What is the probability that there will be more than two failures in the first year?
(*c*) What is the expected number of failures per year?

4.25 Demonstrate that Eq. 4.100 satisfies Eq. 4.98.

4.26 The MTBF for punctures of truck tires is 150,000 miles. A truck with 10 tires carries 1 spare.
(*a*) What is the probability that the spare will be used on a 10,000-mile trip?
(*b*) What is the probability that more than the single spare will be required on a 10,000-mile trip?

4.27 In Exercise 4.14, suppose that there are three watchmen on duty every night for 8 hr.

(*a*) How many flashlight failures would you expect in one year?

(*b*) Assuming that the failures are not caused by battery or bulb wear-out (these are replaced frequently), how many spare flashlights would be required to be on hand at the beginning of the year, if the probability of running out of spares is to be less than 10%?

4.28 Widgets have a constant failure rate with MTTF = 5 days. Ten widgets are tested for one day.

(*a*) What is the expected number of failures during the test?

(*b*) What is the probability that *more than one* will fail during the test?

(*c*) For how long would you run the test if you wanted the expected number of failures to be five?

4.29 The reliability of a cutting tool is given by

$$R(t) = \begin{cases} (1 - 0.2t)^2, & 0 \leq t \leq 5, \\ 0, & t > 5, \end{cases}$$

where t is in hours.

(*a*) What is the MTTF?

(*b*) How frequently should the tool be changed if failures are to be held to no more than 5%?

(*c*) Is the failure rate decreasing or increasing? Justify your result.

CHAPTER 5

Reliability Testing

5.1 INTRODUCTION

We turn now to the problem of reliability testing. Reliability estimates and the testing needed to verify them are important considerations during each phase of the life cycle discussed in Chapter 1. Reliability objectives and the determination of component reliability requirements to meet those objectives enter the earliest conceptual design and definition of a system. The parts count method of the preceding chapter and similar techniques may be used to estimate reliability from the known values of standard components. A comparison of similar existing systems and a number of other qualitative techniques may also be employed. During the detailed design phase prototypes may be built, tested, and analyzed for failure modes in order to improve reliability through redesign.

During the manufacturing and construction phase, qualification tests and acceptance testing become important to ensure that the delivered product meets the standards for which it is designed. Through improvement in quality control, defects in the manufacturing process can be eliminated. Finally, the collection of reliability data throughout the operational life of a system is an important task, not only for the correction of defects that may become apparent only with extensive field service, but also for the setting and optimization of maintenance schedules, parts replacement, and warranty policies.

Although the details differ with the product under consideration, reliability testing at any point in the life cycle is often severely limited by both money and time. Unless the subject of the test is a very inexpensive mass-produced component, it is costly to devote enough units to testing to make the sample size as large as one would like, particularly when the test is likely to cause wear and even destruction of the test units. The time over which the test units must be operated in order to obtain sufficient failure data also may be severely restricted by the date at which the design must be frozen, the manufacture commenced, or the product delivered. Finally, there is a premium attached to having reliability information early in the life cycle when there are few test prototypes available. The later design, manufacture, or operating modifications are made, the more expensive they are likely to be.

In some situations the analysis of test data by standard statistical methods may be problematical. If only a few prototypes are built, for example, the sample size may be too small to apply many of the standard statistical methods for estimating reliability. As the prototypes are operated and failures occur, however, the analysis of the failure modes provides valuable qualitative information for improving the reliability through modifying the design, the manufacturing process, or operational procedures. Such a process of test-fix-test-fix on even a single prototype can lead to significant growth in reliability through refinement in design or manufacturing procedures.

The binomial pass–no pass sampling, discussed in Chapter 2, finds extensive use, particularly at the acceptance testing stage of the life cycle. It is valuable in destructive testing or in other procedures that test a system's strength or other properties from which reliability may be inferred. This sampling has a disadvantage, however, in that little information is gained in understanding the time-dependent behavior of reliability. In binomial sampling all the units are operated for a fixed length of time or a fixed number of cycles, and then the failed fraction is tabulated.

A more informative procedure is life testing, in which the time to failure of each unit is recorded. As we shall see, this information allows the time dependence of the reliability and of the failure rate to be estimated. Although the determination of the time to failure of each unit is possible in laboratory tests, this is less likely to be the case for reliability data collected from field experience. From field data, for example, we may not be able to determine the exact time to failure, but only the number of failures during the first month, during the second month, and so on. When data are grouped in the process of collection, some information is lost. Nevertheless, it also may be used as a base for estimating the time dependence of reliability and failure rates.

A further word is in order concerning the relative merits of laboratory and field data. Both have their uses. Laboratory data are likely to provide more information per sample unit, both in the precise time to failure and in the mechanism by which the failures occur. Moreover, such data, because they are collected earlier, provide timely information for modifying design and manufacturing processes. On the other hand, the sample size for field data is likely to be much larger, allowing more precise statistical estimates to be made, even if the data are gathered in grouped form. Equally as important, laboratory testing may not adequately represent the environmental condition of the field, even though attempts are made to do so. The exposures to dirt, temperature, humidity, and other environmental loading encountered in practice may be difficult to predict and simulate in the laboratory. Similarly, the quality of maintenance provided by field crews or consumers is unlikely to equal that performed by laboratory technicians.

In this chapter we examine several ways of interpreting data from life testing. There are powerful statistical methods that may be applied to glean the maximum amount of information out of such test data. For our part, we shall rely primarily on graphical techniques that are accessible to the engineer with no statistical background beyond that provided in the preceding chapters. From this it should not be inferred, however, that more advanced methods are to be avoided. When the expense of doing life testing or of collecting field data has been authorized, the best technique available is likely to be justified for the analysis.

Although our concentration is on life testing in which time is the random variable, it is important to realize that with minor modification both the parametric and nonparametric methods discussed herein are often applicable to the analysis of data from other random variables: the strength of material, the intensity of mechanical loadings, and so on. We pursue these topics further in Chapter 6.

In the following section we examine the nonparametric analysis of complete data. By complete data we mean that the test is run to completion and all the failure times are utilized. In nonparametric analysis the results are not fitted to a particular distribution, obviating the problems of deciding which distribution best describes the data and of estimating the distribution parameters.

In Section 5.3 we examine some of the practical problems that arise both in the length of time required for life testing to be performed and in the complications caused by the several modes of failure. In considering censored data, we examine tests that are stopped before all the units are failed or from which some of the units are removed before failure. In accelerated testing we examine techniques for increasing the rate at which failures occur, thereby decreasing the amount of time necessary to complete the test.

In Section 5.4 we demonstrate graphical methods for chosing the distribution that best describes $f(t)$, the distribution of times to failure, and for determining the parameters inherent in this distribution. In Section 5.5 we examine in some detail the sampling methods for estimating the MTTF when the constant failure rate model is known to be applicable; in Section 5.6 Bayesian analysis of constant failure rates is discussed.

Finally, in Section 5.6 we depart from conventional life testing to examine the measurement of reliability growth. Reliability is said to grow when design is modified to prevent the failures that occur during prototype testing. The test-fix-test-fix process normally carried out on prototypes typically leads to a refined design with a longer MTTF.

5.2 NONPARAMETRIC METHODS

In nonparametric methods the data from testing or operating experience are plotted directly, without any attempt to fit them to a paticular distribution. Such methods may be applied to two types of data, ungrouped and grouped. Ungrouped data is obtained from reliability testing, and in some cases from field experience. It consists of a series of specific times at which the individual equipment failures took place. Table 5.1 is an example of ungrouped data. Grouped data are obtained most frequently from field experience, and in the simplest form, consist of the number of items failed within each of a number of equal time periods. No information is available on the specific times within the intervals at which failures took place. Table 5.2 is typical of grouped data. Both tables are examples of complete data; all the units are failed before the test is determined.

Ungrouped Data

Some field data and virtually all data from planned reliability tests are obtained in ungrouped form. That is, the results consist of a series of failure times $t_1, t_2, \ldots, t_i, \ldots, t_N$ for the N units in the test. In statistical nomenclature the t_i are referred to as the rank statistics of the test. Suppose that we estimate the reliability at t_i simply as the fraction of units surviving. The number of units surviving at t_i is just $n = N - i$. Consequently,

$$\hat{R}(t_i) = 1 - \frac{i}{N}, \tag{5.1}$$

and correspondingly, the CDF estimate is

$$\hat{F}(t_i) = 1 - \hat{R}(t_i) = \frac{i}{N}. \tag{5.2}$$

TABLE 5.1 Failure Times[a]

i	t_i	i	t_i
0	0.0	5	1.00
1	0.41	6	1.08
2	0.58	7	1.17
3	0.75	8	1.25
4	0.83	9	1.35

[a] Data from H. G. Martz and R. A. Waller, *Bayesian Reliability Analysis*, Wiley, New York, 1982.

TABLE 5.2 Grouped Failure Data[a]

Time interval	*Number of failures*
$0 \leq t < 3$	21
$3 \leq t < 6$	10
$6 \leq t < 9$	7
$9 \leq t < 12$	9
$12 \leq t < 15$	2
$15 \leq t < 18$	1

[a] Data from H. G. Martz and R. A. Waller, *Bayesian Reliability Analysis,* Wiley, New York, 1982.

Consistent with the preceding chapters, we use the caret to indicate estimates as opposed to true values.

Although Eq. 5.1 may be used to estimate reliability, it has some shortcomings when N is not a large number, say less than 10 or 15, which often happens. In particular, we find that reliability is zero for times after the Nth failure. If a much larger test were run, for example using $10 \times N$ units, it is highly likely that several of these units would fail at times greater than t_N. Therefore, Eq. 5.1 may seriously underestimate the reliability. The estimate is somewhat improved by arguing that if a very large number of units were to be tested, roughly equal numbers of failures would occur in each of the intervals between the t_i, and the number of failures after t_N would probably be about equal to the number within any one interval. From this argument we may estimate the fraction of units not to have failed by t_i to be

$$\hat{F}(t_i) = \frac{i}{N + 1}, \tag{5.3}$$

and since $R = 1 - F$,

$$\hat{R}(t_i) = \frac{N + 1 - i}{N + 1}. \tag{5.4}$$

Other more advanced statistical arguments have been used to estimate the fraction of failures occurring before t_i.* One of the more widely used is

$$\hat{F}(t_i) = \frac{i - 0.3}{N + 0.4}. \tag{5.5}$$

In practice, the randomness and limited amounts of data introduce more uncertainty than the particular form that is used to estimate $\hat{F}$ or $\hat{R}$. For large values of N, they yield nearly identical results for $F(t)$ after the

* See, for example, W. Nelson, *Applied Life Data Analysis,* Wiley, New York, 1982.

first few failures. For the most part we shall use Eq. 5.4 as a reasonable compromise between computational ease and accuracy.

Equation 4.10 may be used to estimate the values of the PDF between the t_i. We approximate the derivative on the right-hand side of Eq. 4.10 by a simple difference formula

$$\hat{f}(t) = -\frac{\hat{R}(t_{i+1}) - \hat{R}(t_i)}{t_{i+1} - t_i}, \qquad t_i < t < t_{i+1}. \tag{5.6}$$

With the reliability given by Eq. 5.4, we have for the PDF,

$$\hat{f}(t) = \frac{1}{(t_{i+1} - t_i)(N + 1)}, \qquad t_i < t < t_{i+1}. \tag{5.7}$$

We may also estimate the failure rate. However, since $\hat{f}(t)$ is discontinuous at the t_i, we cannot simply evaluate Eq. 4.14 at $t = t_i$. Instead, we note from Eq. 4.11 that $\lambda(t)\,\Delta t$, is a conditional probability for failure in the interval between t and $t + \Delta t$, given that the system is operational at t. Thus, combining Eqs. 4.11 through 4.13, we have

$$\lambda(t)(t_{i+1} - t_i) = \frac{f(t)(t_{i+1} - t_i)}{R(t_i)}, \qquad t \leq t \leq t_{i+1}, \tag{5.8}$$

provided that we take $t = t_i$ and $\Delta t = ti+1 - t_i$. Finally, replacing $f(t)$ and $R(t_i)$ by the estimates $\hat{R}(t_i)$ and $\hat{f}(t)$ given by Eqs. 5.4 and 5.7, we obtain

$$\hat{\lambda}(t) = \frac{1}{(t_{i+1} - t_i)(N + 1 - i)}, \qquad t_i < t < t_{i+1}. \tag{5.9}$$

The use of these ungrouped data estimators for $R(t)$, $f(t)$, and $\lambda(t)$ are best understood with an example.

EXAMPLE 5.1

From the data in Table 5.1 construct graphs for the reliability, failure PDF, and failure rate as a functin of time.

Solution The necessary calculations are carried out in Table 5.3 using Eqs. 5.4, 5.7, and 5.9. The results are plotted in Fig. 5.1. Note that from these results there are indications of an increasing failure rate and therefore of wear or aging effects.

The estimate of the MTTF or variance of the failure distribution for ungrouped data is straightforward. We simply adopt the unbiased point estimators discussed in chapter 3. The mean is given by Eq. 3.111,

$$\hat{\mu} = \frac{1}{N}\sum_{i=1}^{N} t_i, \tag{5.10}$$

TABLE 5.3 Ungrouped Data Computations

i	t_i	$t_{i+1} - t_i$	$\hat{R}(t_i)$	$\hat{f}(t)$	$\hat{\lambda}(t)$
0	0	0.41	1.0	0.213	0.213
1	0.41	0.17	0.9	0.588	0.658
2	0.58	0.17	0.8	0.588	0.735
3	0.75	0.08	0.7	1.25	1.785
4	0.83	0.17	0.6	0.588	0.98
5	1.00	0.08	0.5	1.25	2.50
6	1.08	0.09	0.4	1.11	2.78
7	1.17	0.08	0.3	1.25	4.17
8	1.25	0.10	0.2	1.00	5.00
9	1.35	—	0.1	—	—

and for the variance Eq. 3.124 becomes

$$\hat{\sigma}^2 = \frac{N}{N-1}\left[\frac{1}{N}\sum_{i=1}^{N} t_i^2 - \hat{\mu}^2\right]. \tag{5.11}$$

Grouped Data

Suppose that we want to estimate the reliability, failure rate, or other properties of a failure distribution from data such as those given in Table 5.2. We begin with the reliability. The test is begun with N items. The number of surviving items is tabulated at the end of each of the M time intervals into which the data are grouped: $t_1\, t_2, \ldots, t_i, \ldots t_M$. The number of surviving items at these times is found to be $n_1, n_2, \ldots, n_i, \ldots$. Since the reliability $R(t)$ is defined as the probability that a system will operate successfully for time t, we estimate the reliability at time t_i to be

$$\hat{R}(t_i) = \frac{n_i}{N}, \qquad i = 1, 2, \ldots, M, \tag{5.12}$$

which is a straightforward generalization of Eq. 5.1. Since the number of failures is generally significantly larger for grouped than for ungrouped

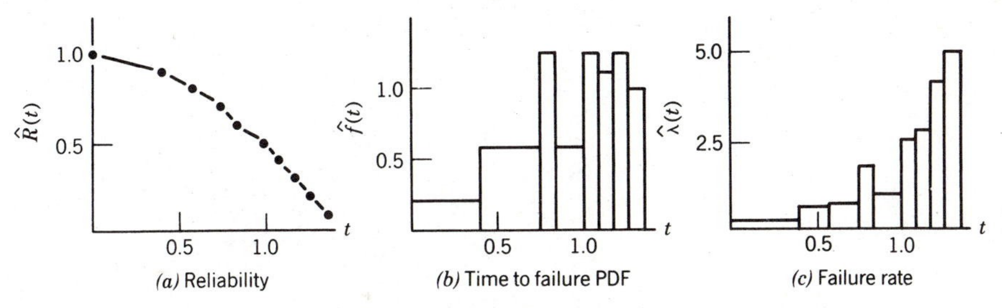

FIGURE 5.1 Nonparametric estimates from ungrouped life data.

data, it usually is not meaningful to derive more precise estimates analogous to Eqs. 5.4 and 5.5. Knowing the values of the reliability at the t_i, we approximate the values for $t_i < t < t_{i+1}$ by linear interpolation. With the reliability given by Eq. 5.12, we may once again use Eq. 5.6 to obtain the PDF:

$$\hat{f}(t) = \frac{n_i - n_{i+1}}{(t_{i+1} - t_i)N}, \qquad t_i < t < t_{i+1}. \tag{5.13}$$

Similarly, Eqs. 5.12 and 5.13 may be inserted into Eq. 5.8 to estimate the failure rate:

$$\hat{\lambda}(t) = \frac{n_i - n_{i+1}}{(t_{i+1} - t_i)n_i}, \qquad t_i < t < t_{i+1}. \tag{5.14}$$

Theses estimation procedures are illustrated in the following example.

EXAMPLE 5.2

From the data in Table 5.2 estimate the reliability, the PDF for the time to failure, and the failure rate. Is the failure rate increasing or decreasing?

Solution The necessary calculations, from Eq. 5.12, 5.13, and 5.14, are indicated in Table 5.4. The resulting values for the three quantities are plotted in Fig. 5.2. For $R(t)$ linear interpolation is employed between the values of t_i. It is clear that the failure rate once again increases with time.

In addition to obtaining plots of the results for grouped data, we may estimate the mean, variance, or other properties of the failure distribution. To do this, we simply approximate $f(t)$ by a histogram, as indicated in Table 5.2. Let f_i be the value calculated by Eq. 5.13 for $t_i < t < t_{i+1}$. Then the integral in Eq. 3.15 is estimated as

$$\hat{\mu} = \sum_{i=0}^{M-1} \bar{t}_i f_i \, \Delta_i, \tag{5.15}$$

TABLE 5.4 Grouped Computations

i	t_i	n_i	$\hat{R}(t_i)$	$\hat{f}(t_i)$	$\hat{\lambda}(t_i)$
0	0	50	1.00	0.14	0.14
1	3	29	0.58	0.07	0.11
2	6	19	0.38	0.05	0.12
3	9	12	0.24	0.06	0.25
4	12	3	0.06	0.01	0.22
5	15	1	0.02	0.01	0.33
6	18	0	0.00		

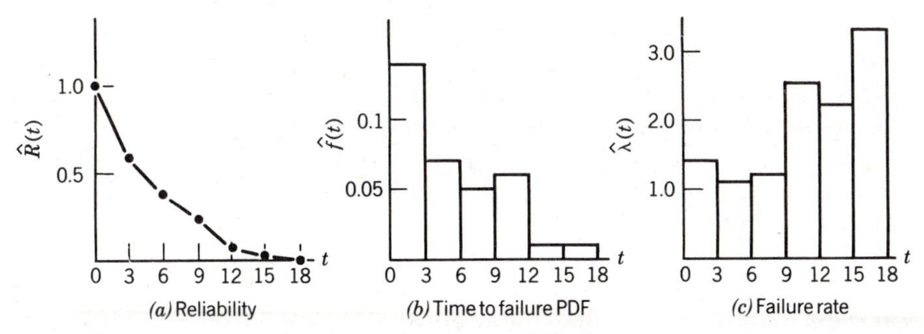

FIGURE 5.2 Nonparametric estimates from grouped life data.

where time is evaluated at the midpoint of the interval

$$\bar{t}_i = \frac{1}{2}(t_i + t_{i+1}), \tag{5.16}$$

and the width of the interval is

$$\Delta_i = t_{i+1} - t_i. \tag{5.17}$$

The variance, given by Eq. 3.22, is approximated as

$$\hat{\sigma}^2 = \sum_{i=0}^{M-1} \bar{t}_i^2 f_i \, \Delta_i - \hat{\mu}^2. \tag{5.18}$$

5.3 CENSORING AND ACCELERATION

In this section we first consider censoring, a major and frequent complication of testing. Censoring is said to occur if the data are incomplete, either because specimens are removed during the test or beacuse the test is not run to completion. We then examine accelerated life testing, a group of techniques by which conditions are modified to decrease the time required to complete reliability tests.

Singly Censored Data

In many reliability tests the data are not complete, either because the test has been stopped before all the specimens have failed or because intermediate results are needed before all the specimens have failed. The data are then said to be singly censored, or censored on the right, since most data are plotted with time on the horizontal axis.

With censored grouped data we have available the number of failures for only some of the intervals, say for the first i ($<M$). For ungrouped data there are two types of single censoring. In type I the test is terminated after

some fixed length of time; in type II the test is terminated after some fixed number of failures have taken place. The distinction between the two becomes important when sampling for a particular distribution is considered. For the nonparametric methods used in this section, it is adequate to treat all singly censored ungrouped data as failure-censored; we assume that of N units that begin a test, we are able to obtain the failure times for only the first n ($<N$) failures.

Censoring from the right of either grouped or ungrouped data simply removes that part of the curves in Figs. 5.1 or 5.2 to the right of the time, t_i (grouped) or the failure time (ungrouped) at which the test is terminated. With these limitations the estimates given for the mean and variance in Eqs. 5.10, 5.11, 5.15, and 5.18 are no longer valid. However, the graphical results may still be very useful. Often the early part of the reliability curve is the most important for setting a warrantee period, for determining adequate safety, and for a number of other purposes. Moreover, if early failures are under investigation, the first failures are of primary interest. Even when wearout is of concern, most engineering analysis can be completed without waiting until every last test unit has failed.

Censoring from the right may be deliberately incorporated into a test plan in conjunction with specifying how many units are to be tested. The test engineer may require that a relatively large number of units be tested in order to obtain enough early failures to estimate better the failure rate curve for some specified period of time, say the warrantee period or the design life. If this is the case, many of the units will not fail until well after the time period of interest, and at least a few are likely to survive for very long periods. Thus terminating the test at the end of the period of interest is quite natural.

The value of singly censored data may be further extended by applying the parametric methods discussed in the following sections. If one of the standard PDF's, say the Weibull distribution, can be fitted to the data and the distribution's parameters estimated, the reliability can be extrapolated beyond the end of the test interval. The danger of doing this, of course, is that different failure modes will appear only after longer periods of time, thus invalidating the extrapolation.

Multiply Censored Data

Data are said to be multiply censored if uints are removed at various times during a life test. Such removals are usually required either because a mechanism that is not under study caused failure or because the unit is for some other reason no longer available for testing. Suppose, for example, that records are being kept on a fleet of trucks to determine the time to failure of the transmission. Trucks that are destroyed by severe accidents would be withdrawn from the test, assuming that a transmission failure was not the cause of the accident. Moreover, from time to time some of the trucks might be sold or for other reasons removed from the test population.

When trucks are removed for such reasons, it is easy to pretend that the removed units were not part of the original sample. This would not bias the results, provided that the censored units were representative of the total sample, but it would amount to throwing away valuable data with a concomitant loss in precision of the life-testing results. It is preferable to include the effects of the removed but unfailed units in determining the reliability.

Multiple censoring may be called for even in situations in which all the test units are run to failure, for, in a complex piece of machinery, analysis may indicate two or more different failure modes. Thus, it may prove particularly advantageous to remove units that have not failed from the mode under study in order to describe a particular failure mode with parameters of one of the distributions discussed in the following section.

In what follows we examine the nonparametric analysis of multiply censored data. The techniques that are employed have been developed the most extensively in the biomedical community, but they are also applicable to technological systems. Once the censoring is carried out and the reliability estimate $\hat{R}(t_i)$ is available at each failure or censor time, the substitution $\hat{F}(t_i) = 1 - \hat{R}(t_i)$ may be made and the $\hat{F}(t_i) = 1 - \hat{R}(t_i)$ the $\hat{F}(t_i)$ used as a starting point for the parametric analysis of the following section.

Ungrouped censored data take the form shown in Table 5.5. They consist of a series of times, $t_1, t_2, \ldots, t_i, \ldots, t_N$. Each of these times represents the removal of a unit from the test. The removal may be due to failure, or it may be due to censoring (i.e., removal for any other reason). The convention is to indicate the times associated with censoring removals by placing a plus sign (+) after the number.

A number of techniques have been used to estimate reliabilities from multiply censored data. We employ a modification of the product-limit estimate* which for the case of uncensored data reduces to Eq. 5.4. We begin by deriving a recursive relation for $\hat{R}(t_i)$ in terms of $\hat{R}(t_{i-1})$. First, without

TABLE 5.5 Failure Times

27	39	40+	54	69
85+	93	102	135+	144

* E. L. Kaplan and P. Meier, "Non-parametric Estimation from Incomplete Observations," *J. Am. Stat. Assoc.*, **53,** 457–481 (1958); see also R. G. Miller, Jr., *Survival Analysis,* Wiley, New York, 1981.

censoring, it follows from Eq. 5.4 that

$$\hat{R}(t_{i-1}) = \frac{N + 2 - i}{N + 1}. \tag{5.19}$$

By taking the ratio

$$\frac{\hat{R}(t_i)}{\hat{R}(t_{i-1})} = \frac{N + 1 - i}{N + 2 - i} \tag{5.20}$$

we obtain

$$\hat{R}(t_i) = \frac{N + 1 - i}{N + 2 - i}\hat{R}(t_{i-1}). \tag{5.21}$$

This expression may be interpreted in light of the definition of a conditional probability given by Eq. 2.4. The probability that a unit survives to t_i [i.e., $\hat{R}(t_i)$] is just the product of the probability that it survives to t_{i-1})[i.e. $\hat{R}(t_{i-1})$] multiplied by the conditional probability (i.e. $(N + 1 - i)/(N + 2 - i)$) that it will not fail between t_{i-1} and t_i, given that it is operating at t_{i-1}. Thus, at each t_i at which a failure takes place, we reduce the reliability by using Eq. 5.21.

In the event that a censoring action takes place at t_i, the reliability should not change. Therefore, we take

$$\hat{R}(t_i) = \hat{R}(t_i - 1). \tag{5.22}$$

Equations 5.21 and 5.22 can be combined in the following way. Suppose then we let

$$\delta_i = \begin{cases} 1, & \text{(failure at } t_i\text{)}, \\ 0, & \text{(censor at } t_i\text{)}. \end{cases} \tag{5.23}$$

We may thus write the reliability for multiply censored data as

$$\hat{R}(t_i) = \left(\frac{N + 1 - i}{N + 2 - i}\right)^{\delta_i} \hat{R}(t_{i-1}). \tag{5.24}$$

Or correspondingly, with $R(0) = 1$,

$$\hat{R}(t_i) = \prod_{i'=0}^{i} \left(\frac{N + 1 - i'}{N + 2 - i'}\right)^{\delta_{i'}}. \tag{5.25}$$

In practice this estimate is used to calculate the values of the reliability only at the values of t_i at which failures occur. The time dependence of the reliability between these points may then be interpolated, for instance, by straight-line segments. Once again, the procedure is best illustrated with an example.

TABLE 5.6 Censored Computations

i	t_i	$\dfrac{N+1-i}{N+2-i}$	$\hat{R}(t_i)$
1	27	$\frac{10}{11}$	$R(27) = \frac{10}{11} \times 1.0000 = 0.9\overline{090}$
2	39	$\frac{9}{10}$	$R(39 = \frac{9}{10} \times 0.9\overline{090} = 0.8\overline{181}$
3	40+	$\frac{8}{9}$	—
4	54	$\frac{7}{8}$	$R(54) = \frac{7}{8} \times 0.8\overline{181} = 0.7\overline{1590}$
5	69	$\frac{6}{7}$	$R(69) = \frac{6}{7} \times 0.7\overline{159} = 0.61\overline{36}$
6	85+	$\frac{5}{6}$	—
7	93	$\frac{4}{5}$	$R(93) = \frac{4}{5} \times 0.61\overline{36} = 0.4\overline{909}$
8	102	$\frac{3}{4}$	$R(102) = \frac{3}{4} \times 0.4\overline{909} = 0.36\overline{81}$
9	135+	$\frac{2}{3}$	—
10	144	$\frac{1}{2}$	$R(144) = \frac{1}{2} \times 0.36\overline{81} = 0.18\overline{409}$

EXAMPLE 5.3

Ten motors underwent life testing. Three of these motors were removed from the test and the remaining ones failed. The times in hours are given in Table 5.5. Use the product-limit method to plot the motor reliability versus time.

Solution The necessary calculations are indicated in Table 5.6 and the reliability is plotted in Fig. 5.3.

It should be noted that once the reliability at those values of t_i at which failure takes place has been estimated, Eqs. 5.7 and 5.9 may be used to

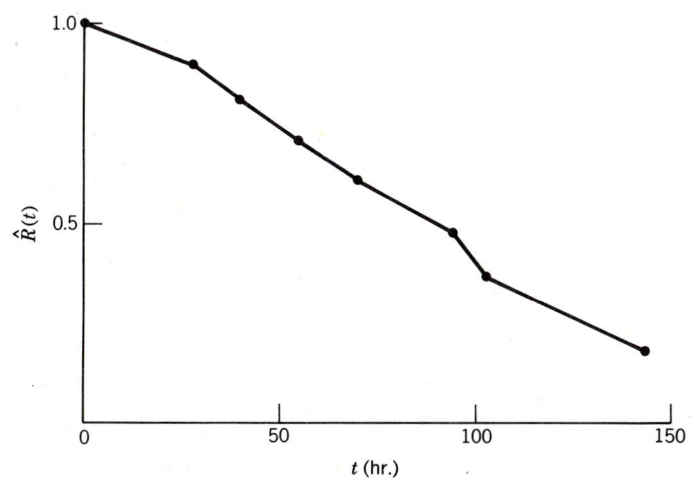

FIGURE 5.3 Reliability estimate from censored life data.

estimate PDF and failure rate, provided that only the values at which failure actually take place are used as t_i and t_{i+1}.

Accelerated Life Testing

Having too little time to complete testing is an ubiquitous problem in making reliability estimates. Frequently, the process of design and manufacture must take place over a much shorter time span than the design life of the product. As a result, there are at least two points in the product life cycle when time deadlines and the need for reliability data are likely to create serious conflicts.

At the completion of the design phase, the time available to perform life testing on prototypes before the design is frozen and mass production is begun is likely to be inadequate. Thus there is a danger that many units will be produced before a design weakness is discovered, creating the difficulty and expense of after-the-fact fixes to the product.

Invariably, some of the first units off the production line are likely to be allocated to reliability testing to ensure that the as-built units meet the reliability requirements placed on the design. Before adequate reliability data can be amassed, however many units may have been produced and shipped, and indeed the production run may have been completed. Thus, if reliability shortcomings are found, they again are likely to affect the large number of units already produced and inspected. To some degree this problem is circumvented by rigorous acceptance testing, provided that tests which do not consume significant amounts of time can be devised to predict reliability properties. Even if the more time-comsuming life tests cannot be completed for timely feedback into the manufacturing process, they still serve some vital functions, such as predicting the necessary quantities of repair parts and providing data for improving the design of subsequent products.

Because of the difficulties encountered in performing life testing with time deadlines, a variety of procedures are used to accelerated life tests. Although none of them is without shortcomings, these procedures nevertheless contribute substantially to the timeliness with which reliability data are obtained. Accelerated tests can be divided roughly into two categories, compressed-time tests and advanced stress tests.

Compressed-Time Testing

In compressed-time testing a product is used more frequently in the life test than it would be in normal use, but the loads and environmental stresses on the product are maintained at the level expected in normal use. Compressed-time testing is applied in its most straightforward fashion when the failure rate—and wearout in particular—depends on the number of cycles that a system goes through rather than on a period of continuous operation.

Consider, as an example, the latch on an automobile door. In field use the door may only be cycled (opened and closed) several times per day. But a compressed-time test can easily be performed in which the open-close cycle is performed a few times per minute. Certain precautions must be adhered to, of course. If the cycle is accelerated too much, the conditions of operation may change, increasing stress levels and thus artificially increasing failure rates. If the latch is worked several times per second, for example, the heat of friction may not have time to dissipate. This, in turn, would cause the latch to overheat, increasing the failure rate and perhaps activating failure mechanisms that would not plague ordinary operation. Nor does this test adequately treat failure caused by corrosion or other failure modes that appear with time. For these effects to be studied, other acceleration techniques of the advanced stress variety must be applied.

Systems that normally operate continuously but intermittently may also be subjected to compressed-time. An automobile engine that normally does not operate more than a few hours per day may be operated continuously until failure occurs. Of course, this will not pick up the cyclical failures caused by the starting and stopping of the engine. For this a separate cycling test is required, or the continuous operation must be interrupted by intervals long enough for the engine to cool to ambient temperatures.

Advanced Stress Testing

Systems that are normally in continuous operation and failures caused by deterioration that occurs even though a unit is inactive present some of the most difficult problems in accelerated testing. In such situations advanced stress testing may be employed to accelerate failures. Generally, if increased loads or harsher environments are applied to a device, the failure rate will increase. Thus, if a decrease in reliability can be related to an increase in stress level, the life tests can be performed at high stress levels, and the reliability at normal stress levels can be inferred.

For example, suppose that the MTTF is estimated at the number of different elevated stress levels. Such stress might typically be temperature, voltage, radiation intensity, mechanical stress, or any number of other variables. The MTTF or other reliability parameter is then plotted versus the stress level, as indicated in Fig. 5.4. A curve is fitted to the data, and the MTTF is estimated at the stress level that the device is expected to experience during normal operation.

Both random failures and aging effects may be the subject of such tests. In the electronics industry components are tested at elevated temperatures to increase the incidence of random failure. In the nuclear industry pressure vessel steels are exposed to extreme levels of neutron irradiation to increase

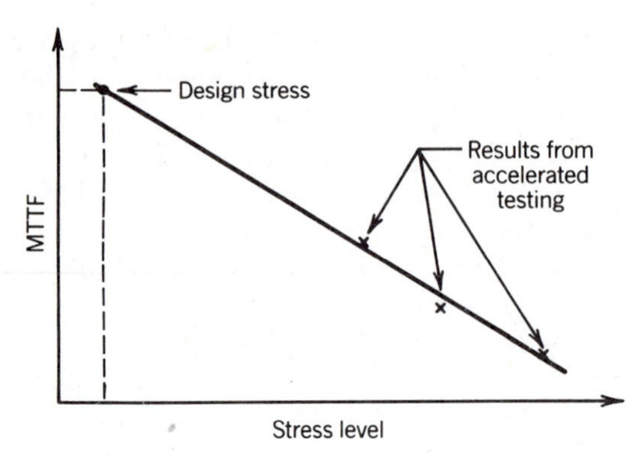

FIGURE 5.4 Estimate of MTTF from accelerated test data.

the rate of embrittlement. Similarly, placing equipment under a high stress level for a short period of time in a proof test may be considered accelerated testing to reveal the early failures from defective manufacture.

Extrapolation from high- to lower-stress conditions, however, is often problematical. It may be possible to run accelerated tests at only one or two stress levels, thus making the relation between stress and reliability difficult to define on the basis of the test data. Even if such a relationship can be defined after carefully accelerated tests have been run at several stress levels, the extrapolation to lower-stress levels may be nonlinear, and failure mechanisms that are important at high-stress levels may no longer be dominant at nominal levels.

Accelerated testing is the most successful when the failure mechanisms are already understood, and can be described quantitatively. For example, a wide variety of chemical reactions, whether they be corrosion of metals, breakdown of lubricants, or diffusion of semiconductor materials, obey the Arrhenius equation*

$$\text{rate of process} = A \exp(-Q/RT), \tag{5.26}$$

where A is a constant, Q is the activation energy, R is the gas constant, and T is the absolute temperature. Thus, for systems in which chemical reactions are responsible for failure, an increase in temperature increases the failure rate in a prescribed manner. For example, if the real time of interest is t, the accelerated test time t_A can be reduced to

$$\frac{t}{t_A} = \exp\left[+\frac{Q}{R}\left(\frac{1}{T} - \frac{1}{T_A}\right)\right], \tag{5.27}$$

* See, for example, M. L. Shooman, *Probabilistic Reliability: An Engineering Approach,* McGraw-Hill, New York, 1968, Chapter 8.

where T is the normal temperature, and T_A is the temperature at which the accelerated test is performed.

Other time-scaling laws are also available. For example, tests of electric equipment are often accelerated by increasing the voltage at which the test units are operated. Here the scaling law is*

$$\frac{t}{t_A} = \left(\frac{V_A}{V}\right)^3, \tag{5.28}$$

where V_A and V are the accelerated test and normal rated voltage, respectively. Other more empirical relations that apply to humidity and other environmental factors are also available.

Accelerated testing is useful, but it must be carried out with great care to ensure that results are not erroneous. We must know for sure that the phenomena for which the acceleration factor (t/t_A) has been calculated are the failure mechanisms. Experience gained with similar products and a careful comparison of the failure mechanisms occuring in accelerated and real time tests will help determine whether we are testing the correct phenomena.

5.4 PARAMETRIC METHODS

Plotting the reliability or other quantites versus time as in the preceding sections often yields valuable information. In general, however, it is more desirable to fit the reliability data to some particular distribution, such as the exponential, normal, or Weibull. For if this can be accomplished, a great deal more can often be determined about the nature of the failure mechanisms, and the resulting model can be used more readily in the analytical techniques discussed in the following chapters.

In order to obtain parametric models for failure distributions, we must first determine what distribution will adequately represent the data and then determine the parameters. There are a variety of advanced statistical methods for determining the goodness of fit of data to a particular distribution, for estimating the parameters for the distribution, and for calculating confidence levels for each parameter.† In what follows, however, we confine our attention to relatively simple graphical methods. Such techniques allow us to evaluate the goodness of fit visually, without using advanced mathematics, and at the same time to estimate the parameters that define the distribution.

In general, the procedure that we follow consists of choosing a distri-

* C. M. Ryerson, "Acceptance Testing," in *Reliability Handbook,* W. G. Ireson (ed.), McGraw-Hill, New York, 1966.

† See, for example, J. F. Lawless, *Statistical Models and Methods for Lifetime Data,* Wiley, New York, 1982; also Nelson, op. cit.

bution and then plotting ungrouped failure data on the appropriate graph paper for this distribution. If the data are described by the distribution, the data points will be clustered along a straight line. The parameters are then estimated from the slope and intercept of the line. For the exponential, Weibull, normal and lognormal distributions we derive the relationships necessary to understand the construction of the graph and to compute the parameters from the resulting line. Similar techniques are available for other distributions, such as the extreme-value distributions discussed in Chapter 6. A blank sheet of the graph for each distribution is provided in Appendix D; these may be reproduced by the reader for the purpose of working the exercises at the end of the chapter.

Exponential Distribution

Often the exponential distribution or constant failure rate model is the first to be used when we attempt to parameterize data. In addition to being the only distribution for which only one parameter must be estimated, it provides a reasonable starting point for considering other two or three-parameter distributions. For as will be seen, the distribution of the data may indicate whether the failure rate is increasing or decreasing, and this in turn may provide insight whether another distribution should be considered.

To plot data, we begin by taking the natural logarithm of Eq. 4.25,

$$\ln R = -\lambda t; \tag{5.29}$$

or using $\ln(1/R) = -\ln R$, we have

$$\ln(1/R) = \lambda t. \tag{5.30}$$

It is customary to construct graph paper in terms of $F = 1 - R$, the CDF. Thus we have

$$\ln\left(\frac{1}{1-F}\right) = \lambda t. \tag{5.31}$$

An exponential distribution probability paper is shown in Fig. 5.5. The numerical values labeled on the vertical axis are those of $\hat{F}(t_i)$, which may be obtained from Eq. 5.3,

$$\hat{F}(t_i) = \frac{i}{N+1}, \tag{5.32}$$

where N is the number of test units. It will be noted that $\lambda t = 1$ when $1 - F = e^{-1}$ or $F = 0.632$. Thus the value of $1/\lambda$ is equal to the time at which $F = 0.632$. The data through which the straight line is drawn on Fig. 5.5 come from the following example.

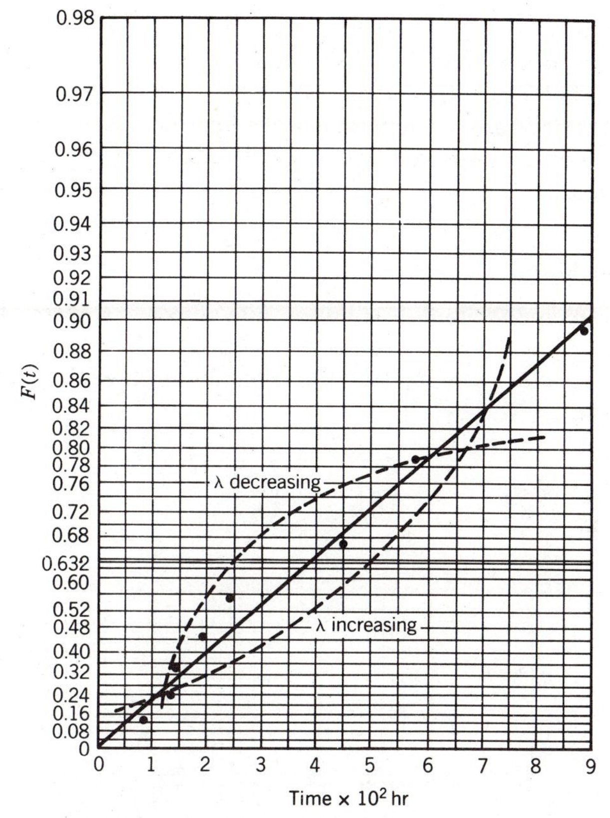

FIGURE 5.5 Graphical parameter estimation for the exponential distribution.

EXAMPLE 5.4

The following are the failure times from eight control circuits in hours: 80, 134, 148, 186, 238, 450, 581, and 890. Estimate the failure rate by making a plot on exponential distribution probability paper.

Solution The calculations are carried out in Table 5.7. From Fig. 5.5 we see that $F = 0.632$ when $t \approx 400$ hr. Therefore we estimate $\lambda = 0.0025/\text{hr}$. ◀

TABLE 5.7 Exponential Calculations

i	t_i	$\frac{i}{N+1}$	i	t_i	$\frac{i}{N+1}$
1	80	0.111	5	238	0.555
2	134	0.222	6	450	0.666
3	148	0.333	7	581	0.777
4	186	0.444	8	890	0.888

The following is an important feature of plotting failure times on logarithmic paper. If the failure rate is not constant, the curvature of the data may indicate whether the failure rate is increasing or decreasing. The dotted lines on Fig. 5.5 indicate the general pattern that the data would follow were the failure rate increasing (concave upward) or decreasing (concave downward) with time.

Weibull Distribution

The two-parameter Weibull distribution may also be estimated by plotting failure times on specially constructed graph paper. To arrange the Weibull data on a straight line, we first take the logarithm of the Weibull expression for the reliability given by Eq. 4.58:

$$\left(\frac{t}{\theta}\right)^m = \ln\left(\frac{1}{R}\right). \tag{5.33}$$

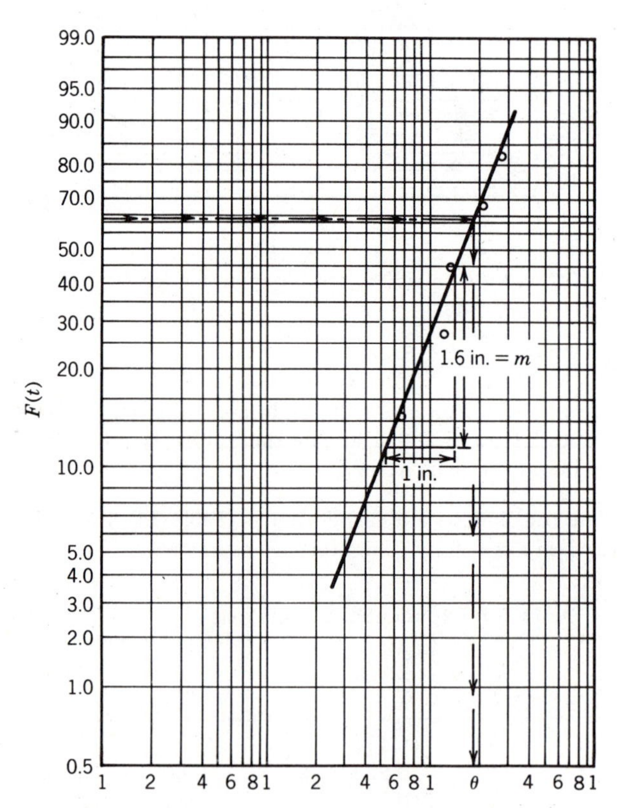

FIGURE 5.6 Graphical parameter estimation for the Weibull distribution.

Then, taking the logarithm again, we obtain

$$\ln t = \frac{1}{m} \ln\left[\ln\left(\frac{1}{R}\right)\right] + \ln \theta. \tag{5.34}$$

If we rewrite this equation as

$$\ln\left[\ln\left(\frac{1}{R}\right)\right] = m \ln t - m \ln \theta, \tag{5.35}$$

we see that it has the form $y = mx + b$, where the ordinate is $\ln[\ln(1/R)]$ and the abscissa is $\ln t$. Such graph paper is shown in Fig. 5.6. Once again it is the convention to number the vertical axis values of the CDF or $F = 1 - R$ rather than R. Thus, using Eq. 5.35, we plot

$$\ln\left[\ln\left(\frac{1}{1-F}\right)\right] = m \ln t - m \ln \theta, \tag{5.36}$$

where $\hat{F}(t_i)$ is given by Eq. 5.32 for the ith failure of the N test units, and the failure times are plotted on the horizontal axis. The two Weibull parameters are estimated directly from the straight line. As shown, the slope m is obtained by drawing a right triangle with a horizontal side of length one; the length of the vertical side is then the slope. The value of θ is estimated by noting that the left-hand side of Eq. 5.36 vanishes when $F = 0.632$ yielding $t = \theta$. The data and line of Fig. 5.6 result from the following problem.

EXAMPLE 5.5
Estimate the Weibull parameters for the life test data: 67, 120, 130, 220, and 290 hr.

Solution The corresponding values of $\hat{F}(t_i)$ are $i/(N + 1) = 0.166, 0.33, 0.500, 0.666$, and 0.833. The data are plotted in Fig. 5.6. The parameters are estimated to be m = 1.6 ◀ and θ = 1.9 hr. ◀

The three-parameter Weibull distribution is very useful when no failures occur before some threshold time. If such data are plotted on two-parameter Weibull paper, the result is a concave downward line as indicated in Fig. 5.7.

One possible way to correct for the curvature is to select a threshold, t_o equal to the first failure time t_1. This step has often been found in practice to overcompensate for the curvature and turn it upward. A more fruitful procedure is to choose a smaller threshold of the form*

$$t_o = a t_1, \tag{5.37}$$

* K. C. Kapur and L. R. Lamberson, *Reliability in Engineering Design,* Wiley, New York, 1977.

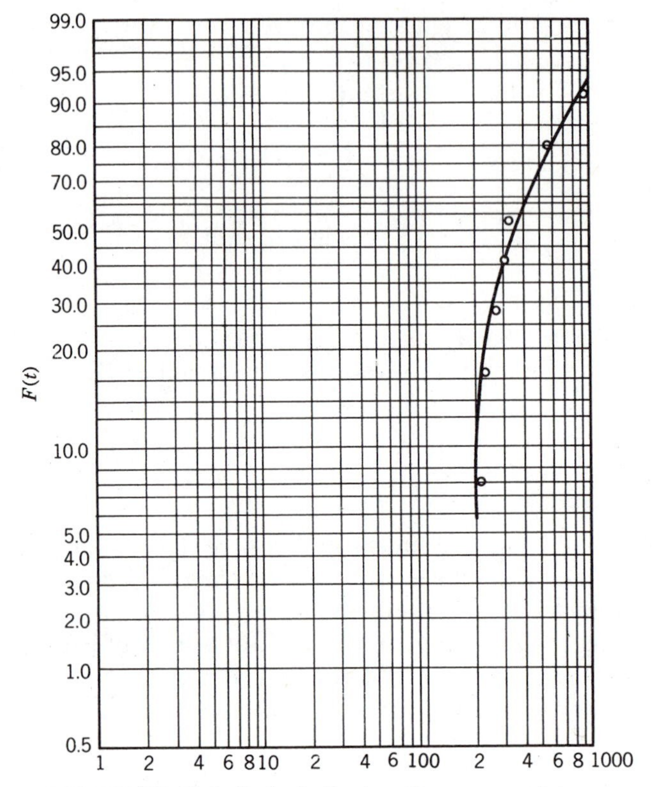

FIGURE 5.7 Weibull plot indicative of a nonzero minimum life.

where a is a number less than one. Some trial and error is usually necessary, until an approximately linear plot is obtained. For the present data a value of $a = 0.89$ is used to construct a three-parameter Weibull distribution. Instead of Eq. 5.36, we now obtain from Eq. 4.65,

$$\ln\left[\ln\frac{1}{1 - F}\right] = m \ln(t - t_o) - m \ln \theta \tag{5.38}$$

to be plotted. In other words we simply replace t by $t - t_o$ on the horizontal axis. The slope and therefore m is determined in the same manner as before. The graph is shown in Fig. 5.8.

Normal Distribution

Graphical methods may also be used to determine whether a sequence of failure times or other data may be approximated by a normal or a lognormal

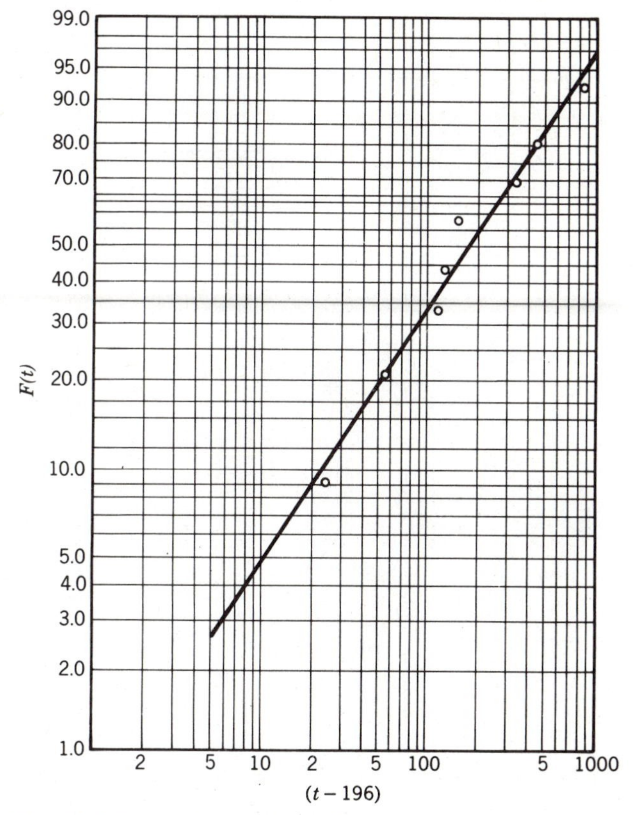

FIGURE 5.8 Weibull plot adjusted for a nonzero minimum life.

distribution. We begin with the standardized CDF for the normal distribution given by Eq. 4.43,

$$F(t) = \Phi\left(\frac{t - \mu}{\sigma}\right), \tag{5.39}$$

where μ is the mean and σ is the standard deviation. Normal paper is based on inverting Eq. 5.39 to obtain

$$\Phi^{-1}(F) = \frac{1}{\sigma}t - \frac{\mu}{\sigma}. \tag{5.40}$$

Here the inverse of the standardized normal distribution, $\Phi^{-1}(F)$, is plotted on the vertical axis and time is plotted on the horizontal axis. If the failure data are normally distributed, the line will be straight. Such normal plotting paper is shown in Fig. 5.9. Since $F(\mu) = 0.5$ for a normal distribution, the mean, μ, is determined as the time for which $F = 0.5$, as shown. Similarly, since $F(\mu + \sigma) = 0.84$, the value of σ is just the horizontal distance between

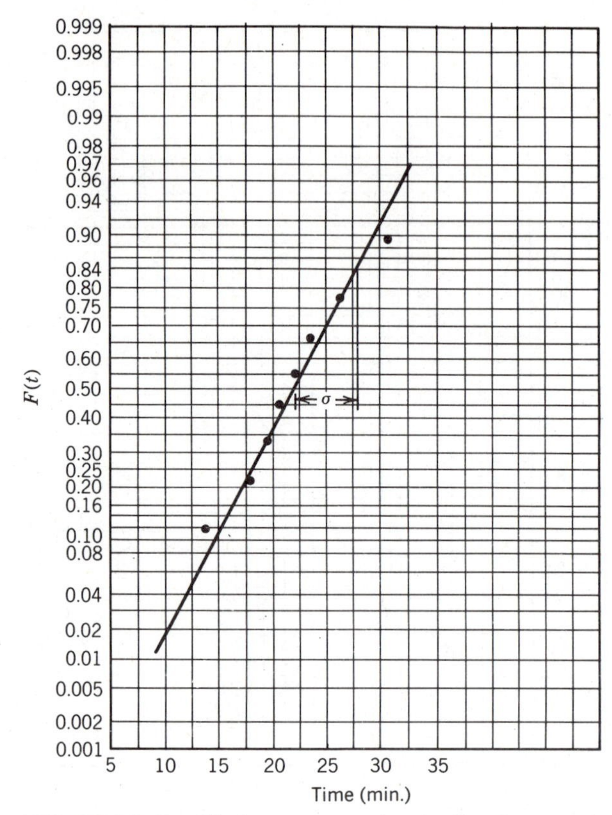

FIGURE 5.9 Graphical parameter estimation for the normal distribution.

the points $F = 0.84$ and $F = 0.5$, as shown. The data plotted in Fig. 5.9 are from the following problem.

EXAMPLE 5.6

The following times were recorded for the wearout time (in minutes) for the cutting edge on a machine tool:

18.1, 23.2, 26.5, 20.6, 19.3, 22.3, 13.7, 31.0.

Plot the data on normal distribution paper and (*a*) estimate μ and σ from the graph and (*b*) from Eqs. 5.10 and 5.11.

Solution (*a*) First reorder the data and calculate $\hat{F}(t_i) = \dfrac{i}{N+1} = \dfrac{i}{9}$:

i	1	2	3	4	5	6	7	8
t_i	13.7	18.1	19.3	20.6	22.3	23.2	26.5	31.0
$\hat{F}(t_i)$	1/9	2/9	3/9	4/9	5/9	6/9	7/9	8/9

The results are plotted in Fig. 5.9. We obtain from the graph

$$\hat{\mu} = 22 \text{ min}, \blacktriangleleft \qquad \sigma = 28 - 22 = 6.0 \text{ min}. \blacktriangleleft$$

(*b*) From Eq. 5.10 we have

$$\hat{\mu} = \frac{1}{N}\sum_i t_i = \frac{174.7}{8} = 21.8 \text{ min}. \blacktriangleleft$$

From Eq. 5.11 we have

$$\hat{\sigma}^2 = \frac{N}{N-1}\left(\frac{1}{N}\sum_i t_i^2 - \hat{\mu}^2\right).$$

With

$$\frac{1}{N}\sum_i t_i^2 = \frac{4011}{8} = 501.4$$

we obtain

$$\hat{\sigma}^2 = \frac{8}{7}(501.4 - 21.8^2) = 29.9$$

or

$$\hat{\sigma} = 5.5. \blacktriangleleft.$$

Lognormal Distribution

The lognormal distribution paper is constructed very similarly to normal paper. From Eq. 4.50 we have

$$F(t) = \Phi\left[\frac{1}{s}\ln\left(\frac{t}{t_0}\right)\right], \tag{5.41}$$

where $\ln t_0$ is the log mean and s is the log standard deviation. We again invert the standardized normal distribution to obtain

$$\Phi^{-1}(F) = \frac{1}{s}\ln t - \frac{1}{s}\ln t_0. \tag{5.42}$$

Equations 5.40 and 5.42 are seen to be analogous; lognormal paper is the same as normal paper except that the logarithm of time instead of time is plotted on the horizontal axis. As indicated in Fig. 5.10, with the data from the following example problem, the value of the slope, $1/s$, is determined from a right triangle, and the value to t_0 corresponds to $F = 0.5$. Once t_0 and s are known, the mean and variance, μ and σ^2, may be determined from Eqs. 4.51 and 4.52, respectively.

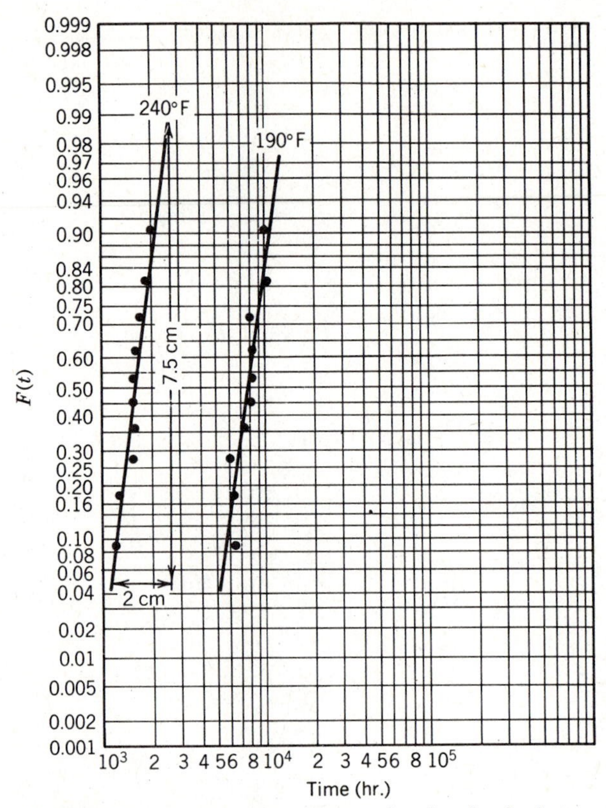

FIGURE 5.10 Graphical parameter estimation for the lognormal distribution.

EXAMPLE 5.7

Twenty specimens of a new electric insulation are life-tested, half at 190°F and the other half at 240°F. The failure times are as follows (in hours).*

At 190°F: 7228, 7228, 7228, 8448, 9167, 9167, 9167, 9167, 10511, 10511.

At 240°F: 1175, 1175, 1521, 1567, 1617, 1665, 1665, 1713, 1761, 1953.

Some equal failure times occur, because failure was determined by periodic inspection. The times given are the midpoints of the period during which failure occurred. Plot the results on lognormal paper and estimate the parameters.

Solution For each temperature there are 10 points. Therefore, the successive times should be plotted at

$$\hat{F}(t_i) = \frac{i}{N+1} = \frac{i}{11}.$$

* Data from Nelson, op. cit.

The results are shown in Fig. 5.10.

For 190°F: $t_0 \approx 7800$ hr, $s \approx 2.0$ cm/7.5 cm $= 0.27$. ◀

For 240°F: $t_0 \approx 1600$ hr, $s \approx 2.0$ cm/7.5 cm $= 0.27$. ◀

Thus the increased temperature strongly decreases the median time to failure and also the MTTF, but it hardly affects the logstandard deviation of the result at all.

5.5 FAILURE RATE ESTIMATES

In this and the following section we examine in more detail the testing procedures for determining the MTTF when the data are exponentially distributed. This is justified both because the exponential distribution (i.e., the constant failure rate model) is the most widely applied in reliability engineering, and because it provides insight into the problems of parameter estimation that are indicative of those encountered with other distributions.

We must, of course, determine whether the constant failure rate model is applicable to the test at hand. At least four approaches to this problem may be taken. The exponential distribution may be assumed, based on experience with equipment of similar design. It may be identified by using one of the standard statistical goodness-of-fit criteria or by plotting the times to failure on exponential paper, as described in the preceding section, and examining the results visually for the required straight-line behavior. Finally, it may be argued from the failure mode whether the failures are random, as opposed to early or aging failures. If defective products or aging effects are identified as causing some of the failures, the data may be censored appropriately.

Given that we have decided on a model, in this case the exponential distribution, there are two approaches to the parameter estimation, sampling and Bayesian analysis. In sampling we use only the test data to estimate the parameter and to establish a confidence interval on the estimate. In this case the discussion centers on how to obtain the best estimates at the least cost, and in particular on what types of deliberate censoring may be beneficial. In the Bayesian analysis discussed in Section 5.6, we assume that there already exists some estimate of the range within which the parameter should lie. The test data—however meager—are used to upgrade that estimate. The relative merits of sampling and of Bayesian analysis are discussed at length elsewhere.*

The exponential distribution has only a single parameter to be estimated, the failure rate λ. Rather than estimate the failure rate directly, most sampling schemes are cast in terms of the MTTF, denoted by MTTF $\equiv \mu = 1/\lambda$. For uncensored data the value of μ may be estimated from Eq. 5.10. Moreover, when N, the number of test specimens, is sufficiently large, the central limit theorem, which was discussed in Chapter 3, may be used to

* H. G. Martz and R. A. Waller, *Bayesian Reliability Analysis,* Wiley, New York, 1982.

estimate a confidence interval. In particular, the 69% confidence interval is given by $\hat{\mu} \pm \sigma/\sqrt{N}$, where σ^2 is the variance of the distribution being used. Since for the exponential distribution $\sigma = \mu$, we may estimate the 69% confidence interval from $\hat{\mu} \pm \hat{\mu}/\sqrt{N}$.

Censoring on the Right

It is clear from the foregoing expressions that for a precise estimate a large sampling size is required. Using many test specimens is expensive, but, more important, a very long time is required to complete the test. As N becomes large, the last failure is likely to occur only after several MTTFs have elapsed. Moreover, the analysis of the failures that occur after long periods of time is problematic for two reasons. First, a design life is normally less than the MTTF, and it is often not possible to hold up final design, production, or operation while tests are carried out over many design lives. Equally important, many of the last failures are likely to be caused by aging effects. Thus they must be removed from the data by censoring if a true picture of the random failures is to be gained.

Type I and type II censoring from the right are attractive alternatives to uncensored sampling. By limiting the period of the test while increasing the number of units tested, we can eliminate most of the aging failures, and estimate more precisely the time-independent failure rate. Within this framework four different test plans may be used. With the assumption that the test is begun with N test units, these plans may be distinguished as follows. If the test is terminated at some specified time, say t_*, then type I censoring is said to take place. If the test is terminated immediately after a particular number of failures, say n, then type II censoring is said to take place. With either type I or type II censoring, we may run the test in either of two ways. In the nonreplacement method each unit is removed from the test at the time of failure. In the replacement method each unit is immediately repaired or replaced following failure so that there are always N units operating until the test is terminated.

The choice between type I and type II censoring involves the following trade-off. Type I censoring is more convenient because the duration of the test t_* can be specified when the test is planned. The time t_n of the nth failure, at which a test with type II censoring is terminated, however, cannot be predicted with precision at the time the test is planned, for t_n is a random variable. Conversely, the precision of the measurement of the MTTF for the exponential distribution is a function of the number of failures rather than of the test time. Therefore, it is often considered advisable to wait until some specified number of failures have occurred before concluding the test.

A number of factors also come into play in determining whether nonreplacement or replacement tests are to be used. In laboratory tests the cost of the test units compared with the cost of the apparatus required to perform the test may be the most significant factor. Consider two extreme examples.

First, if jet engines are being tested, nonreplacement is the likely choice. When a specified number of engines are available, more will fail within a given length of time if they are all started at the same time than if some of them are held in reserve to replace those that fail. The same is true of any other expensive piece of equipment that is to be tested as a whole.

Conversely, suppose that we are testing fuel injectors for large internal-combustion engines. The supply of fuel injectors may be much larger than the number of engines upon which to test them. Therefore, it would make sense to keep all the engines running for the entire length of the test by immediately replacing each fuel injector following failure, provided that the replacement can be carried out swiftly and at minimum cost. Minimizing cost is an important provision, for generally the personnel costs are larger with replacement tests; in nonreplacement tests personnel or instrumentation is required only to record the failure times. In replacement tests personnel and equipment must be available for carrying out the repairs or replacements within a short period of time.

The situation is likely to be quite different when the data are to be accumulated from actual field experience with breakdowns. Here, in the normal course of events, equipment is likely to be repaired or replaced over a time span that is short compared to the MTTF. Conversely, records may indicate only the number of breakdowns, not when they occurred. The number of breakdowns might be inferred, for example, from spare parts orders or from numbers of service calls. In these circumstances replacement testing describes the situation. Moreover, unlike nonreplacement testing, the MTTF estimation does not require that the times of failures be recorded.

One last class of test remains to be mentioned. Sometimes referred to as percentage survival, it is a simple count of the fraction (or percentage) of failed units. From the properties of the exponential distribution, we infer the MTTF. This test procedure requires no surveillance, for failed equipment does not need to be replaced or times of failure recorded. Not surprisingly, the estimate obtained is less precise. The method is normally not recommended, unless failures are not apparent at the time they take place and can only be determined by destructive testing or other invasive techniques following the conclusion of the test.

MTTF Estimates

With the exception of the percentage survival technique, the same estimator may be shown to be valid for all the test procedures described:*

$$\hat{\mu} = \frac{T}{n},$$

$$T = \text{total operational time of all test units}, \tag{5.43}$$

$$n = \text{number of failures}.$$

* I. Bazovsky, *Reliability Theory and Practice,* Prentice-Hall, Englewood Cliffs, NJ, 1961.

For each class of test, however, the total operating time T is calculated differently.

Consider first nonreplacement testing with type I censoring (i.e., the test is terminated at some predetermined time t_*). If $t_1, t_2, \ldots, t_n$ are the times of the n failures, the total operational time T for the N units tested is

$$T = \sum_{i=1}^{n} t_i + (N - n)t_*, \tag{5.44}$$

since $N - n$ units operate for the full time t_*.

EXAMPLE 5.8

A 30-day nonreplacement test is carried out on 20 rate gyroscopes. During this period of time 9 units fail; examination of the failed units indicates that none of the failures is due to defective manufacture or to wear mechanisms. The failure times (in days) are 27.4, 13.5, 10.5, 20.0, 23.6, 29.1, 27.7, 5.1, 14.4. Estimate the MTTF.

Solution From Eq. 5.44 with $N = 20$ and $n = 9$,

$$T = \sum_{i=1}^{9} t_i + (20 - 9) \times 30$$

$$= 171.3 + 11 \times 30 = 501.3$$

$$\hat{\mu} = \frac{T}{n} = \frac{501.3}{9} = 55.7 \text{ days.} \blacktriangleleft$$

For type II censoring the test is stopped at t_n, the time of the nth failure. Thus, if there is no replacement of test units, the total operating time is calculated from

$$T = \sum_{i=1}^{n} t_i + (N - n)t_n, \tag{5.45}$$

since the unfailed $(N - n)$ units are taken out of service at the time of the nth failure. Note that in the event that some of the units, say k of them, are removed from the test because they fail from another mechanism, such as aging, then T is still calculated by Eq. 5.44 or Eq. 5.45. Now, however, the estimate is obtained by dividing only by the number $n - k$ of random failures:

$$\hat{\mu} = \frac{T}{n - k}. \tag{5.46}$$

EXAMPLE 5.9

The engineer in charge of the test in the preceding problem decides to continue to test until 10 of the 20 rate gyroscopes have failed. The tenth failure occurs at 41.2 days, at which time the test is terminated. Estimate the MTTF.

Solution From Eq. 5.45 with $N = 20$ and $n = 10$,

$$T = \sum_{i=1}^{10} t_n + (20 - 10)41.2$$

$$T = (171.3 + 41.2) + 10 \times 41.2 = 624.5$$

$$\hat{\mu} = \frac{T}{n} = \frac{624.5}{10} = 62.4 \text{ days.} \blacktriangleleft$$

In replacement testing all N units are operated for the entire length of the test. Thus, for type I censoring, we have $T = Nt_*$, where t_* is the specified test time. Hence

$$\hat{\mu} = \frac{Nt_*}{n}. \tag{5.47}$$

For type II censoring, we have $T = Nt_n$, where t_n is the time at which the nth unit fails. Thus $T = Nt_n$ or

$$\hat{\mu} = \frac{Nt_n}{n}. \tag{5.48}$$

EXAMPLE 5.10

A chemical plant has 24 process control circuits. During 5000 hr of plant operation the circuits experience 14 failures. After each failure the unit is immediately replaced. What is the MTTF for the control circuits?

Solution From Eq. 5.47,

$$T = Nt_* = 24 \times 5000 = 120{,}000$$

$$\hat{\mu} = \frac{T}{n} = \frac{120{,}000}{14} = 8571 \text{ hr.} \blacktriangleleft$$

EXAMPLE 5.11

Six units of a new high-precision pressure monitor are placed on an industrial furnace. After each failure the monitor is immediately replaced. However, the eighth failure occurs after only 840 hours of service. It is decided that the high-temperature environment is too severe for the instruments to function reliably, and the furnace is shut down to replace the pressure monitors with a more reliable, and expensive, design. Assuming that the failures are random, estimate the MTTF of the monitors.

Solution From Eq. 5.48,

$$T = Nt_8 = 6 \times 840 = 5040 \text{ hr}$$

$$\hat{\mu} = \frac{T}{n} = \frac{5040}{8} = 630 \text{ hr.} \blacktriangleleft$$

As alluded to earlier, the MTTF may also be estimated from the percentage survival method. We begin by first estimating the reliability at the end of the test, time t_0 as $R(t_0) = 1 - n/N$. With an exponential distribution, however, the reliability is given by

$$R(t_0) = \mathrm{Exp}(-t_0/\mu). \tag{5.49}$$

Thus, combining these equations, we estimate MTTF from

$$\mu = \frac{t_0}{\ln[1/(1 - n/N)]}. \tag{5.50}$$

EXAMPLE 5.12

A National Guard unit is supplied with 20,000 rounds of ammunition for a new model rifle. After 5 years, 18,200 rounds remain unused. From these 200 rounds are chosen randomly and test-fired. Twelve of them misfire. Assuming that the misfires are random failures of the ammunition caused by storage conditions, estimate the MTTF.

Solution In Eq. 5.50 take $n = 12$, $N = 200$, and $t_0 = 5$ years. We have

$$\hat{\mu} = \frac{5}{\ln\{1/[1 - 12/200]\}} = 81 \text{ years.} \blacktriangleleft$$

Confidence Intervals

We next consider the precision of the MTTF estimates made with Eq. 5.43. The confidence limits for both replacement and nonreplacement tests may be expressed in terms of $\hat{\mu}$ and the number of failures by using the χ^2 distribution.* The results are given conveniently by the curves shown in Fig. 5.11. We consider type II censoring first.

Let $U_{\alpha/2,n}$ and $L_{\alpha/2,n}$ be the upper and lower limits for the $100 \times (1 - \alpha)$ percent confidence interval for type II censoring. The two-sided confidence interval states that if the test is stopped after the nth failure, there is a $1 - \alpha$ probability that the true value of n lies between $L_{\alpha/2,n}$ and $U_{\alpha/2,n}$:

$$P\{L_{\alpha/2,n} \leq \mu < U_{\alpha/2,n}\} = 1 - \alpha. \tag{5.51}$$

It turns out that the ratios $L_{\alpha/2,n}/\hat{\mu}$ and $U_{\alpha/2,n}/\hat{\mu}$ are independent of the operating time T. Therefore, they can be plotted as functions of α and n, the number of failures. The plot is shown in Fig. 5.11. Thus, if $\hat{\mu}$ has been estimated from one of the forms of Eq. 5.43, the confidence interval can be read from Fig. 5.11. This is best illustrated by examples.

* Bazovsky, op. cit.

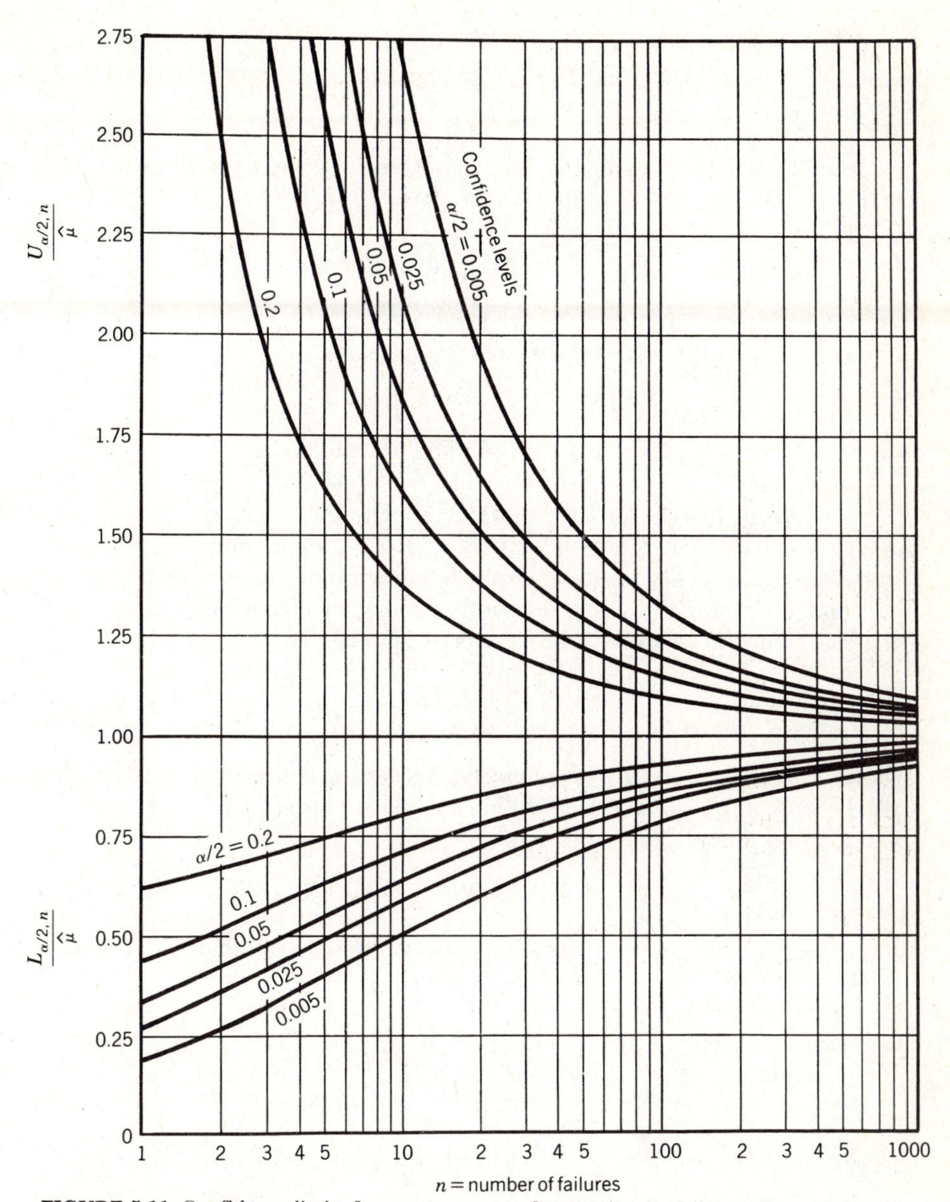

FIGURE 5.11 Confidence limits for measurement of mean-time-to-failures. (From Igor Bazovsky, *Reliability Theory and Practice*, © 1961, p. 241, with permission from Prentice-Hall, Englewood Cliffs, NJ.)

EXAMPLE 5.13

What is the 90% confidence interval for the rate gyroscopes tested in Example 5.9, taking the failure at 41.2 days into account?

Solution For a 90% confidence interval we have $100(1 - \alpha) = 90$, or $\alpha = 0.1$ and $\alpha/2 = 0.05$. For $n = 10$ failures we find from Fig. 5.11 that

$$\frac{L_{0.05,10}}{\hat{\mu}} \approx 0.65, \qquad \frac{U_{0.05,10}}{\hat{\mu}} \approx 1.82.$$

Therefore, using $\hat{\mu} = 62.4$ days from Example 5.9:

$$L_{0.05,10} \approx 0.65 \times 62.4 = 41 \text{ days}, \blacktriangleleft$$

$$U_{0.05,10} \approx 1.82 \times 62.4 = 114 \text{ days}, \blacktriangleleft$$

$$41 < \mu < 114 \text{ days} \blacktriangleleft \text{ with 90\% confidence.}$$

With slight modifications the results of Fig. 5.11 may also be applied to type I censoring, where the test is ended at some time t_*. Using the properties of the χ^2 distribution, it may be shown that the upper confidence limit and $\hat{\mu}$ remain the same. The lower confidence limit, in general, decreases. It may be related to the results in Fig. 5.11 by

$$\frac{L^*_{\alpha/2,n}}{\hat{\mu}} = \frac{n}{n+1}\frac{L_{\alpha/2,(n+1)}}{\hat{\mu}}, \tag{5.52}$$

where L^* is the value for type I censoring, and L is the plotted value for type II censoring. Again, the confidence limits are applicable to both nonreplacement and replacement testing.

EXAMPLE 5.14

During the first year of operation a demineralizer suffers seven shutdowns. Estimate the MTBF and the 95% confidence interval.

Solution From Eq. 5.43,

$$\hat{\mu} = \text{MTBF} = \frac{T}{n} = \frac{12 \text{ months}}{7} = 1.71 \text{ months.} \blacktriangleleft$$

For a 95% confidence interval $\alpha = 0.05$ and $\alpha/2 = 0.025$. From Fig. 5.11,

$$\frac{L^*_{0.025,n}}{\hat{\mu}} = \frac{n}{n+1}\frac{L_{0.025,n+1}}{\hat{\mu}} = \frac{7}{8}\frac{L_{0.025,8}}{\hat{\mu}} \approx \frac{7}{8} \times 0.57 = 0.50$$

$$L_{0.025,7} = 0.50 \times 1.71 = 0.86 \text{ month}, \blacktriangleleft$$

$$U_{0.025,7} = 2.5 \times 1.71 = 4.27 \text{ months.} \blacktriangleleft$$

Thus

$$0.86 \text{ months} < \mu < 4.27 \text{ months} \blacktriangleleft$$

with 95% confidence.

In some situations, particularly in setting specifications, we are not interested in the MTBF, but only in assuring that it be greater than some specified value. If the MTBF must be greater than the specified value at a confidence level of $\alpha/2$, we estimate $L_{\alpha/2,n}/\hat{\mu}$ or $L^*_{\alpha/2,n}/\hat{\mu}$ from Fig. 5.11 and determines the value of $\hat{\mu}$ with an appropriate form of Eq. 5.43.

EXAMPLE 5.15

A computer specification calls for an MTBF of at least 100 hr with 90% confidence. If a prototype fails for the first time at 210 hr, can these test data be used to demonstrate that the specification has been met?

Solution $\hat{\mu} = T/n = 210/1 = 210$ hr. For the 90% one-sided confidence interval $\alpha/2 = 0.1$. From Fig. 5.11,

$$L_{0.1,1}/\hat{\mu} \approx 0.44,$$

$$L_{0.1,1} = 0.44 \times 210 = 93 \text{ hr.} \blacktriangleleft$$

The test is inadequate, ◀ since the lower confidence limit is smaller than the specified value of 100 hr.

A word is in order concerning the percentage survival test discussed earlier. It is a form of binomial sampling, with the ratio n/N being the estimate of the failure probability of failure. Consequently, the method discussed in Chapter 2 can be used to estimate the confidence interval of the failure probability, and from this the confidence interval on the MTTF can be estimated. The uncertainty is greater than that obtained from testing in which the actual failure times are recorded.

EXAMPLE 5.16

Estimate the 90% confidence interval for the National Guard ammunition problem, Example 5.12.

Solution Since, in 5 years, 12 of 200 rounds fail, the 5-year failure probability may be calculated from Eq. 2.91 to be

$$\hat{p} = \frac{n}{N} = \frac{12}{200} = 0.06 = 1 - \hat{R}.$$

Since this test is a form of binomial sampling, we can look up the 90% confidence interval on p from Appendix B. We obtain for $n = 12$, $0.01 < p < 0.31$. For a constant failure rate we have

$$p = 1 - e^{-t/\mu} \qquad \text{or} \qquad \mu = -t/\ln(1 - p).$$

Therefore, with $t = 25$ years,

$$\frac{-25}{\ln(1 - 0.31)} < \mu < \frac{-25}{\ln(1 - 0.01)}$$

$$67 \text{ years} < \mu < 2487 \text{ years.} \blacktriangleleft$$

with 90% confidence.

5.6 BAYESIAN ANALYSIS

Reliability engineering is frequently faced with a paucity of available test data. If only a very few units can be tested to failure because of expense or production deadlines, the foregoing graphical methods are no longer very helpful for estimating parameters. Similarly, if more advanced statistical calculations are employed, the resulting confidence intervals are likely to be so wide that the estimates will be of limited value.

In such situations engineers may turn to the alternative of using reliability models and parameters from similar equipment. These may take the form of the generic estimates that are often available in handbooks. Similarly, the experienced engineers may extrapolate values from previous equipment models with which they are familiar. This indeed is frequently done, particularly if the constant failure rate or some other standard model can be assumed applicable. Nevertheless, it would be unfortunate if test results on the system under consideration—however few—could not be factored into the reliability estimates.

Bayesian analysis makes this possible,* for given an estimate of a parameter such as a failure rate, the test results can be used to upgrade that estimate in a systematic way. More specifically, the engineer utilizes handbooks, expert opinion, and previous experiences to formulate a probability distribution expressing the uncertainty in the true value of a parameter. This is referred to as the prior distribution. The best point estimate of the parameter would normally appear as the mean or median of this distribution. With Bayesian analysis the test data are used to modify the distribution, yielding the so-called posterior distribution. Since the posterior distribution represents the new state of knowledge, its mean or median represents an improved point estimate, given the availability of the test results.

Bayesian analysis may be applied to upgrading estimates for a wide variety of reliability problems. In order to limit the amount of additional mathematical apparatus that must be brought into play, we confine our attention here to two illustrations, both involving upgrading the estimate of a constant failure rate with data from a reliability test of fixed duration. They involve, respectively, discrete and continuous probability distributions.

Discrete Distribution

The Bayesian formula stems from the fact that the intersection of two probabilities can be written in terms of two different conditional probabilities; for example, as in Eqs. 2.4 and 2.6. Replacing X by X_i and combining these relationships, we obtain the most elementary form of the Bayes equation:

$$P\{X_i|Y\} = \frac{P\{Y|X_i\}P\{X_i\}}{P\{Y\}}. \tag{5.53}$$

* Martz and Waller, op. cit.

We may give the following interpretations to these probabilities: $P\{X_i\}$ is our estimate of a probability that $\mathbf{X}$ has a value of X_i, and Y represents the outcome of an experiment. The probability $P\{X_i|Y\}$ is our upgraded estimate, given the outcome of the experiment. To evaluate this result, we must be able to estimate the probability of an experimental outcome Y given that $\mathbf{X}$ has a value X_i. Finally, $P\{Y\}$ is determined as follows.

Suppose that X_1 $X_2, \ldots, X_n$ are the only possible values that $\mathbf{X}$ may take on. Since $\mathbf{X}$ can have only one value, the events X_i are mutually exclusive, and therefore,

$$\sum_{i=1}^{n} P\{X_i\} = 1, \tag{5.54}$$

and

$$\sum_{i=1}^{n} P\{X_i|Y\} = 1. \tag{5.55}$$

If we combine this equation with Eq. 5.53, it follows that

$$P\{Y\} = \sum_{i=1}^{n} P\{Y|X_i\}P\{X_i\}. \tag{5.56}$$

Therefore, the Bayes equation, Eq. 5.53, may be written as

$$P\{X_i|Y\} = \frac{P\{Y|X_i\}P\{X_i\}}{\sum_{i'=1}^{n} P\{Y|X_{i'}\}P\{X_{i'}\}}. \tag{5.57}$$

The use of the Bayes equation is best understood through a simple example.

EXAMPLE 5.17

An engineer calls in two experts to estimate the MTTF of a new process computer. Expert 1 estimates 30 months and expert 2 estimates 12 months. Since the engineer gives their opinions equal weight, he estimates the MTTF to be

$$\text{MTTF} = 0.5 \times 30 + 0.5 \times 12 = 21 \text{ months}.$$

Subsequently, a 6-month test is run, and the prototype for the new computer does not fail. In the light of these test results, (*a*) how should the experts' opinions be weighed, and (*b*) how should the estimated MTTF be upgraded?

Solution Let $P\{X_1\} = P\{X_2\} = 0.5$ be the prior probabilities that the MTTF estimates of experts 1 and 2 are correct. If the experts' opinions are correct, the probability of 6-month operation without failure is

$$P\{Y|X_i\} = \exp(-t/\text{MTTF}_i),$$

assuming that the constant failure rate model is adequate. Thus

$$P\{Y|X_1\} = e^{-6/30} = 0.819,$$

$$P\{Y|X_2\} = e^{-6/12} = 0.607.$$

Thus, according to Eq. 5.57, the revised probabilities that each of the experts are correct are

$$P\{X_1|Y\} = \frac{0.819 \times 0.5}{0.819 \times 0.5 + 0.607 \times 0.5} = 0.574,$$

$$P\{X_2|Y\} = \frac{0.607 \times 0.5}{0.819 \times 0.5 + 0.607 \times 0.5} = 0.425.$$

With these weights the upgraded estimate is

$$\text{MTTF} = 0.574 \times 30 + 0.426 \times 12 = 22.3 \text{ months.} \blacktriangleleft$$

Continuous Distribution

In order to carry out the failure rate estimate with a continuous distribution, we shall require the form of Eq. 5.53 in which **x** is a continuous and **y** is a discrete random variable. The necessary conditional probability equations are Eqs. 3.61 and 3.62; from them we obtain

$$f_{\mathbf{x}}(x|y_n) = \frac{f_{\mathbf{y}}(y_n|x)f_{\mathbf{x}}(x)}{f_{\mathbf{y}}(y_n)}. \tag{5.58}$$

The PMF $f_{\mathbf{y}}(y_n)$ on the right-hand side may be eliminated by inserting this equation into Eq. 3.63. We have

$$\int_{-\infty}^{\infty} f_{\mathbf{x}}(x|y_n)\,dx = \frac{1}{f_{\mathbf{y}}(y_n)}\int_{-\infty}^{\infty} f_{\mathbf{y}}(y_n|x)f_{\mathbf{x}}(x)\,dx = 1, \tag{5.59}$$

or simply

$$f_{\mathbf{y}}(y_n) = \int_{-\infty}^{\infty} f_{\mathbf{y}}(y_n|x)f_{\mathbf{x}}(x)\,dx. \tag{5.60}$$

The desired Bayesian equation is then found from Eq. 5.58 to be

$$f_{\mathbf{x}}(x|y_n) = \frac{f_{\mathbf{y}}(y_n|x)f_{\mathbf{x}}(x)}{\int_{-\infty}^{\infty} f_{\mathbf{y}}(y_n|x')f_{\mathbf{x}}(x')\,dx'}. \tag{5.61}$$

This equation may now be given a more concrete interpretation. Suppose that we are trying to estimate some parameter x appearing in a distribution. This might be a failure rate, the mean strength of a structure, or any number of other quantities. Since we are uncertain of the value of x, we treat it as a random variable **x**, where the PDF $f_{\mathbf{x}}(x)$ expresses our uncertainty. In our problem we shall be estimating a failure rate. Suppose that we characterize our uncertainty by the normal distribution. Thus $x \rightarrow \lambda$, and

$$f_{\lambda}(\lambda) = \frac{1}{\sqrt{2\pi}\,\sigma_0}\exp\left[-\frac{1}{2\sigma_0^2}(\lambda - \lambda_0)^2\right], \tag{5.62}$$

where λ_0 is our best point estimate. The variance indicates our degree of uncertainty; we are 69% certain that the true value lies between $\lambda_0 - \sigma_0$ and $\lambda + \sigma_0$. This is our prior distribution; note that it is assumed implicitly that the constant failure rate model is valid.

The quantity $f_{\mathbf{y}}(y_n|x)$ is the probability that an experiment that may result in several different discrete values of **y** will have a particular result y_n, given that the parameter takes on a value x. In our illustration, we shall assume that N units are tested for some time t_0, and that, at the end of the test, the number n that have failed are counted. Thus we take $y_n \rightarrow n$ as the discrete variable. With the assumption that the tests are independent, the results are given by the binomial distribution:

$$f_{\mathbf{n}}(n|\lambda) = C_n^N p(\lambda)^n [1 - p(\lambda)]^{N-n}. \tag{5.63}$$

Here $p(\lambda)$ is the failure probability for a unit operating for time t_o with a failure rate λ, or simply

$$p(\lambda) = 1 - e^{-\lambda t_0}. \tag{5.64}$$

Thus we have

$$f_{\mathbf{n}}(n|\lambda) = C_n^N [1 - e^{-\lambda t_0}]^n \exp[-(N - n)\lambda t_0]. \tag{5.65}$$

With these changes of variables (i.e., $x \rightarrow \lambda$ and $y_n \rightarrow n$), we may write Eq. 5.61 as

$$f_{\lambda}(\lambda|n) = \frac{[1 - e^{-\lambda t_0}]^n \exp[-(N - n)\lambda t_0] f_{\lambda}(\lambda)}{\int_{-\infty}^{\infty} [1 - e^{-\lambda' t_0}]^n \exp[-(N - n)\lambda' t_0] f_{\lambda}(\lambda')\, d\lambda'}, \tag{5.66}$$

where $f_{\lambda}(\lambda)$ is given by Eq. 5.62 or any other PDF expressing our uncertainty in the failure rate.

The PDF $f_{\lambda}(\lambda|n)$ is our estimate of the uncertainty in λ, modified to take into account the test data that have become available. If the test had taken a different form, say the actual failure times of the N units, Bayesian estimates could be used to refine the PDF further, but then the mathematics required would be substantially more extensive, and we therefore do not pursue it here.

Once the upgraded PDF is known, we may also upgrade our point estimate of λ. Let λ_1 be the upgraded mean value

$$\lambda_1 = \int_{-\infty}^{\infty} \lambda f_{\lambda}(\lambda|n)\, d\lambda, \tag{5.67}$$

which can be compared to our original estimate λ_0. Similarly, to see if we have gained any additional precision from the test data, we may calculate the variance of the new distribution

$$\sigma_1^2 = \int_{-\infty}^{\infty} \lambda^2 f_{\lambda}(\lambda|n)\, d\lambda - \lambda_1^2, \tag{5.68}$$

and compare it to σ_0.

These ideas are made more concrete by using specific data, as in the following example.

EXAMPLE 5.18

With long experience a reliability engineer is confident that a given class of power supplies can be represented by a constant failure rate model. After studying the design of a new model, she compares it to similar models for which data are available, consults several handbooks that give approximate failure rate data, and decides that with her current state of knowledge the failure rate estimate can be represented by a normal distribution with mean $\lambda_0 = 0.04$/day and standard deviation $\sigma_0 = 0.01$/day. Three prototype units are subjected to the company's standard 30-day field test, and one of them fails. Use Bayesian methods (*a*) to find an updated PDF for λ, and (*b*) to upgrade the point estimate of λ.

Solution (*a*) From the initial estimate

$$f_\lambda(\lambda) = \frac{100}{\sqrt{2\pi}} \exp[-5 \times 10^3(\lambda - 0.04)^2]$$

and with one failure $n = 1$ and $t_0 = 30$ we obtain from Eq. 5.66

$$f_\lambda(\lambda|\mathbf{n} = 1) = \frac{1}{I}(e^{-60\lambda} - e^{-90\lambda}) \exp[-5 \times 10^3(\lambda - 0.04)^2], \blacktriangleleft$$

where

$$I = \int_{-\infty}^{\infty} (e^{-60\lambda} - e^{-90\lambda}) \exp[-5 \times 10^3(\lambda - 0.04)^2]\, d\lambda,$$

which can be evaluated analytically by using the completion of squares as in Chapter 6, or by numerical integration.
(*b*) Substituting the expression just given for $f_\lambda(\lambda|\mathbf{n} = 1)$ into Eq. 5.67 yields

$$\lambda_1 = \frac{1}{I}\int_{-\infty}^{\infty} \lambda(e^{-60\lambda} - e^{-90\lambda}) \exp[-5 \times 10^3(\lambda - 0.04)^2]\, d\lambda.$$

The analytical evaluation of this integral is possible but tedious. We evaluate it approximately by simple numerical integration between $0 \leq \lambda < 0.1$ to obtain $\lambda_1 = 0.036$. ◀

This lowering of the failure rate estimate is to be expected, in view of the test results. If $\lambda_0 = 0.04$ had been accurate, the failure probability for any unit would have been

$$p(\lambda_0) = 1 - e^{-\lambda_0 t_0} = 1 - e^{-0.04 \times 30} = 0.7,$$

meaning that the expected number of failures would have been about two had the test been repeated many times.

5.7 RELIABILITY GROWTH TESTING

As discussed earlier, predictions of reliability during design and development are extremely valuable, for they are available at a time when design modifications or other corrections can be made at much less expense than is the case later in the product life cycle. At the same time, there may not be enough prototypes available to obtain useful results from the life-testing procedures discussed in the preceding sections. Thus reliability engineers must often depend on theoretical predictions, such as the parts count method, or employ whatever scant prototype data are available to do Bayesian upgrades of theoretical or judgmental estimates.

The manner in which engineering development is normally carried out presents an additional difficulty in conventional life testing. Typically, after one or more prototypes are constructed, they are tested to failure. The failure is analyzed. The prototypes are then fixed by modifying the design or by altering operating procedures, and the test is repeated. In this way the design is refined as the engineering staff progresses through a learning curve, gaining information obtainable only through the testing experience.

Such prototype testing is an indispensable part of the design process, but the results cannot be used by the reliability engineer in a standard analysis of a life test. Each time the prototype is fixed by design modification, a new system results; consequently, it is not valid to treat the data as repeated failures of the same system. Moreover, as the bugs are worked out of the design, the periods of time between failures tend to become longer, and the reliability of the system is thus said to grow as the design is modified through this test-fix-test-fix process. What is needed is a technique for using such prototype data to estimate MTBF or other reliability parameters for the final system design. Such a technique was originated by Duane* and has been refined and widely applied.

Suppose that we define the following:

T = total operation time accumulated on all the prototypes,

$n(T)$ = number of failures from the beginning of testing through time T.

If more than one prototype is tested, then T is calculated in the same way as in Section 5.5, depending on whether replacement or nonreplacement testing is used. Finally, we assume that as failures occur, the design is modified to eliminate the failure modes.

Duane observed that if $n(T)/T$ is plotted versus T on log-log paper, the result, as indicated in Fig. 5.12, tends to be a straight line, whatever the type of electromechanical equipment under consideration. From such empirical

* J. J. Duane, "Learning Curve Approach to Reliability Modeling," *IEEE Trans. Aerospace,* **2,** 563–566 (1964).

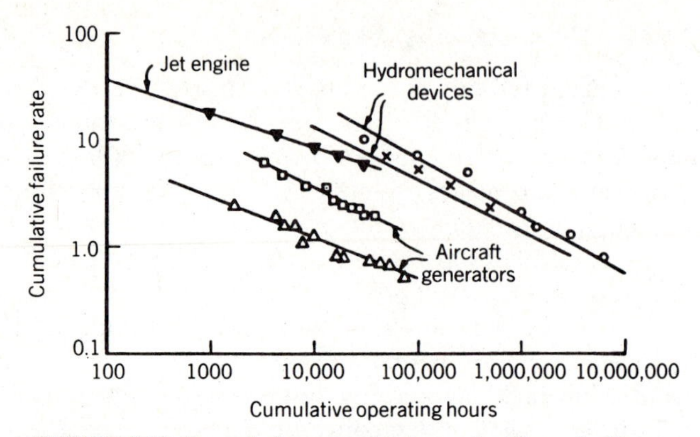

FIGURE 5.12 Duane's data on a log-log scale. [From L. H. Crow, "On Tracking Reliability Growth," *Proceedings 1975 Reliability and Maintainability Symposium*, 438–443 (1975).]

relationships, referred to as Duane plots, we may estimate the growth of the MTBF with time and therefore also extrapolate how much reliability is likely to be gained from further development work.*

To analyze reliability growth, we must first relate $n(T)$ to the failure rate and generalize the definition of the MTBF. The failure rate $\lambda(t)$ may be interpreted as just the number of failures per unit operating time. Thus

$$\frac{d}{dT} n(T) = \lambda(T). \tag{5.69}$$

We may then integrate this equation to obtain the cumulative failure rate

$$n(T) = \int_0^T \lambda(t)\, dt. \tag{5.70}$$

Generalizing the constant failure rate definition of MTBF given by Eq. 4.26, we may define an instantaneous MTBF to be

$$\mu_i(T) \equiv \frac{1}{\lambda(T)} = \frac{1}{\dfrac{d}{dT} n(T)}, \tag{5.71}$$

and then define the corresponding cumulative MTBF to be

$$\mu_c(T) \equiv \frac{1}{(1/T)\displaystyle\int_0^T \lambda(t)\, dt} = \frac{T}{n(T)}. \tag{5.72}$$

* See A. E. Green, *System Safety Reliability*, Wiley, New York, 1983, Chapter 5; also A. H. Kling and J. E. Arsenault, "Reliability Testing" in *Reliability and Maintainability of Electronic Systems*, J. E. Arsenault and J. A. Roberts (eds.), Computer Science Press, Potomac, MD, 1980.

Our objective is to obtain estimates of $\lambda(t)$ and the instantaneous MTBF from the data.

We note that since Duane plots are straight lines, we may write, in general,

$$\ln[n(T)/T] = a - \alpha \ln T. \tag{5.73}$$

Solving for $n(T)$, we obtain

$$n(T) = KT^{1-\alpha}, \tag{5.74}$$

where $K = e^a$. The instantaneous failure rate may then be determined from Eq. 5.69 to be

$$\lambda(T) = (1 - \alpha)KT^{-\alpha}. \tag{5.75}$$

Correspondingly, from Eq. 5.71, the instantaneous MTBF is

$$\mu_i(T) = \frac{1}{(1 - \alpha)K} T^{\alpha}. \tag{5.76}$$

Since α is a positive number, usually about one-half, this equation illustrates the growth of the MTBF, and therefore of the reliability, with accumulated test time.

A widely used variant of the Duane plot consists of plotting $T/n(T)$ versus T on log-log paper. This has a number of attractive features. From Eq. 5.72 we see that the ordinate is now just the cumulative MTBF. Moreover, since it follows from Eqs. 5.72 and 5.74 that

$$\mu_c(T) = \frac{1}{K} T^{\alpha}, \tag{5.77}$$

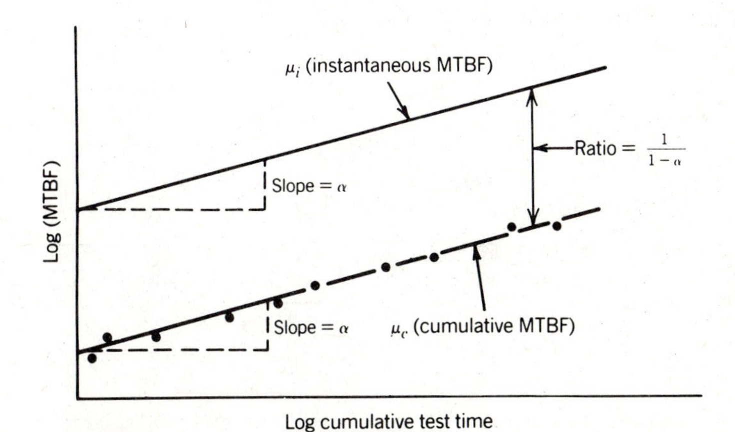

FIGURE 5.13 Duane's plot showing the determination of instantaneous MTBF.

we can measure α directly as the slope of the curve. Finally, we may determine the instantaneous MTBF quite simply. Since the combination of Eqs. 5.76 and 5.77 yields

$$\mu_i(T) = \frac{1}{1 - \alpha} \mu_c(T), \tag{5.78}$$

the estimated instantaneous MTBF may also be plotted as a straight line. This is illustrated in Fig. 5.13.

The resemblance between Eq. 5.75 and the Weibull distribution cannot be escaped, for if we compare Eqs. 5.75 and 4.55, we see that they are the same provided that $m = 1 - \alpha$. Thus the Duane plots appear to yield Weibull distributions that are characteristic of wearin, since $\alpha > 0$ and therefore $m < 1$. The two situations should not be confused, however, for they are quite different. In wearin the failed units are simply removed, and the test continued, without making any improvements to the remaining test units. In the test-fix-test-fix process after each failure the design is modified on all the test units to improve their reliability. Thus the failure rate decreases because the design has been improved and not simply because weaker units have been removed.

The methods discussed here for estimating reliability growth of course have limitations. When failures may no longer be attributed to removable design defects, the growth in reliability may no longer be significant. Moreover, refining the design to reduce random failures, or those caused by wear, may eventually become prohibitively expensive or may unacceptably limit the performance of the system. Similarly, the reliability obtained for mass-produced items may be lower than that of the tested prototypes, and that obtained in field service may be lower than that of laboratory tests. To deal with these problems, we must concentrate on production quality control and realistically anticipate field conditions.

Bibliography

Ang, A. H-S., and W. H. Tang, *Probability Concepts in Engineering Planning and Design,* Vol. 1, Wiley, New York, 1975.

Barlow, R. E., and F. Proschan, *Statistical Theory of Reliability and Life Testing Probability Models,* Holt, Rinehart and Winston, New York, 1975.

Bazovsky, I., *Reliability Theory and Practice,* Prentice-Hall, Englewood Cliffs, NJ, 1961.

Green, A. E., *Safety Systems Analysis,* Wiley, New York, 1983.

Kapur, K. C., and L. R. Lamberson, *Reliability in Engineering Design,* Wiley, New York, 1977.

Lawless, J. F., *Statistical Models and Methods for Lifetime Data,* Wiley, New York, 1982.

Mann, N. R., R. E. Schafer, and N. D. Singpurwalla, *Methods for Statistical Analysis of Reliability and Life Data,* Wiley, New York, 1974.

Martz, H. F., and R. A. Walker, *Bayesian Reliability Analysis,* Wiley, New York, 1982.
Miller, R. G., Jr., *Survival Analysis,* Wiley, New York, 1981.
Nelson, W., *Applied Life Data Analysis,* Wiley, New York, 1982.

EXERCISES

5.1 The wearin times of 10 emergency flares in minutes are

17.0, 20.6, 21.3, 21.4, 22.7, 25.6, 26.5, 27.0, 27.7, 29.7.

Use the nonparametric method to make plots of the reliability and failure rate.

5.2 The following are time intervals,* in hours of operation, between successive failures of air-conditioning equipment on a Boeing 720 aircraft. The times appear in the order of occurrence (i.e., the first failure occured at 413 hr, the second at 413 + 14 = 427 hr, and so on).

413, 14, 58, 37, 100, 65, 9, 169, 447, 184, 36, 201, 118, 34, 31, 18, 18, 67, 62, 7, 22, 34.

Use the nonparametric method described in the text to plot the reliability and the failure rate of the air-conditioning equipment.

5.3 Data for the failure times of 309 radio transmitter receivers are given in the following table.†

Time interval, hr	Failures	Time interval, hr	Failures
0–50	41	300–350	18
50–100	44	350–400	16
100–150	50	400–450	15
150–200	48	450–500	11
200–250	28	500–550	7
250–300	29	550–600	11

At 600 hr, 51 of the receiver–transmitters remained in operation. Use the nonparametric method described in the text to plot the reliability and failure rate from this data.

* Data from F. Proschan, "Theoretical Explanation of Observed Decreasing Failure Rate," *Technometrics,* **5,** 375–383 (1963).

† From W. Mendenhall and R. J. Hader, "Estimation of Parameters of Mixed Exponential Distribution Failure Times from Censored Life Test Data," *Biometrika,* **63,** 449–464 (1958).

5.4 The following uncensored grouped data were collected on the failure time of feedwater pumps, in units of 1000 hr:

Interval	Number of failures
$0 \leq t \leq 6$	5
$6 \leq t \leq 12$	19
$12 \leq t \leq 18$	61
$18 \leq t \leq 24$	27
$24 \leq t \leq 30$	20
$30 \leq t \leq 36$	17

Make a nonparametric plot of the reliability and of the failure rate versus time.

5.5 Consider the following multiply censored data* for the field windings for 16 generators. The times to failure and removal times (in months) are

31.7, 39.2, 57.5, 65.0+, 65.8, 70.0, 75.0+, 75.0+, 87.5+, 88.3+, 94.2+, 101.7+, 105.8, 109.2+, 110.0, 130.0+.

Make a nonparametric plot of the reliability.

5.6 Suppose that instead of Eq. 5.3, we use Eq. 5.5 as a starting point for nonparametric analysis. Derive the expressions for $\hat{R}(t_i)$, $\hat{f}(t)$, and $\hat{\lambda}(t)$ that should be used in place of Eqs. 5.4, 5.7, and 5.9.

5.7 Microcircuits undergo accelerated life testing. The analysis is to be carried out using nonparametric methods for ungrouped data.

(*a*) The first test series on six prototype microcircuits results in the following times to failure (in hours): 1.6, 2.6, 5.7, 9.3, 18.2, and 39.6.
Plot a graph of the estimated reliability.

(*b*) The second test series of six prototype microcircuits results in the following times to failure (in hours): 2.5, 2.8, 3.5, 5.7, 10.3, and 23.5. Combine these data with the data from part *a* and plot the reliability estimate on the same graph used for part *a*.

(*c*) Plot the failure rate estimates obtained from the data in part *a* and from the combined data on the same graph.

5.8 At rated voltage a microcircuit has been estimated to have an MTTF of 20,000 hr. An accelerated life test is to be carried out to verify this number. At least 10% of the test circuits must fail before the test is

* From Nelson, op. cit.

terminated if we are to have confidence in the result. If the test must be completed in 30 days, at what percentage of the rated voltage should the circuits be tested?

5.9 Twenty units of a catalytic converter are tested to failure without censoring. The times to failure (in days) are the following:

2.6	3.2	3.4	3.9	5.6
7.1	8.4	8.8	8.9	9.5
9.8	11.3	11.8	11.9	12.7
12.3	16.0	21.9	22.4	24.2

Plot an exponential paper, and determine whether the failure rate is increasing or decreasing with time.

5.10 Plot the results of Exercise 5.9 on Weibull paper and estimate the parameters m and θ.

5.11 Plot the data from Exercise 5.2 on exponential graph paper and determine whether the failure rate increases with time.

5.12 Make a two-parameter Weibull plot of the multiply censored winding data from Exercise 5.5 and estimate m and θ.

5.13 Of a group of 180 transformers, 20 of them fail within the first 4000 hr of operation. The times to failure in hours are as follows:*

10	1046	2096	3200
314	1570	2110	3360
730	1870	2177	3444
740	2020	2306	3508
990	2040	2690	3770

(*a*) Plot the data on normal paper.
(*b*) Estimate μ and σ for the transformers.
(*c*) Estimate how many transformers will fail between 4000 and 8000 hr.

5.14 Plot the data from the Exercise 5.13 on exponential paper to estimate whether the failure rate increases with time.

5.15 (*a*) Plot the data from Exercise 3.24 on normal probability paper.
(*b*) Estimate the mean and the variance, assuming that the distribution is normal.
(*c*) Compare the mean and variance determined from your plot with the values calculated in part *a* of Exercise 3.24.
(*d*) Draw the line on the probability paper that is determined by the mean and variance calculated in part *a* of Exercise 3.24.

* Data from Nelson, op. cit.

5.16 The following times to failure (in days) result from a fatigue test of 10 flanges:

1.66, 83.36, 25.76, 24.36, 334.68, 29.62, 296.82, 13.92, 107.04, 6.26.

(*a*) Plot the data on lognormal paper.
(*b*) Estimate the parameters.
(*c*) Estimate the factor to which the time to failure is known with 90% confidence.

5.17 The times to failure on four compressors are 240, 420, 630, and 1080 hr.
(*a*) Plot this data on lognormal paper.
(*b*) Estimate the most probable time to failure.

5.18 A life test with type II censoring is performed on 50 servomechanisms that are thought to have a constant failure rate. The test is terminated after the twentieth failure. The times to failure (in months) are as follows:

0.10	0.29	0.49	0.51	0.55
0.63	0.68	1.16	1.40	2.24
2.25	2.64	2.99	3.01	3.06
3.15	3.51	3.53	3.99	4.05

The failed servomechanisms are not replaced.
(*a*) Plot the failure times on exponential graph paper and estimate whether the failure rate is constant.
(*b*) Make a point estimate of the MTTF from the appropriate form of Eq. 5.43.
(*c*) Using the MTTF from part *b*, draw a straight line through the data ploted for part *a*.
(*d*) What is the 90% confidence interval on the MTTF?
(*e*) Draw the straight lines on your plot in part *a* corresponding to the confidence limits on the MTTF.

5.19 Suppose that in the preceding problem the life test had to be stopped at 3 months because of a production deadline. Based on a 3-month test, estimate the MTTF and the corresponding 90% confidence interval.

5.20 A nonreplacement reliability test is carried out on 20 high-speed pumps to estimate the value of the failure rate. In order to eliminate wear failures, it is decided to terminate the test after half of the pumps have failed. The times of the first 10 failures (in hours) are

33.7, 36.9, 46.8, 56.6, 62.1, 63.6, 78.4, 79.0, 101.5, 110.2.

(*a*) Estimate the MTTF.
(*b*) Determine the 90% confidence interval for the MTTF.

5.21 A new robot system undergoes test-fix-test-fix development testing. The number of failures during each 100-hr interval in the first 700 hr of operation are recorded. They are 14, 7, 6, 4, 3, 1, and 1.

(*a*) Plot the cumulative MTBF on log-log paper and approximate the data by a straight line.

(*b*) Estimate α from the slope of the line.

(*c*) Estimate the instantaneous MTBF at the end of 700 hr of development testing.

CHAPTER 6

Loads, Capacity, and Reliability

6.1 INTRODUCTION

In the preceding chapters failure rates were used to emphasize the strong dependence of reliability on time. Empirically, these failure rates were found to increase with the complexity of a system and also with the load placed on a system. In this chapter we explore the concepts of loads and capacity and examine their relationship to reliability. This examination allows us both to relate reliability to traditional design approaches using safety factors, and to gain additional insight into the relations between failure rates, random loading, wearin, and wearout.

In a number of engineering fields it has been customary to define safety factors and margins in the following way. Suppose that we define l as the load on a system, structure, or piece of equipment and c as the corresponding capacity. The safety factor is then defined as

$$v = \frac{c}{l}. \tag{6.1}$$

Alternately, the safety margin may be used. It is defined by

$$m = c - l. \tag{6.2}$$

Failure then occurs if the safety factor falls to a value less than one, or if the safety margin becomes negative.

The concepts of load and capacity are employed most widely in structural engineering and related fields, where the load is usually referred to as stress and the capacity as strength. However, they have much wider applicability. For example, if a piece of electric equipment is under consideration, we may speak of electric load and capacity; for a telecommunications system load and capacity may be measured in terms of telephone calls per unit time, and for an energy conversion system thermal units for load and capacity may be used. The point is that a wide variety of applications can be formulated in terms of load and capacity. For a given application, however, l and c must have the same units.

In the traditional approach to design, the safety factor or margin is made large enough to more than compensate for uncertainties in the values of both the load and the capacity of the system under consideration. Thus, although these uncertainties cause the load and the capacity to be viewed as random variables, the calculations are deterministic, using for the most part the best estimates of load and capacity. The probabilistic analysis of loads and capacities necessary for estimating reliability clarifies and rationalizes the determination and use of safety factors and margins. This analysis is particularly useful for situations in which no fixed bound can be put on the loading, for example, with earthquakes, floods and other natural phenomena, or for situations in which flaws or other shortcomings may result in systems with unusually small capacities. Similarly, when economics rather than safety is the primary criteria for setting design margins, the trade-off of performance versus reliability can best be studied by examining the increase in probability of failure as load and capacity approach one another.

The expression for reliability in terms of the random variables $\mathbf{l}$ and $\mathbf{c}$ comes from the notion that there is always some small probability of failure that decreases as the safety factor is increased. We may define the failure probability as

$$p = P\{\mathbf{l} \geqslant \mathbf{c}\}. \tag{6.3}$$

In this context the reliability is defined as the nonfailure probability or

$$r = 1 - p, \tag{6.4}$$

which may also be expressed as

$$r = P\{\mathbf{l} < \mathbf{c}\}. \tag{6.5}$$

In treating loads and capacities probabilistically, we must exercise a great deal of care in expressing the types of loads and the behavior of the capacity. If this is done, we may use the resulting formalism not only to provide a probabilistic relation between safety factors and reliability, but also to gain a better understanding of the relations between loading, capacities, and the time dependence of failure rates as exhibited, for example, in the bathtub curve.

In Section 6.2 we develop expressions for reliability for a single loading and then, in Section 6.3, relate the results to safety factors. Because both loads and capacities are frequently expressed in terms of extreme-value distributions, we discuss such distributions and relate them to reliability in Section 6.4. In Section 6.5 we consider repetitive loading and discuss the circumstances under which failure rates are constant in systems with known time-independent capacity. In the concluding section these restrictions on the capacity are removed to demonstrate how wearin and wearout effects appear.

6.2 RELIABILITY WITH A SINGLE LOADING

In this section we derive the relations between load, capacity, and reliability for systems that are loaded only once. The resulting reliability does not depend on time, for the reliability is just the probability that the system survives the application of the load. Nevertheless, before the expressions for the reliability can be derived, the restrictions on the nature of the loads and capacity must be clearly understood.

Load Application

In referring to the load on a system, we are in fact referring to the maximum load from the beginning of application until the load is removed. Figure 6.1 indicates the time dependence of several loading patterns that may be treated as single on loading l, provided that appropriate restrictions are met.

Figure 6.1*a* represents a single loading of finite duration. Missiles dur-

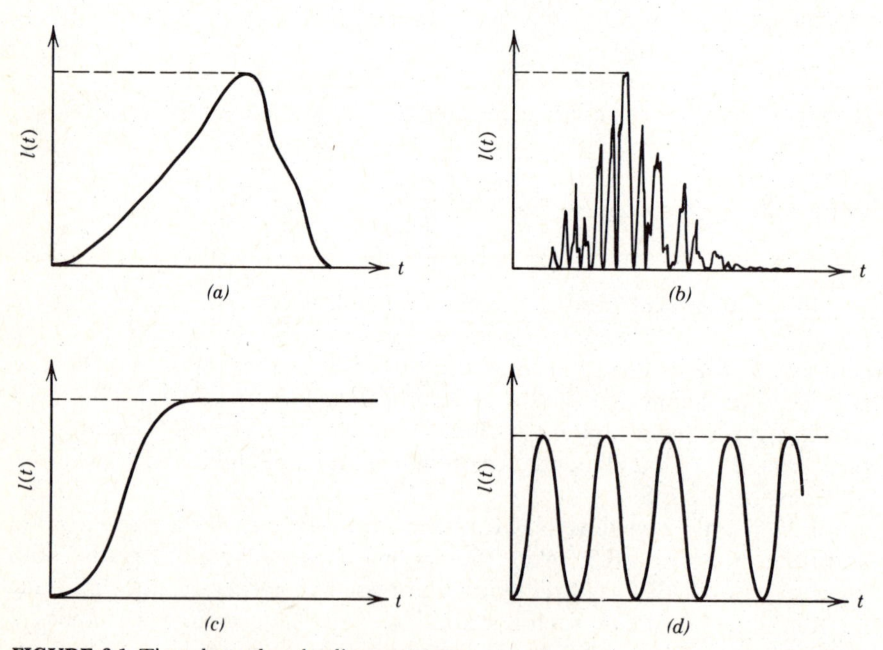

FIGURE 6.1 Time-dependent loading patterns.

ing launch, flashbulbs, and any number of other devices that are used only once have such loadings. Such one-time-only loads are also a ubiquitous feature of manufacturing processes, occurring for instance when torque is applied to a bolt or pressure is applied to a rivet. Loading often is not applied in a smooth manner, but rather as a series of shocks, as shown in Fig. 6.1*b*. This behavior would be typical of the vibrational loading on a structure during an earthquake and of the impact loading on an aircraft during landing. In many situations, the extreme value of many short-time loadings may be treated as a single loading provided that there is a definite beginning and end to the disturbance giving rise to it.

The duration of the load in Figs. 6.1*a* and *b* is short enough that no weakening of the system capacity takes place. If no decrease in system capacity is possible, the situations shown in Figs. 6.1*c* and *d* may also be viewed as single loadings, even though they are not of finite duration. The loading shown in Fig. 6.1*c* is typical of the dead loads from the weight of structures; these increase during construction and then remain at a constant value. This formulation of the loading is widely used in structural analysis when the load-bearing capacity not only may remain constant, but may in some instance increase somewhat with time because of the curing of concrete or the work-hardening of metals.

Subject to the same restrictions, the patterns shown in Fig. 6.1*d* may be viewed as a single loading. Provided that the peaks are of the same magnitude, the system will either fail the first time the load is applied or will not fail at all. Under such cyclic loading, however, the assumption that the system capacity will not decrease with time should be suspect. Metal fatigue and other wear effects are likely to weaken the capacity of the system gradually, as described in Section 6.6. Similarly, if the values of peak magnitudes vary from cycle to cycle, we must consider the time dependence of reliability explicitly, as in Section 6.5.

Thus far we have assumed that a system is subjected to only one load and that reliability is determined by the capacity of the system as a whole to resist this load. In reality a system is invariably subjected to a variety of different loads; if it does not have the capacity to sustain any one of these, it will fail. An obvious example is a piece of machinery or other equipment, each of whose components are subjected to different loads; failure of any one component will make the system fail. A more monolithic structure, such as a dam, is subject to static loads from its own weight, dynamic loads from earthquakes, flood loadings, and so on. Nevertheless, the considerations that follow remain applicable, provided that the loads are considered in terms of the probability of a particular failure mode or of the loading of a particular component. If the failure modes can be assumed to be approximately independent of one another, the reliability of the overall system can be calculated as the product of the failure mode reliabilities, as discussed in Chapter 4.

Definitions

To derive an expression for the reliability, we must first define independent PDFs for the load, **l**, and for the capacity, **c**. Let

$$f_{\mathbf{l}}(l)\,dl = P\{l \leqslant \mathbf{l} \leqslant l + dl\} \tag{6.6}$$

be the probability that the load is between l and $l + dl$. Similarly, let

$$f_{\mathbf{c}}(c)\,dc = P\{c \leqslant \mathbf{c} < c + dc\} \tag{6.7}$$

be the probability that the capacity has a value between c and $c + dc$. Thus $f_{\mathbf{l}}(l)$ and $f_{\mathbf{c}}(c)$ are the necessary PDFs; we include the subscripts to avoid any possible confusion between the two. The corresponding CDFs may also be defined. They are

$$F_{\mathbf{c}}(c) = \int_0^c f_{\mathbf{c}}(c')\,dc', \tag{6.8}$$

$$F_{\mathbf{l}}(l) = \int_0^l f_{\mathbf{l}}(l')\,dl'. \tag{6.9}$$

We consider first a system with a known capacity c and a distribution of possible loads, as shown in Fig. 6.2*a*. For fixed c, the reliability of the system is just the probability that $\mathbf{l} < c$, the shaded area in the figure. Thus

$$r(c) = \int_0^c f_{\mathbf{l}}(l)\,dl. \tag{6.10}$$

The reliability, therefore, is just $F_{\mathbf{l}}(c)$, the CDF of the load evaluated at c. Clearly, for a system of known capacity, the reliability is equal to one as $c \to \infty$, and to zero as $c \to 0$.

Now suppose that the capacity also involves uncertainty; it is described by the PDF $f_{\mathbf{c}}(c)$. The expected value of the reliability is then obtained from

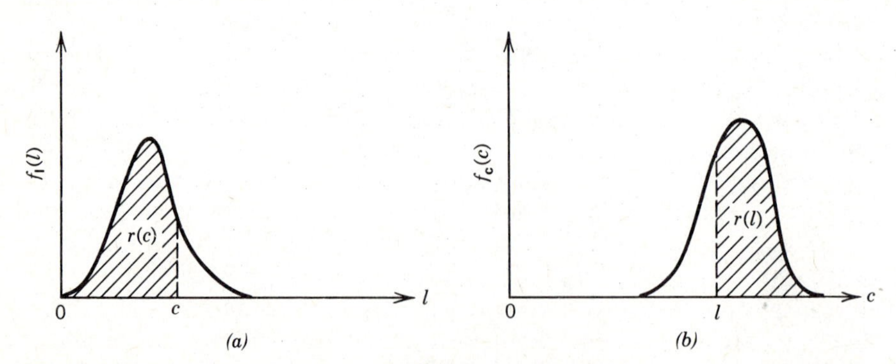

FIGURE 6.2 Area interpretation of reliability: (*a*) variable load, fixed capacity; (*b*) variable capacity, fixed load.

averaging over the distribution of capacities:

$$r = \int_0^\infty r(c) f_c(c)\, dc. \tag{6.11}$$

Substituting in Eq. 6.10, we have

$$r = \int_0^\infty \left[\int_0^c f_l(l)\, dl \right] f_c(c)\, dc. \tag{6.12}$$

The failure probability may then be determined from Eq. 6.4 to be

$$p = 1 - \int_0^\infty \left[\int_0^c f_l(l)\, dl \right] f_c(c)\, dc. \tag{6.13}$$

Alternately, we may substitute the condition on the load PDF,

$$\int_0^c f_l(l)\, dl = 1 - \int_c^\infty f_l(l)\, dl, \tag{6.14}$$

into Eq. 6.12. Then, using the condition

$$\int_0^\infty f_c(c)\, dc = 1, \tag{6.15}$$

we obtain for the failure probability

$$p = \int_0^\infty \left[\int_c^\infty f_l(l)\, dl \right] f_c(c)\, dc. \tag{6.16}$$

As shown in Fig. 6.3, the probability of failure is loosely associated with the

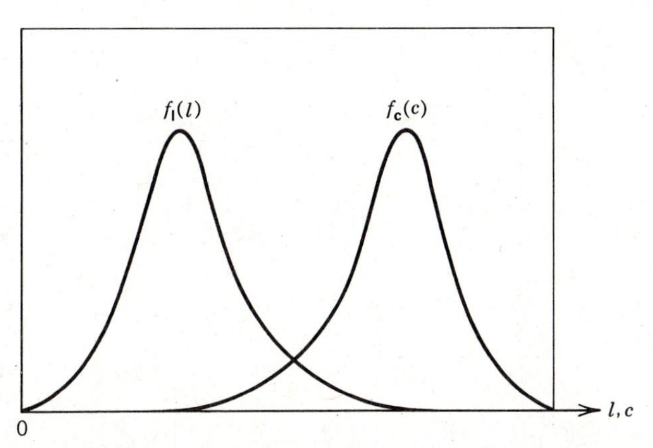

FIGURE 6.3 Graphical reliability interpretation with variable load and capacity.

overlap of the PDFs for load and capacity in the sense that if there is no overlap, the failure probability is zero and $r = 1$.

EXAMPLE 6.1

The bending moment on a match stick during striking is estimated to be distributed exponentially. It is found that match sticks of a given strength break 20% of the time. Therefore, the manufacturer increases the strength of the matches by 50%. What fraction of the strengthened matches are expected to break as they are struck?

Solution Assume that the strength (capacity) is known; then for the standard matches we have

$$0.8 = r = \int_0^c f_l(l)\, dl = \int_0^c \lambda e^{-\lambda l}\, dl = 1 - e^{-\lambda c}.$$

Therefore, $e^{-\lambda c} = 0.2$ or $\lambda c = -ln(0.2)$, where λ is the unknown parameter of the exponential loading distribution. For the strengthened matches

$$r' = \int_0^{1.5c} f_l(l)\, dl = \int_0^{1.5c} \lambda\, e^{-\lambda l}\, dl = 1 - e^{-1.5\lambda c},$$

$$p' \equiv 1 - r' = \exp[+1.5 \times ln(0.2)] = 0.2^{1.5} = 0.089.$$

Thus about 9% ◀ of the strengthened matches are expected to break.

Another derivation of r and p is possible. Although the derivation may be shown to yield results that are identical to Eqs. 6.12 and 6.13, the intermediate results are useful for different sets of circumstances. To illustrate, let us consider a system with known load but uncertain capacity represented by the distribution $f_c(c)$. The reliability for this system with known load is then given by the shaded area in Fig. 6.2*b*.

$$r(l) = \int_l^\infty f_c(c)\, dc, \tag{6.17}$$

or equivalently,

$$r(l) = 1 - \int_0^l f_c(c)\, dc. \tag{6.18}$$

For a system in which the load is also represented by a distribution, the expected value of the reliability is obtained by averaging over the load distribution,

$$r = \int_0^\infty f_l(l) r(l)\, dl, \tag{6.19}$$

or more explicitly

$$r = \int_0^\infty f_l(l) \left[\int_l^\infty f_c(c)\, dc \right] dl. \tag{6.20}$$

Similarly, we may consider the variation of the capacity first in deriving an expression for the failure probability. For a system with a fixed load the failure probability will be the unshaded area under the curve in Fig. 6.2*b*:

$$p(l) = \int_0^l f_c(c)\, dc. \tag{6.21}$$

Then, averaging over the distribution of loads, we have

$$p = \int_0^\infty f_l(l)\left[\int_0^l f_c(c)\, dc\right] dl. \tag{6.22}$$

It is easily shown that Eqs. 6.12 and 6.20 are the same. First write Eq. 6.12 as the double integral

$$r = \int_0^\infty \left[\int_0^c f_c(c) f_l(l)\, dl\right] dc, \tag{6.23}$$

where the shaded domain of integration appears in Fig. 6.4. If we reverse the order of integration, taking the c integration first, we have

$$r = \int_0^\infty \left[\int_l^\infty f_c(c) f_l(l)\, dc\right] dl. \tag{6.24}$$

Putting $f_l(l)$ outside the integral over c, we obtain Eq. 6.20.

To recapitulate, Eqs. 6.12 and 6.20 may be shown to be identical, as may Eqs. 6.16 and 6.22. However, the intermediate results for $r(c)$, $p(c)$, $r(l)$, and $p(l)$ are useful when considering systems whose capacity varies little compared to their load, or vice versa.

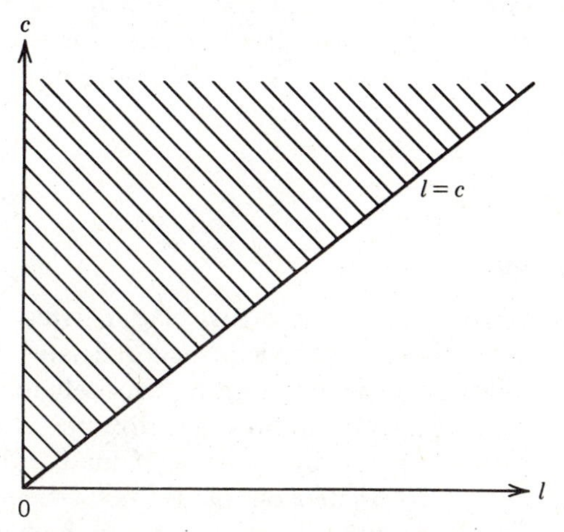

FIGURE 6.4 Domain of integration for reliability calculation.

6.3 RELIABILITY AND SAFETY FACTORS

In the preceding section reliability for a single loading is defined in terms of the independent PDFs for load and capacity. Similarly, it is possible to define safety factors in terms of these distributions. Two of the most widely accepted definitions are as follows. In the central safety factor the values of load and capacity in Eq. 6.1 are taken to be the mean values

$$\bar{l} = \int_{-\infty}^{\infty} l f_l(l)\, dl, \tag{6.25}$$

$$\bar{c} = \int_{-\infty}^{\infty} c f_c(c)\, dc. \tag{6.26}$$

Thus the safety factor is

$$\upsilon = \bar{c}/\bar{l}. \tag{6.27}$$

There is a second alternative if we express the safety factor in terms of the most probable values l_0 and c_0 at the load and capacity distributions. The safety factor in Eq. 6.1 is then

$$\upsilon = c_0/l_0. \tag{6.28}$$

These definitions are naturally associated with loads and capacities represented in terms of normal or of lognormal distributions, respectively. Then the reliability can be expressed in terms of the safety factor along with measures of the uncertainty in load and capacity. Other distributions may also be used in relating reliability to safety factors. Such is the case with the extreme-value distribution discussed in the following section. With such analysis the effects of design changes and quality control can be evaluated. For design determines the mean, $\bar{c}$, or most probable value, c_0, of the capacity, whereas the degree of quality control in manufacture or construction influences primarily the variance of $f_c(c)$ about the mean. Similarly, the conditions under which operations take place determine the load distribution $f_l(l)$ as well as the mean value $\bar{l}$.

Normal Distributions

The normal distribution is widely used for relating safety factors to reliability, particularly when small variations in materials and dimensional tolerances and the inability to determine loading precisely make capacity and load uncertain. The normal distribution is appropriate when variability in loads, capacity, or both is caused by the sum of many effects, no one of which is dominant. An appropriate example is the load and capacity of an elevator large enough to carry several people. Since the load is the sum of the weights of the people, the variability of the weight is likely to be very close to a normal distribution for the reasons discussed in Chapter 3. The variability in the weight of any one person is unlikely to have an overriding effect on the total load. Similarly, if the elevator cable is made up of many independent strands of wire, its capacity will be the sum of the strengths of the individual strands. Since the variability in strength of any one strand will not have much effect on the cable capacity, the normal distribution may be used to model the cable capacity.

Suppose that the load and capacity are represented by normal distributions,

$$f_l(l) = \frac{1}{\sqrt{2\pi}\sigma_l} \exp\left[-\frac{1}{2} \frac{(l - \bar{l})^2}{\sigma_l^2} \right] \tag{6.29}$$

and

$$f_c(c) = \frac{1}{\sqrt{2\pi}\sigma_c} \exp\left[-\frac{1}{2} \frac{(c - \bar{c})^2}{\sigma_c^2} \right], \tag{6.30}$$

where the mean values of the load and capacity are denoted by $\bar{l}$ and $\bar{c}$, and the corresponding standard deviations are σ_l and σ_c. Substituting these expressions into Eq. 6.12, we obtain for the reliability

$$\begin{aligned} r = &\int_{-\infty}^{\infty} \frac{1}{\sqrt{2\pi}\sigma_c} \exp\left[-\frac{1}{2} \frac{(c - \bar{c})^2}{\sigma_c^2} \right] \\ &\times \left\{ \int_{-\infty}^{c} \frac{1}{\sqrt{2\pi}\sigma_l} \exp\left[-\frac{1}{2} \frac{(l - \bar{l})^2}{\sigma_l^2} \right] dl \right\} dc. \end{aligned} \tag{6.31}$$

This expression* for the reliability may be reduced to a much simpler form involving only a single normal integral. To accomplish this, however, involves a significant amount of algebraic manipulation. We begin by transforming variables to the dimensionless quantities

$$x = (c - \bar{c})/\sigma_c, \tag{6.32}$$

$$y = (l - \bar{l})/\sigma_l. \tag{6.33}$$

Equation 6.31 may then be rewritten as

$$r = \frac{1}{2\pi} \int_{-\infty}^{\infty} \left\{ \int_{-\infty}^{(\sigma_c x + \bar{c} - \bar{l})/\sigma_l} \exp[-\tfrac{1}{2}(x^2 + y^2)]\, dy \right\} dx. \tag{6.34}$$

This double integral may be viewed geometrically as an integral over the shaded part of the $x - y$ plane shown in Fig. 6.5. The line demarking the

* Note that we have extended the lower limits on the integrals to $-\infty$ in order to accommodate the use of normal distributions. The effect on the result is negligible for $\bar{c} >> \sigma_c$ and $\bar{l} >> \sigma_l$.

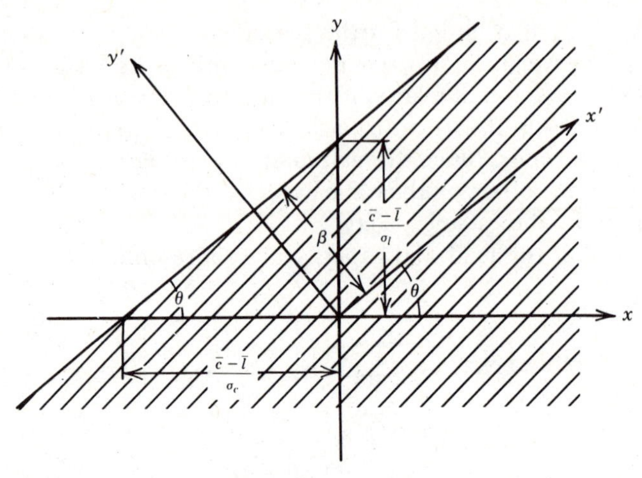

FIGURE 6.5 Domain of integration for normal load and capacity.

edge of the region of integration is determined by the upper limit of the y integration in Eq. 6.34:

$$y = \frac{1}{\sigma_l}(\sigma_c x + \bar{c} - \bar{l}). \tag{6.35}$$

By rotating the coordinates through the angle θ, we may rewrite the reliability as a single standardized normal function. To this end we take

$$x' = x \cos\theta + y \sin\theta \tag{6.36}$$

and

$$y' = -x \sin\theta + y \cos\theta. \tag{6.37}$$

It may then be shown that

$$x^2 + y^2 = x'^2 + y'^2 \tag{6.38}$$

and

$$dx\, dy = dx'\, dy', \tag{6.39}$$

allowing us to write the reliability as

$$r = \frac{1}{2\pi}\int_{-\infty}^{\infty}\left\{\int_{-\infty}^{\beta} \exp[-\tfrac{1}{2}(x'^2 + y'^2]\, dy'\right\} dx'. \tag{6.40}$$

The upper limit on the y' integration is just the distance β shown in Fig. 6.5. With elementary trigonometry, β may be shown to be a constant given by

$$\beta = \frac{\bar{c} - \bar{l}}{(\sigma_c^2 + \sigma_l^2)^{1/2}}. \tag{6.41}$$

The quantity β is referred to as the safety or reliability index. Since β is a constant, the order of integration may be reversed. Then, since

$$\frac{1}{\sqrt{2\pi}}\int_{-\infty}^{\infty} e^{-\frac{1}{2}x'^2}\,dx' = \Phi(\infty) = 1, \tag{6.42}$$

the remaining integral, in y, may be written as a standardized normal CDF to yield the reliability in terms of the safety index β:

$$r = \Phi(\beta). \tag{6.43}$$

The results of this equation may be put in a more graphic form by expressing them in terms of the safety factor, Eq. 6.27. A standard measure of the dispersion about the mean is the coefficient of variation, defined as the standard deviation divided by the mean:

$$\rho = \sigma/\mu. \tag{6.44}$$

Thus we may write

$$\rho_c = \sigma_c/\bar{c} \tag{6.45}$$

and

$$\rho_l = \sigma_l/\bar{l}. \tag{6.46}$$

With these definitions we may express the safety index in terms of the central safety factor and the coefficients of variation:

$$\beta = \frac{\upsilon - 1}{(\rho_c^2\upsilon^2 + \rho_l^2)^{1/2}}. \tag{6.47}$$

In Fig. 6.6 the standardized normal distribution is plotted. The area under the curve to the left of β is the reliability r; the area to the right is the failure probability p. In Fig. 6.6*b* the CDF for the normal distribution is plotted. Thus, given a value of β, we can calculate r and p. Conversely,

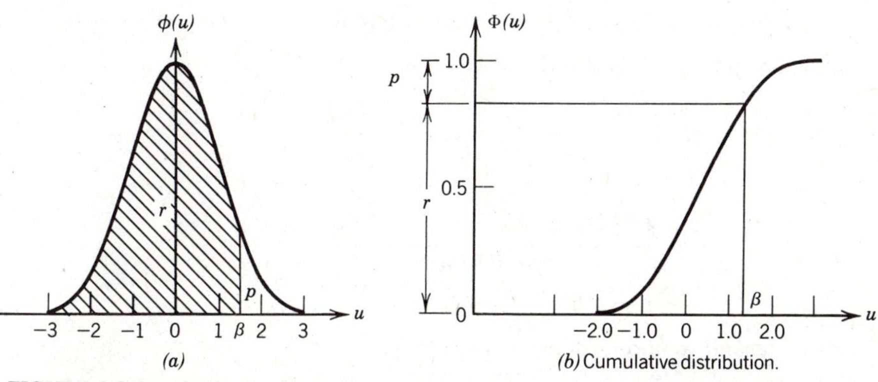

FIGURE 6.6 Standard normal distribution: (*a*) probability density function PDF, (*b*) cumulative distribution function (CDF).

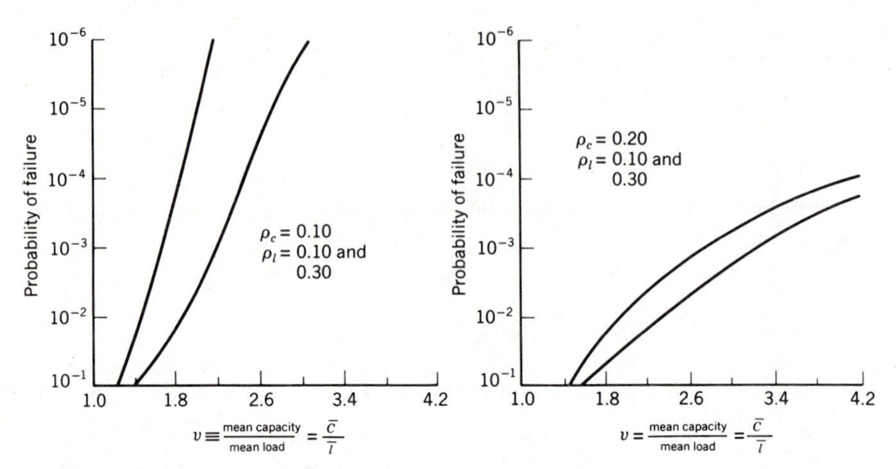

FIGURE 6.7 Probability of failure for normal load and capacity (From Gary C. Hart, *Uncertainty Analysis, Loads, and Safety in Structural Engineering*, © 1982, p. 107, with permission from Prentice-Hall, Englewood Cliffs, NJ.)

if the reliability is specified and the coefficients of variation are known, we may determine the value of the safety factor. In Fig. 6.7 the relation between safety factor and probability of failure is indicated for some representative values of the coefficients of variation.

EXAMPLE 6.2

Suppose that the coefficients of variation are $\rho_c = 0.1$ and $\rho_l = 0.15$. If we assume normal distributions, what safety factor is required to obtain a failure probability of no more than 0.005?

Solution $p = 0.005$; $r = 0.995$; $r = \Phi(\beta) = 0.995$. Therefore, from Appendix C, $\beta = 2.575$. We must solve Eq. 6.47 for υ. We have

$$\beta^2(\rho_c^2\upsilon^2 + \rho_l^2) = (\upsilon - 1)^2 \quad \text{or} \quad (1 - \beta^2\rho_c^2)\upsilon^2 - 2\upsilon + (1 - \beta^2\rho_l^2) = 0.$$

Solving this quadratic equation in υ, we have

$$\upsilon = \frac{2 \pm [4 - 4(1 - \beta^2\rho_l^2)(1 - \beta^2\rho_c^2)]^{1/2}}{2(1 - \beta^2\rho_c^2)}$$

or

$$\upsilon = \frac{2 \pm 2(1 - 0.8508 \times 0.9337)^{1/2}}{2 \times 0.9336} = \frac{1 \pm 0.4534}{0.9337}$$

$$= 1.56, \blacktriangleleft \quad 0.5853,$$

since the second solution will not satisfy Eq. 6.47.

In using Eqs. 6.43 and 6.47 to estimate reliability, we assume that the load and capacity are normally distributed and that the means and variances can be estimated. In practice, the paucity of data often does not allow us to say with any certainty what the distributions of load and capacity are. In these situations, however, the sample mean and variance can often be obtained. They can then be used to calculate the reliability index defined by Eq. 6.47; often the reliability can be estimated from Eq. 6.43. Such approaches are referred to as second-moment methods, since only the zero and second moments of the load and capacity distributions, as defined by Eq. 3.18, need to be estimated.

Second-moment methods* have been widely employed, for they represent the logical next step beyond the simple use of safety factors in that they also account for the variance of the distributions. Such methods must be employed with care, however, for when the distributions deviate greatly from normal distributions, the resulting formulas may be in serious error. This may be seen from the different expressions for reliability when lognormal or extreme-value distributions are employed.

Lognormal Distributions

The lognormal distribution is useful when the uncertainty about the load, or capacity, or both, is relatively large. Often it is expressed as having 90% confidence that the load or the capacity lies within some factor, say two, of the best estimates l_0 or c_0. In Chapter 3 the properties of the lognormal distribution were presented. As indicated there, the lognormal distribution is most appropriate when the value of the variable is determined by the product of several different factors. For load and capacity, we rewrite Eq. 3.95 for the PDFs as

$$f_l(l) = \frac{1}{\sqrt{2\pi}s_l l}\exp\left\{-\frac{1}{2s_l^2}\left[\ln\left(\frac{l}{l_0}\right)\right]^2\right\}, \qquad 0 < l \leqslant \infty, \tag{6.48}$$

and

$$f_c(c) = \frac{1}{\sqrt{2\pi}s_c c}\exp\left\{-\frac{1}{2s_c^2}\left[\ln\left(\frac{c}{c_0}\right)\right]^2\right\}, \qquad 0 < c \leqslant \infty. \tag{6.49}$$

* C. A. Cornell, "Structural Safety Specifications Based on Second-Moment Reliability," *Symposium of the International Association of Bridge, and Structural Engineers*, London, 1969; see also A. H.-S. Ang and W. H. Tang, *Probability Concepts in Engineering Planning and Design*, Vol. 2, Wiley, New York, 1984.

If Eqs. 6.48 and 6.49 are substituted into Eq. 6.12, the resulting expression for the reliability is

$$r = \int_0^\infty \frac{1}{\sqrt{2\pi}s_c c} \exp\left\{-\frac{1}{2s_c^2}\left[\ln\left(\frac{c}{c_0}\right)\right]^2\right\}$$
$$\times \left(\int_0^c \frac{1}{\sqrt{2\pi}s_l l} \exp\left\{-\frac{1}{2s_l^2}\left[\ln\left(\frac{l}{l_0}\right)\right]^2\right\} dl\right) dc. \tag{6.50}$$

Note, however, that with the substitutions

$$y = \frac{1}{s_l}\ln\left(\frac{l}{l_0}\right) \tag{6.51}$$

and

$$x = \frac{1}{s_c}\ln\left(\frac{c}{c_0}\right), \tag{6.52}$$

we obtain

$$r = \frac{1}{2\pi}\int_{-\infty}^{\infty}\left\{\int_{-\infty}^{(1/s_l)[s_c x + \ln(c_0/l_0)]} \exp[-\tfrac{1}{2}(x^2 + y^2)]\, dy\right\} dx. \tag{6.53}$$

The forms of the reliability in Eq. 6.34 and in this equation are identical if in the upper limit of the y integration we substitute s_l and s_c for σ_l and σ_s, respectively, and replace $\bar{c} - \bar{l}$ with $\ln(c_0/l_0)$. Thus the reliability still has the form of a standardized normal distribution given by Eq. 6.43. Now, however, the argument β is given by

$$\beta = \frac{\ln(c_0/l_0)}{(s_c^2 + s_l^2)^{1/2}}. \tag{6.54}$$

EXAMPLE 6.3

Suppose that both the load and the capacity on a device are known within a factor of two with 90% confidence. What value of the safety factor, c_0/l_0, must be used if the failure probability is to be no more than 0.1%?

Solution For $\Phi(\beta) = r = 1 - p = 0.99$ we find from Appendix C that $\beta = 2.33$. From Eq. 3.105 for 90% confidence with a factor of $n = 2$ uncertainty, we have for both load and capacity $s_c = s_l = s = (1/1.645)\ln(n) = (1/1.645)\ln(2) = 0.4214$. Solve Eq. 6.54 for c_0/l_0:

$$\frac{c_0}{l_0} = \exp[\beta(s_c^2 + s_l^2)^{1/2}] = \exp(\beta\sqrt{2}s)$$
$$= \exp(2.33 \times 1.414 \times 0.4214) = 4.01. \blacktriangleleft$$

6.4 EXTREME-VALUE DISTRIBUTIONS

A salient feature of failure probabilities is indicated clearly by Fig. 6.3. In situations in which safety is a significant factor, the probability of failure depends strongly on the lower tail of the capacity distribution and on the upper tail of the load distribution. Normal and lognormal distributions are useful in representing these tails when there are many contributions, no one of which is dominant. But these tails are not described well by normal or lognormal distributions when the load or capacity is not determined by either the sum or the product of many relatively small contributions. In contrast, it may be the extreme of many contributions that governs the load or the capacity. For example, it is not the sum of the accelerations but rather the extreme value that determines the primary earthquake loading on a structure. Similarly, it is not the sum of the flaw contributions, but rather the largest of many flaws that may limit the capacity of a pressure vessel. Extreme-value distributions have proved to be very useful in the analysis of reliability problems of this nature.*

In this section we introduce the use of maximum extreme-value distributions for the treatment of loads, and minimum extreme-value distributions for capacity determination. We begin by introducing the extreme over a sequence of some finite number N of random variables, for, although the number of practical problems that can be treated is limited, the concepts are more transparent. We then proceed to the standard asymptotic extreme-value distributions for large N. These have proved to be exceedingly useful in treating a variety of reliability problems for which large numbers of random variables must be considered. We conclude with a discussion of data gathering for extreme-value distributions.

Maximum Extreme Values

Suppose that we have a number N of random variables, say loads, $\mathbf{x}_1$, $\mathbf{x}_2$, . . . , $\mathbf{x}_n$, . . . , $\mathbf{x}_N$, each with a CDF $F_{\mathbf{x}_n}(x)$. If we then define the variable $\mathbf{y}$ as the maximum of the $\mathbf{x}_n$, it is also a random variable. The CDF of $\mathbf{y}$ is expressed in terms of those of the $\mathbf{x}_n$ by realizing that

$$F_{\mathbf{y}}(y) \equiv P\{\mathbf{y} \leq y\} = P\{\mathbf{x}_1 \leq y) \cap (\mathbf{x}_2 \leq y) \cap \cdots \cap (\mathbf{x}_N \leq y)\}. \quad (6.55)$$

Therefore, if the $\mathbf{x}_n$ are independent,

$$F_{\mathbf{y}}(y) = P\{\mathbf{x}_1 \leq y\}\, P\{\mathbf{x}_2 \leq y\} \cdots P\{\mathbf{x}_N \leq y\}. \quad (6.56)$$

Since the probabilities on the right-hand side are just the CDFs of the $\mathbf{x}_n$, evaluated at $\mathbf{x}_n = y$,

$$F_{\mathbf{y}}(y) = F_{\mathbf{x}_1}(y)\, F_{\mathbf{x}_2}(y) \cdots F_{\mathbf{x}_N}(y). \quad (6.57)$$

* E. J. Gumbel, *Statistics of Extremes,* Columbia University Press, New York, 1958.

In the most frequently appearing case, all the $\mathbf{x}_n$ are identically distributed with CDF, $F_{\mathbf{x}}(y)$. Hence

$$F_{\mathbf{y}}(y) = [F_{\mathbf{x}}(y)]^N. \tag{6.58}$$

The PDF for $\mathbf{y}$ may be obtained, as in Eq. 3.4, by differentiating

$$f_{\mathbf{y}}(y) = \frac{d}{dy} F_{\mathbf{y}}(y) = N[F_{\mathbf{x}}(y)]^{N-1} \frac{d}{dy} F_{\mathbf{x}}(y), \tag{6.59}$$

or

$$f_{\mathbf{y}}(y) = N[F_{\mathbf{x}}(y)]^{N-1} f_{\mathbf{x}}(y), \tag{6.60}$$

where $f_{\mathbf{x}}(y)$ is the PDF of each of the x_n.

In these expressions $F_{\mathbf{x}}(x)$ is referred to as the parent distribution and $F_{\mathbf{y}}(y)$ is the maximum extreme-value distribution. The extreme-value behavior may be illustrated by using a known distribution for the parent. Suppose, for example, that we use the exponential distribution

$$f_{\mathbf{x}}(x) = \alpha e^{-\alpha x}, \qquad F_{\mathbf{x}}(x) = 1 - e^{-\alpha x}, \tag{6.61}$$

where $\mathbf{x}$ might represent the impact loading on some system. The CDF for

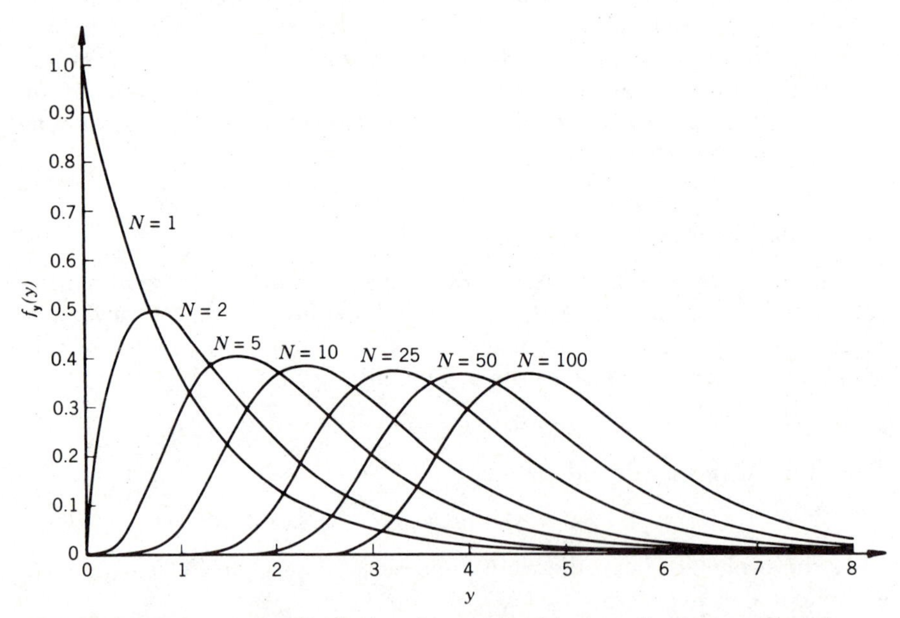

FIGURE 6.8 Extreme-value distributions for exponential parent distribution. (From A. H-S. Ang and W. H. Tang, *Probability Concepts in Engineering Planning and Design*, Vol. 2, p. 189. Copyright © 1984, by John Wiley and Sons, New York. Used with permission.)

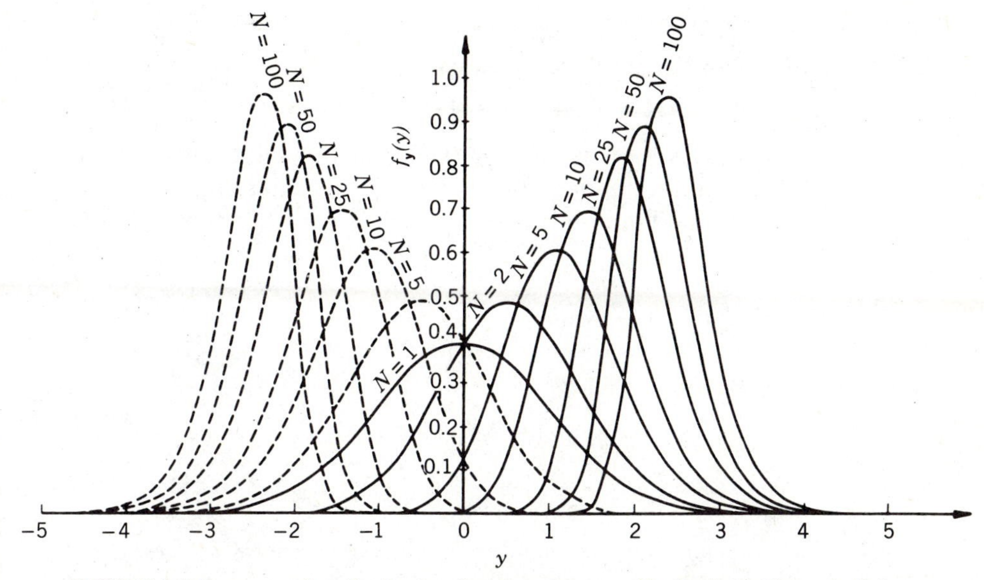

FIGURE 6.9 Extreme-value distribution for normal parent distributions. (From A. H-S. Ang and W. H. Tang, *Probability Concepts in Engineering Planning and Design*, Vol. 2, p. 195. Copyright © 1984, by John Wiley and Sons, New York. Used with permission.)

the maximum impact in N independent loadings would then be given according to Eq. 6.58,

$$F_y(y) = (1 - e^{-\alpha y})^N, \tag{6.62}$$

with a corresponding PDF of

$$f_y(y) = N\alpha(1 - e^{-\alpha y})^{N-1}\, e^{-\alpha y}. \tag{6.63}$$

The relation between the maximum extreme-value distribution and its parent, when the parent is an exponential distribution, is shown in Fig. 6.8.

Extreme-value distributions can be formed for other parent distributions as well. For example, if the parent distribution is the normal distribution, in terms of the reduced variate u, the maximum extreme-value distribution appears as the solid lines in Fig. 6.9.

Maximum extreme-value distributions are useful in calculating reliability when the load distribution appearing in the reliability formulas consists of the maximum of N loadings. We recall from Eq. 6.10 that for a system with known capacity c, the reliability is just the CDF of the load evaluated at c. Thus, if the load is given by an extreme-value distribution F_l, we have simply

$$r(c) = F_l(c) = [F_{l'}(c)]^N, \tag{6.64}$$

where $F_{l'}$ is now understood to be the CDF of the parent distribution, that is, for any one of the N loadings to which the system is subjected. The corresponding failure probability, $1 - r(c)$, is

$$p(c) = \tilde{F}_l(c) = 1 - [1 - \tilde{F}_{l'}(c)]^N, \tag{6.65}$$

where $\tilde{F}_l = 1 - F_l$ and $\tilde{F}_{l'} = 1 - F_{l'}$ are the CCDF of **l** and **l**′, respectively.

If the capacity is large enough that $\tilde{F}_{l'}(c) << 1$, a simple approximation can be made by expanding the last term in Eq. 6.65 with binomial coefficients,

$$p(c) = 1 - 1 + N\tilde{F}_{l'}(c) - \frac{N(N-1)}{2}\tilde{F}_{l'}(c)^2 + \cdots, \tag{6.66}$$

so that for $N\tilde{F}_{\mathbf{x}}(c) << 1$ we have

$$p(c) \approx N\tilde{F}_{l'}(c). \tag{6.67}$$

EXAMPLE 6.4

The final assembly of a large but delicate piece of machinery requires that it be moved eight times between work stations before it is installed in a protective casing; there is impact loading on the machinery each time it is set down. The impulse loading is described by Eq. 6.62 with $\alpha = 0.02$ sec/kg · m. (*a*) If, in the initial design, an impulse of more than 250 kg · m/sec will cause the machinery to be damaged (i.e., failed), what is the failure probability for the assembly procedure? (*b*) If the machinery is to be redesigned so that no more than 0.5% of the units will be damaged in assembly, what must the capacity for impact loading be?

Solution (*a*) The capacity for impact loading is $c = 250$ kg · m/sec. The failure probability is

$$p(c) = 1 - F_l(c) = 1 - (1 - e^{-0.02 \times 250})^8$$

$$p(250) = 0.053 = 5.3\%. \blacktriangleleft$$

(*b*) We must have $p(c) \leqslant 0.005$ or $F_l(c) \geqslant 0.995$. From Eq. 6.62 we have

$$0.995 \leqslant (1 - e^{-0.02c})^8$$

$$c \geqslant \frac{-1}{0.02}\ln[1 - (0.995)^{1/8}]$$

$$c \geqslant 369 \text{ kg} \cdot \text{m/sec}. \blacktriangleleft$$

Minimum Extreme Values

The distribution of minimum extreme values may be derived in a manner analogous to that just discussed. Our primary use for them is in describing system capacity distributions that depend on a weakest component or member. Again we assume that there are N random variables $\mathbf{x}_1, \mathbf{x}_2, \ldots, \mathbf{x}_n, \ldots,$

$\mathbf{x}_N$, each with a CDF $F_{\mathbf{x}_n}(x)$. If $\mathbf{z}$ is the smallest of the $\mathbf{x}_n$, it is also a random variable. Its CDF is given by

$$F_{\mathbf{z}}(z) = P\{\mathbf{z} \leq z\} = 1 - P\{\mathbf{z} > z\}. \tag{6.68}$$

However, if $\mathbf{z}$ is the minimum value of the $\mathbf{x}_n$,

$$P\{\mathbf{z} > z\} = P\{(\mathbf{x}_1 > z) \cap (\mathbf{x}_2 > z) \cap \cdots \cap (\mathbf{x}_N > z)\}. \tag{6.69}$$

If the $\mathbf{x}_n$ are mutually independent, this expression becomes

$$P\{\mathbf{z} > z\} = P\{\mathbf{x}_1 > z\}P\{\mathbf{x}_2 > z\} \cdots P\{\mathbf{x}_N > z\}. \tag{6.70}$$

Finally, since the CDF of $\mathbf{x}_n$ is defined by

$$F_{\mathbf{x}}(x) = P\{\mathbf{x}_n \leq x\} = 1 - P\{\mathbf{x}_n > x\}, \tag{6.71}$$

we may combine the foregoing equations to obtain

$$F_{\mathbf{z}}(z) = 1 - [1 - F_{\mathbf{x}_1}(z)][1 - F_{\mathbf{x}_2}(z)] \cdots [1 - F_{\mathbf{x}_N}(z)]. \tag{6.72}$$

If we assume that all the $\mathbf{x}_n$ are identically distributed, we may simplify this result by replacing the $F_{\mathbf{x}_n}$ by $F_{\mathbf{x}}$. We thus obtain

$$F_{\mathbf{z}}(z) = 1 - [1 - F_{\mathbf{x}}(z)]^N. \tag{6.73}$$

Then, using Eq. 3.4 to relate the PDF to the CDF, we obtain

$$f_{\mathbf{z}}(z) = N[1 - F_{\mathbf{x}}(z)]^{N-1}f_{\mathbf{x}}(z). \tag{6.74}$$

To illustrate, in Fig. 6.9 the dotted lines represent minimum extreme-value distributions, with the parent distribution once again the normal distribution written in terms of the reduced variate u.

The application of minimum extreme values to reliability calculations is often referred to as the chain rule, since a chain (system) is no stronger than its weakest link (component). For example, suppose that we are to calculate the reliability $r(l)$ of a system subject to a constant load. From Eqs. 6.17 and 6.18 we recall that at constant load the reliability is just the CCDF of the system capacity. If the capacity in Eq. 6.18 is given by $F_{\mathbf{c}}$, the minimum extreme-value distribution for the capacity, we have, using Eq. 6.73,

$$r(l) = 1 - F_{\mathbf{c}}(l) = [1 - F_{\mathbf{c}'}(l)]^N, \tag{6.75}$$

where $F_{\mathbf{c}'}$ is understood to represent the capacity of the parent distribution, that is, of any one of N identical components (links) of a nonredundant system (chain).

For a load that is sufficiently small, we may take $F_{\mathbf{c}'}(l) \ll 1$. Then writing the right-hand side of Eq. 6.75 in terms of binomial coefficients as

$$r(l) = 1 - NF_{\mathbf{c}'}(l) + \frac{N(N-1)}{2} F_{\mathbf{c}'}(l)^2 - \cdots \tag{6.76}$$

so that for $NF_{c'}(l) \ll 1$, we obtain

$$r(l) \approx 1 - NF_{c'}(l). \tag{6.77}$$

EXAMPLE 6.5

One hundred-foot lengths of chain must have a failure probability of no more than 0.001 under a prescribed load. A number of 10-ft lengths of the chain are tested under the same load. What is the largest failure probability that is acceptable for the 10-ft lengths?

Solution Let r_{10} and r_{100} be the probability of failure for 10 and 100-ft lengths, respectively. In addition, let N and $10N$ be the number of links in the respective lengths of chain. We assume independent failure of the links. From Eq. 6.75:

$$r_{10} = [1 - F_{c'}(l)]^N, \qquad r_{100} = [1 - F_{c'}(l)]^{10N}.$$

Eliminating $F_{c'}(l)$ yields

$$r_{10}^{1/N} = r_{100}^{1/10N}$$

Therefore, $r_{10} = r_{100}^{1/10} = (0.999)^{1/10} = 0.99990$, and the largest acceptable probability of failure is

$$p_{10} = 1 - r_{10} = 10^{-4}. \blacktriangleleft$$

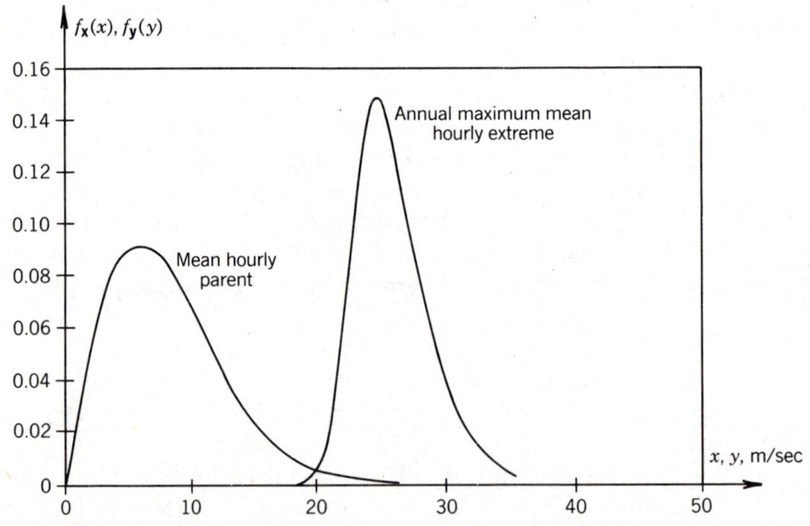

FIGURE 6.10 Distributions of parent and annual extreme wind speed for Lerwick. (From P. Thoft-Christe and M. J. Baker, *Structural Reliability and Its* Application, © 1982, p. 224, with permission from Springer-Verlag, Heidelberg, Germany.)

Asymptotic Extreme-Value Distributions

The extreme-value distributions discussed in the preceding subsections serve to illustrate, in a simple way, the effect of maximum extreme values of the loading on $r(c)$, the reliability at known capacity, and the effect of minimum extreme values in capacity on $r(l)$, the reliability with constant load. In practice, however, the use of Eqs. 6.58 or 6.73 for the CDFs may become cumbersome. Often N, the number of variables over which the extremes in the foregoing equations are taken, is very large, if it is known at all. Similarly, the assumption that all the $\mathbf{x}_n$ are identically distributed may not be valid, and even the assumption that the variables are independent may be suspect.

As an illustration of these complications, consider Fig. 6.10 in which is shown both a PDF for average hourly wind velocity and the annual maximum of the average hourly wind speed. It would be difficult to obtain the maximum PDF directly from the hourly average owing to the large value of $N = 365.25 \times 24 = 8766$. More important, the distributions for the hourly

TABLE 6.1 Extreme-Value Distributions

Distributions of the largest value y	
Type I	
$F_{\mathbf{y}}(y) = \exp[-e^{-(y-u)/\theta}]$	$-\infty \leqslant y \leqslant \infty$, $\theta > 0$
Type II	
$F_{\mathbf{y}}(y) = \exp\left[-\left(\frac{y}{\theta}\right)^{-m}\right]$,	$y \geqslant 0$, $\theta > 0,\ m > 0$
Type III	
$F_{\mathbf{y}}(y) = \exp\left[-\left(\frac{u-y}{\theta}\right)^{m}\right]$,	$y \leqslant u$, $\theta > 0,\ m > 0$
Distributions of the smallest value z	
Type I	
$F_{\mathbf{z}}(z) = 1 - \exp[-e^{(z-u)/\theta}]$,	$-\infty \leqslant z \leqslant \infty$
Type II	
$F_{\mathbf{z}}(z) = 1 - \exp\left[-\left(\frac{-z}{\theta}\right)^{-m}\right]$,	$-\infty \leqslant z \leqslant 0$, $\theta > 0,\ m > 0$
Type III	
$F_{\mathbf{z}}(z) = 1 - \exp\left[-\left(\frac{z-u}{\theta}\right)^{m}\right]$,	$u \leqslant z < \infty$, $\theta > 0,\ m > 0$

averages vary both with the time of day and with the season. As a result, in many situations the extreme-value distribution is measured directly, without an attempt to measure the parent distribution. Instruments are installed to measure directly maximum wind gusts, maximum ground accelerations from earthquakes, and so on.

Data on extreme values are likely to be scarce, particularly if maximum (or minimum) events occurs at considerable intervals, for example, the maximum annual flow of a river or the maximum annual acceleration from earthquakes. It is therefore beneficial if few-parameter analytic distributions can be formulated for modeling distributions of extreme values. Methods similar to those used in the reliability testing discussed in Chapter 5 can then be employed to fit the available data to a model. The model, in turn, may be used to extrapolate to higher and less probable loads, or to lower and less probable capacities.

There are three classes of asymptotic extreme-value distributions,* the CDFs for which are given in Table 6.1. They may be shown to arise when N, the number of variables over which the extreme is taken, becomes large, with remarkably few restrictions on the forms of the parent distributions. The distributions differ both in the domain of the extreme-value variable and in the form of the upper or lower tail of the parent distributions.

Type I

We shall concentrate our attention first on the type I or Gumbel distributions. The maximum extreme-value distribution results from letting N become large in Eq. 6.60 when for large x the parent distribution has the form

$$F_{\mathbf{x}}(x) \approx 1 - e^{-g(x)}, \tag{6.78}$$

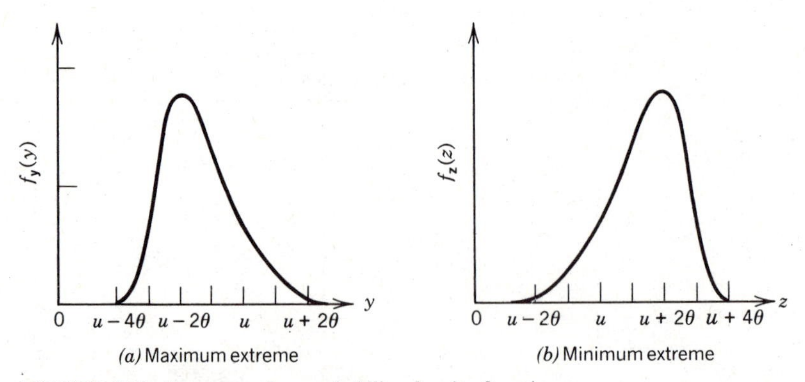

FIGURE 6.11 Extreme-value probability density functions.

* E. J. Gumbel op. cit.

and $g(x)$ is an increasing function of x. Thus normal and exponential parent distributions both give rise to the type I maximum extreme-value distribution when N becomes large.

Equation 6.60 may be used to show that the PDF corresponding to the CDF in Table 6.1 for the type I maximum extreme-value distribution is

$$f_{\mathbf{y}}(y) = \frac{1}{\theta} e^{-(y-u)/\theta} \exp[-e^{-(y-u)/\theta}], \qquad -\infty \leqslant y \leqslant \infty. \tag{6.79}$$

The parameter u is the value of y for which the maximum in $f_{\mathbf{y}}(y)$ occurs, whereas θ is a measure of the dispersion. The mean and variance are given by

$$\mu_{\mathbf{y}} = u + \gamma\theta \tag{6.80}$$

and

$$\sigma_{\mathbf{y}}^2 = \frac{\pi^2}{6} \theta^2, \tag{6.81}$$

where $\gamma = 0.5772157 \cdots$ is Euler's constant. The PDF for this maximum extreme-value distribution is plotted in Fig. 6.11*a*.

As with the normal distribution, it is often useful to express the extreme-value distribution in terms of a reduced variate in order to express the PDF in a parameter-free standardized form. If we define the reduced variate

$$\mathbf{w} = \frac{\mathbf{y} - u}{\theta}, \tag{6.82}$$

we have

$$F_{\mathbf{w}}(w) = e^{-e^{-w}}. \tag{6.83}$$

Because of the form of the CDF, the type I distribution is frequently referred to as the double exponential distribution. This quantity is tabulated in Appendix C.

Analogously, we may show that the type I minimum extreme-value distribution results as $N \to \infty$ for a parent distribution with exponential-like lower tails. The type I PDF corresponding to $F_{\mathbf{z}}(z)$ in Table 6.1 is then

$$f_{\mathbf{z}}(z) = \frac{1}{\theta} e^{(z-u)/\theta} \exp[-e^{(z-u)/\theta}], \qquad -\infty \leqslant z \leqslant \infty \tag{6.84}$$

with a mean of

$$\mu_{\mathbf{z}} = u - \theta\gamma \tag{6.85}$$

and a variance of

$$\sigma_{\mathbf{z}}^2 = \frac{\pi^2}{6} \theta^2. \tag{6.86}$$

The PDF for this minimum extreme-value distribution is plotted in Fig. 6.11*b*. The table in Appendix C for the largest extreme variable may also be used for the smallest value. Because of the antisymmetry of the two distributions, with

$$\mathbf{w} = (u - \mathbf{z})/\theta, \tag{6.87}$$

we again obtain Eq. 6.83 as the CDF for **w**.

Type II

The type II distributions differ from the type I in a number of respects, the most important of which is that the independent variable **x** of the parent distribution is limited on the left by zero. The maximum extreme-value distribution arises from parent distributions that for large x have the form

$$F_{\mathbf{x}}(x) \simeq 1 - \beta\left(\frac{1}{x}\right)^m. \tag{6.88}$$

The PDF corresponding to the type II distribution in Table 6.1 is

$$f_{\mathbf{y}}(y) = \frac{m}{\theta}\left(\frac{\theta}{y}\right)^{m+1} e^{-(\theta/y)^m} \tag{6.89}$$

with a mean of

$$\mu_{\mathbf{y}} = \theta\Gamma(1 - 1/m), \qquad m > 1 \tag{6.90}$$

and variance

$$\sigma_{\mathbf{y}}^2 = \frac{\Gamma(1 - 2/m)}{\Gamma^2(1 - 1/m)} - 1, \qquad m > 2. \tag{6.91}$$

Inspection of Eq. 6.89 reveals that the type II $F_{\mathbf{y}}(y)$ is just a two-parameter Weibull distribution, Eq. 4.56, as discussed in Chapter 4, with m replaced by $-m$. The relation between type I and type II distributions is analogous to that between normal and lognormal distributions. Specifically, if **y** has a type II distribution, then ln **y** has a type I distribution. Thus, if we define the reduced variate

$$\mathbf{w} = m \ln(\mathbf{y}/\theta), \tag{6.92}$$

we again obtain the parameter-free form Eq. 6.83 as the CDF for **w**. The corresponding type II distribution for the smallest value can also be found; however, it is applied very infrequently.

Type III

Type III extreme-value distributions arise from parent distributions in which the tail falls off about some finite value u of x. Thus, for the maximum distribution,

$$F_{\mathbf{x}}(x) \simeq 1 - C(u - x)^m, \qquad x \leqslant u, \tag{6.93}$$

and for the minimum distribution

$$F_{\mathbf{x}}(x) \simeq C(x - u)^m, \qquad x \geqslant u. \tag{6.94}$$

It will be noted that the more widely used minimum type III extreme-value distribution is identical to the three-parameter Weibull distribution discussed in Chapter 4; simply set $t = x$ and $t_0 = u$.

Combined Distributions

In general, it is difficult to evaluate analytically the expressions given for reliability in Section 6.2 when the load and capacity are given by different distributions. However, when the load or capacity is given by an extreme-value distribution of type I and the other by a normal distribution, both analytical results and some insight can be obtained.

Consider first a system whose capacity is approximated by the minimum extreme-value distribution of type I, but about whose loading there is only a small amount of uncertainty. This situation is depicted in Fig. 6.12*a*. We assume that $\bar{l}$, the mean value of the load, is much smaller than the mean, $\bar{c} \equiv u - \theta\gamma$, of the minimum extreme-value distribution that represents the capacity: $\bar{l} \ll \bar{c}$. For known loading the reliability is given by Eq. 6.18. Combining $F_{\mathbf{y}}(z)$ of type I minimum extreme value from Table 6.1, we have

$$r(l) = \exp[-e^{(l-u)/\theta}], \tag{6.95}$$

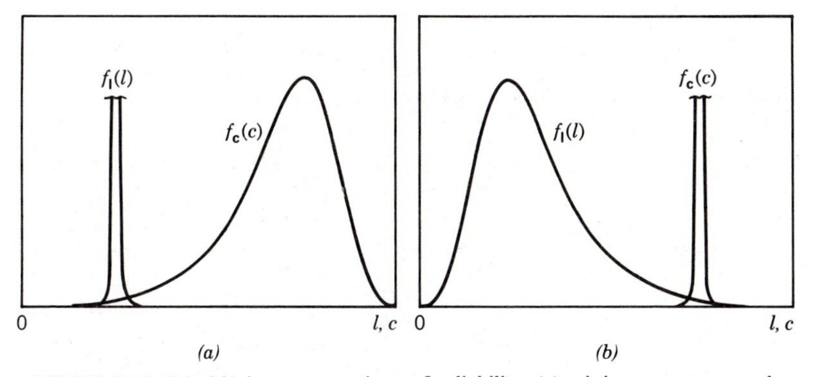

FIGURE 6.12 Graphical representations of reliability: (*a*) minimum extreme-value distribution for capacity, (*b*) maximum extreme-value distribution for loading.

which for small enough values of l (i.e., $l \ll u$) becomes

$$r(l) \approx 1 - \exp\left(\frac{l - u}{\theta}\right). \tag{6.96}$$

Now suppose that we want to take into account some natural variation in the loading on the system. If this is represented by a distribution with small variance of the load about the mean, Eq. 6.19 may be employed to express the reliability as

$$r = 1 - \int_0^\infty f_l(l) \exp\left(\frac{l - u}{\theta}\right) dl. \tag{6.97}$$

Again, it must be assumed that the variance of the load is not large, $\sigma_l \ll \bar{c} - \bar{l}$, so that the expansion, Eq. 6.96, is valid over the entire range of l where $f_l(l)$ is significantly greater than zero. Suppose that we represent the load distribution by the normal distribution, Eq. 6.29. The situation would then appear as in Fig. 6.12*a*. To evaluate the reliability, we first combine Eqs. 6.29 and 6.97 to obtain

$$r = 1 - \frac{1}{\sqrt{2\pi}\sigma_l} \int_{-\infty}^{\infty} \exp\left[-\frac{(l - \bar{l})^2}{2\sigma_l^2} + \frac{l - u}{\theta}\right] dl. \tag{6.98}$$

The integral* may be evaluated analytically by completing the square of the argument of the exponential and then writing the result in terms of a standardized normal distribution. To do this, we first pull the terms that do not depend on l outside of the integral,

$$r = 1 - \frac{1}{\sqrt{2\pi}\sigma_l} \exp\left(-\frac{\bar{l}^2}{2\sigma_l^2} - \frac{u}{\theta}\right) \int_{-\infty}^{\infty} \exp[-g(l)]\, dl, \tag{6.99}$$

where the remaining argument of the exponential inside the integral is now

$$g(l) = \frac{l^2}{2\sigma_l^2} - \left(\frac{\bar{l}}{\sigma_l^2} + \frac{1}{\theta}\right) l. \tag{6.100}$$

To complete the square, we rewrite $g(l)$ as

$$g(l) = \frac{1}{2\sigma_l^2}(l - l_0)^2 - \frac{l_0^2}{2\sigma_l^2}, \tag{6.101}$$

where

$$l_0 \equiv \bar{l} + \frac{\sigma_l^2}{\theta}. \tag{6.102}$$

* The lower limit on the integral is extended to $-\infty$ to accommodate the use of the normal distribution. The effect on the result is negligible when $\bar{l} \gg \sigma_l$.

The reliability is then

$$r = 1 - \exp\left(-\frac{\bar{l}}{2\sigma_l^2} - \frac{u}{\theta} + \frac{l_0^2}{2\sigma_l^2}\right)\int_{-\infty}^{\infty}\frac{1}{\sqrt{2\pi}\sigma_l}\exp\left[-\frac{1}{2\sigma_l^2}(l - l_0)^2\right]dl. \tag{6.103}$$

We immediately recognize that the integral on the right-hand side may be rewritten as the standardized normal integral $\Phi(\infty)$, which has a value of one. If we now use Eq. 6.102 to eliminate l_0 from the argument of the remaining exponential, we obtain for the reliability

$$r = 1 - \exp\left[\frac{1}{2}\left(\frac{\sigma_l}{\theta}\right)^2\right]\exp\left(\frac{\bar{l} - u}{\theta}\right), \tag{6.104}$$

where $u \equiv \bar{c} + \theta\gamma \gg \bar{l}$ and γ is Euler's constant.

In the converse situation the capacity has only a small degree of uncertainty, whereas the loading is represented by a maximum extreme-value distribution, again with the stipulation that $\bar{c} \gg \bar{l}$. This situation is depicted in Fig. 6.12*b*. The reliability at known capacity is first obtained by substituting the type I maximum distribution from Table 6.1 into Eq. 6.10,

$$r(c) = F_y(c) = \exp[-e^{-(c-u)/\theta}], \tag{6.105}$$

or for large c,

$$r(c) \approx 1 - e^{-(c-u)/\theta}. \tag{6.106}$$

Thus, from Eq. 6.11, we have

$$r = \int_0^{\infty} f_c(c)\left[1 - \exp\left(\frac{u - c}{\theta}\right)\right]dc, \tag{6.107}$$

provided that the variance in $f_c(c)$ is small enough that Eq. 6.106 is valid.

We utilize the normal distribution, with small standard deviation, to represent the distribution of capacities in Eq. 6.107. From Eq. 6.30 we obtain

$$r = 1 - \frac{1}{\sqrt{2\pi}\sigma_c}\int_{-\infty}^{\infty}\exp\left[-\frac{(c - \bar{c})^2}{2\sigma_c^2} - \frac{c - u}{\theta}\right]dc. \tag{6.108}$$

The integral on the right-hand side may be evaluated analytically in a manner completely analogous to that used for the minimum extreme-value distribution discussed earlier. The resulting reliability is

$$r = 1 - \exp\left[-\frac{1}{2}\left(\frac{\sigma_c}{\theta}\right)^2\right]\exp\left(\frac{u - \bar{c}}{\theta}\right). \tag{6.109}$$

where $u \equiv \bar{l} - \theta\gamma \ll \bar{c}$ and γ is Euler's constant.

Data Collection

With extreme-value distributions, as with other distributions, empirical data or other means must be available to estimate the distribution parameters before reliability estimates can be made. Although a detailed discussion of the techniques of data collection and analysis may be found in more advanced texts,* some general remarks are in order. A simple illustration can be provided using graphical methods similar to those in Chapter 5.

A collection of extreme-value data typically consists of a series of peak values. If the peaks may be approximated to be independent, a number of techniques may be used to estimate the parameters. The classic example of such an analysis is in predicting the extreme annual flow rates in rivers. For this purpose we may use simple graphical methods quite analogous to those discussed in Chapter 5 for the analysis of life data. Now, however, the random variable is the magnitude of the load for which the system must be designed rather than time, as in the life-testing case. The CDF for the loading is transformed so that the distribution will represent a straight line on a specially constructed graph paper. The data are plotted on that paper, and

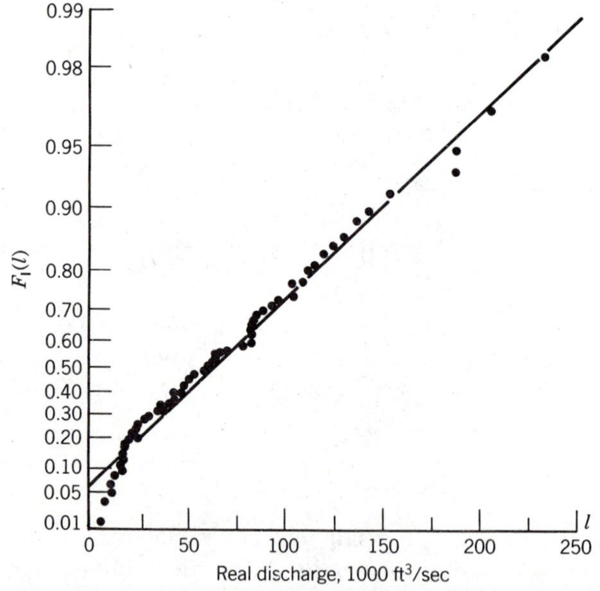

FIGURE 6.13 Annual flood illustration on extreme-value paper (Data from Feather River, Oroville Calif. Source: J. R. Benjamin and C. A. Cornell, Probability, Statistics, and Decisions for Civil Engineers, McGraw Hill Book Co. New York 1970, p. 493. Reprinted by permission.)

* See, for example, J. C. Benjamin and C. A. Cornell, *Probability, Statistics, and Decisions for Civil Engineers*, McGraw-Hill, New York, 1970; see also A. H-S. Ang and W. H. Tang, op. cit.

the parameters are estimated from the slope and intercept of the line. The resulting straight-line extrapolation then permits the return period for loads of greater magnitude to be estimated.

To obtain the linear representation, we begin with the type I extreme-value CDF from Table 6.1:

$$F_1(l) = \exp[-e^{-(l-u)/\theta}]. \tag{6.110}$$

Taking the natural logarithm of this expression, we have

$$-\ln F_1(l) = e^{-(l-u)/\theta}. \tag{6.111}$$

Then taking the ln once again, we obtain

$$[-\ln[-\ln F_1(l)] = \frac{1}{\theta}l - \frac{1}{\theta}u. \tag{6.112}$$

As in Chapter 5, we have obtained a linear equation of the form $y = mx + b$, provided that we plot $[-\ln[-\ln F_l(l)]$ on the ordinate and l on the abscissa. The value of θ is determined by noting that $l = u$ when $F_1(l) = e^{-1} = 0.368$.

To illustrate the use of extreme-value graph paper, suppose that we have available the flood magnitudes for N years. We rank these from the smallest to the largest, $l_1 < l_2 < l_3 \cdots < l_N$, just as we did with the failure times in Chapter 5. The CDF is estimated from the analog of Eq. 5.3:

$$F_1(l_n) \approx \frac{n}{N+1}. \tag{6.113}$$

Thus for each value of l_n plotted on the horizontal axis, the corresponding value of $n/(N + 1)$ is plotted on the nonlinear vertical axis. Some typical annual flood data are plotted in this way in Fig. 6.13. Taking the slope of the line, we obtain $\theta \approx 1.1 \times 10^3$ ft^3/sec. The load corresponding to $F = 0.368$ yields $u = 45 \times 10^3$ ft^3/sec.

6.5 FAILURE RATES AND REPETITIVE LOADING

A number of expressions are derived in the preceding sections for the reliability of a system and the probability that it will fail under a single loading. These quantities are represented, respectively, by the lowercase letters r and p. In this section we examine the reliability of a system under repeated loadings of random magnitude. We restrict our attention to systems with a known capacity that is independent of time, for with these restrictions constant failure rate models are shown to result. In Section 6.6 the inclusion of variability in capacity is shown to be related to the wearin phenomenon of decreasing failure rate, and the deterioration of capacity with time is shown to lead to the increasing failure rates characteristic of wearout.

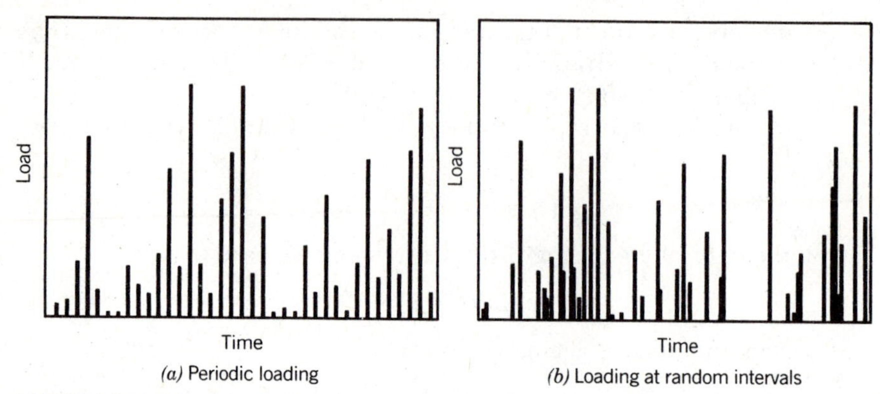

FIGURE 6.14 Repetitive loads of random magnitudes.

Suppose that a system is subjected to repeated loadings, as indicated, for example, by Figs. 6.14*a* and *b*. The two graphs differ in that the loads in the first occur at fixed intervals, whereas those in the second occur at random intervals. For the moment, however, we are more interested in the distributions of magnitudes rather than in their spacing in time. We assume that the load magnitudes are random and independent. Thus each value is drawn from a distribution $f_l(l)$ independent of the values of l that have already occurred.

Such independence of loading magnitudes provides a reasonable approximation for the many diverse circumstances in which the successive loads are very weakly correlated: the maximum annual flood or wind gust, the maximum impact during landing of an aircraft, and the maximum earthquake acceleration, to name only a few.

For a system with fixed, time-independent capacity c, the reliability $r(c)$ for any single loading is independent of the reliabilities from the other loadings. The probability of surviving n such loadings (i.e., the reliability R_n) is obtained by multiplying together the probabilities of surviving the individual loadings:

$$R_n(c) = r(c)^n. \tag{6.114}$$

To convert this expression to $R(t|c)$, the reliability as a function of time, we must determine how frequently the loadings occur. Two idealized cases are considered, periodic loading, and loading at random Poisson-distributed time intervals.

Periodic Loading

For loading at fixed time intervals we proceed by first taking the natural logarithm of Eq. 6.114; then exponentiating, we obtain

$$R_n(c) = \exp[\ln r(c)^n]. \tag{6.115}$$

This may also be written as

$$R_n(c) = \exp[n \ln r(c)]. \tag{6.116}$$

If the probability of failure during any one loading is small, then $1 - r(c) = p(c) \ll 1$, and we may expand the natural logarithm on the right-hand side of Eq. 6.116 as

$$\ln r(c) = \ln[1 - p(c)] \approx -p(c) \tag{6.117}$$

to obtain with good approximation

$$R_n(c) = \exp[-np(c)]. \tag{6.118}$$

To convert the independent variable from n to t, we must know the interval Δt at which the loadings take place. With Δt known, we can say that at time t there have already been

$$n = \frac{t}{\Delta t} \tag{6.119}$$

loadings. Thus, combining Eqs. 6.118 and 6.119, we find the time dependence of the reliability, given capacity c, to be

$$R(t|c) = \exp\left[-\frac{p(c)}{\Delta t}t\right], \tag{6.120}$$

or simply

$$R(t|c) = e^{-\lambda(c)t}, \tag{6.121}$$

where the capacity-dependent failure rate is given by

$$\lambda(c) = p(c)/\Delta t. \tag{6.122}$$

Periodic phenomena are often discussed in terms of the return period $T(c)$ for a load that exceeds the capacity c, defined by

$$T(c) = \frac{\Delta t}{1 - r(c)}. \tag{6.123}$$

The reliability given by Eq. 6.120 may then be written as

$$R(t|c) = e^{-t/T(c)}. \tag{6.124}$$

Specifically, $T(c)$ is used to represent the frequency at which a loading greater than c may be expected to recur. It is usually applied to natural loadings on a calendar-year basis, such as a 100-year flood or a 50-year blizzard. In a more abstract sense it is also perfectly valid for operational units of time, such as using the thousandth flight landing to estimate the maximum impact loading on an aircraft landing gear.

EXAMPLE 6.6

Historically, a design rule for structures subjected to flooding has been to design for a flood with a return period of twice the design life. If this criterion is used, what is the probability of failure during the design life?

Solution Let T be the design life. Then $T(c) = 2T$ and

$$R(t) = e^{-t/2T}.$$

The probability of failure during design life is

$$1 - R(T) = 1 - e^{-T/2T} = 1 - e^{-1/2} = 0.393. \blacktriangleleft$$

Loading at Random Intervals

We now consider the other extreme from periodic loading. In random loading the time until the next loading occurs is independent of when the last loading occurred. In this situation the Poisson distribution derived in Chapter 4 is applicable. The random events are now taken to be peaks in the loadings, such as indicated in Fig. 6.14*b*.

The probability of there being n loadings during time t is given by the Poisson relationship found in Eq. 4.101:

$$p_n(t) = \frac{(\gamma t)^n}{n!} e^{-\gamma t} \tag{6.125}$$

where γ is the frequency of the loading. Equation 6.114 is the conditional probability that the system will survive, given n loadings and capacity c. Thus the reliability, given capacity c, is obtained from summing over n:

$$R(t|c) = \sum_{n=0}^{\infty} R_n(c) p_n(t). \tag{6.126}$$

Combining Eqs. 6.125 and 6.126 with Eq. 6.114, we have

$$R(t|c) = \sum_{n=0}^{\infty} [r(c)\gamma t]^n \frac{e^{-\gamma t}}{n!}. \tag{6.127}$$

Noting, however, that the exponential may be expanded as

$$\exp[r(c)\gamma t] = \sum_{n=0}^{\infty} \frac{[r(c)\gamma t]^n}{n!}, \tag{6.128}$$

we may rewrite the reliability as

$$R(t|c) = e^{-\lambda(c)t}, \tag{6.129}$$

where with $r(c) = 1 - p(c)$. The constant failure rate is given by

$$\lambda(c) = \gamma p(c). \tag{6.130}$$

As for periodic loading, we once again have obtained an expression for a time-independent failure rate that is dependent on the system capacity. Indeed the close relation between Eq. 6.122 for periodic loading and Eq. 6.130 for loading at random intervals cannot be escaped. If we define τ as the mean time between loads, we have for periodic loading $\tau = \Delta t$. Similarly, if the loading is a Poisson process, the mean time between loading may be shown to be $\tau = 1/\gamma$. Thus, in either case,

$$\lambda(c) = p(c)/\tau. \tag{6.131}$$

This expression is thus valid for loadings at totally correlated time intervals (i.e., periodic) as well as at totally uncorrelated time intervals (Poisson). It is therefore not surprising that empirical data often yield constant failure rates for intermediate cases in which the loading intervals are partially correlated.

The widely observed increase in failure with decreased capacity is clear from the forms of Eqs. 6.122 and 6.130. In both

$$\lambda(c) \propto p(c) = \int_c^\infty f_l(l)\, dl. \tag{6.132}$$

Specifically, if the extreme-value distribution of type I is used, we have

$$\lambda(c) \propto \exp\left(\frac{u - c}{\theta}\right), \qquad c \gg u. \tag{6.133}$$

EXAMPLE 6.7

A telecommunications leasing firm finds that during the one-year warrantee period, 6% of its telephones are returned at least once because they have been dropped and damaged. An extensive testing program earlier indicated that in only 20% of the drops should telephones be damaged. Assuming that the dropping of telephones in normal use is a Poisson process, (*a*) what is the MTBD (mean time between drops)? (*b*) Determine the probabilities that the telephone will not be dropped, will be dropped once, and will be dropped more than once during a year of service. (*c*) If the telephones are redesigned so that only 4% of drops cause damage, what fraction of the phones will be returned with dropping damage at least once during the first year of service?

Solution (*a*) The fraction of telephones not returned is $R = e^{-\gamma p t}$ or $0.94 = e^{-\gamma \times 0.2 \times 1}$. Therefore

$$\gamma = \frac{1}{0.2 \times 1} \ln\left(\frac{1}{0.94}\right) = 0.3094/\text{year},$$

$$\text{MTBD} = \frac{1}{\gamma} = 3.23 \text{ year.} \blacktriangleleft$$

(*b*) From Eq. 6.125 we have

$$p_0(0) = e^{-\gamma \times 1} = e^{-0.3094} = 0.734 \blacktriangleleft \text{(no drops)}.$$

$$p_1(0) = \gamma \cdot 1 \cdot e^{-\gamma \cdot 1} = 0.3094e^{-0.3094} = 0.227 \blacktriangleleft \text{(one drop)}.$$

$$1 - p_0(0) - p_1(1) = 1 - 0.734 - 0.227 = 0.039 \blacktriangleleft \text{(more than one drop)}.$$

(*c*) For the improved design $R = e^{-\gamma pt} = e^{-0.3094 \times 0.04 \times 1} = 0.9877$. Therefore the fraction of the phones returned at least once is

$$1 - 0.9877 = 1.23\%. \blacktriangleleft$$

6.6 TIME-DEPENDENT FAILURE RATES

In the preceding section we discussed loads that are applied repeatedly to a system and have mutually independent magnitudes. Under these circumstances failure rates are constant, provided that two assumptions concerning the system capacity are met. First, the capacity of the system is assumed to be known precisely (i.e., it is not random); second, it is independent of time.

If the first assumption does not hold, the capacity of the system is variable. Some such variability is to be expected from variations in the properties of materials and in dimensional tolerances and from innumerable other variables in the manufacturing and construction processes. Flaws, missing components, assembly errors, or other shortcomings may cause larger decreases in capacity. In either case there will be an initial wearin period of decreasing failure rate, as indicated in the bathtub curve in Fig. 4.1.

Relaxing the second assumption, that the capacity is independent of time, allows us to take into account the wear effects that cause failure rates to increase with time. Such degradation of capacity, which in the aggregate we refer to as wear, is often divided into three categories. If capacity varies only with time, it is referred to as aging. Corrosion, embrittlement, and chemical decomposition are examples of phenomena that may sometimes be independent of the magnitude and frequency of the loading to which a system is subjected. Therefore, for a given environment they depend only on time and hence may be classified as aging. If the capacity of a system decreases with the number of times that it has been loaded, cyclic damage is said to occur. If the capacity decrease depends both on the number of times that loading takes place and on the load magnitudes, the phenomena are referred to as cumulative damage. Metal fatigue is probably the most widely studied form of cumulative damage.

Although cumulative damage is often the most realistic model for wear effects, it is cumbersome to include in simple reliability models. For this reason, as well as the fact that aging often provides a reasonable approximation for wear effects, we discuss wearout in what follows, using an aging model.

Wearin

The results of Section 6.5 assume that the system capacity c has a fixed value. To examine wearin, we now relax this restriction and assume that the capacity is a random variable described by a PDF, $f_c(c)$. This probability distribution may be viewed in two different ways. For mass-produced items it may represent the variability in capacity within the batch of manufactured items. For single or few of a kind systems, such as large structures or industrial plants, the PDF may represent the designer's uncertainty about the as-built capacity of the system. In either case we retain, for now, the assumption that the capacity does not change with time.

The reliability $R(t|c)$ is just a conditional probability, given the capacity c. Therefore, consistent with the rules developed in Chapter 3 for treating two random variables—here **t** and **c**—we may obtain the expected value of the reliability $R(t)$ by averaging over c:

$$R(t) = \int_{-\infty}^{\infty} f_c(c) R(t|c)\, dc, \tag{6.134}$$

Now suppose that we employ the constant failure rate model given by Eq. 6.129 for $R(t|c)$. We have

$$R(t) = \int_{-\infty}^{\infty} f_c(c) \exp[-\lambda(c)t]\, dc. \tag{6.135}$$

Let us consider two cases. In the first we assume that the variation in capacity is small, given by a normal PDF with a small standard deviation. We further assume that the variation of the failure rate over the range of capacities is so small that it can be ignored. Then Eq. 6.135 simply reduces to Eq. 6.129. The second case is of more interest; some fraction, say p_d, of the systems under consideration are flawed in a serious way. There flaws will cause early or wearin failures.

To describe the possibility that systems are flawed, we write the PDF of capacities in terms of Dirac delta functions as

$$f_c(c) = (1 - p_d)\,\delta(c - c_0) + p_d\,\delta(c - c_d), \tag{6.136}$$

where $p_d < 1$ is the probability that the system is defective. The first term on the right-hand side corresponds to the probability that the system will be a properly built system with specified design capacity c_0. By using the Dirac delta function, we are assuming that the capacity variability of the properly built systems can be ignored. The second term corresponds to the probability that the system will be defective and have a reduced capacity $c_d < c_0$. Such a situation might arise, for example, if a critical component were to be left out of a small fraction of the systems in assembly or if, in construction, members were not properly welded together with some probability, p_d.

To see the effect on the failure rate, we first substitute Eq. 6.136 into 6.135:

$$R(t) = (1 - p_d) \exp[-\lambda(c_0)t] + p_d \exp[-\lambda(c_d)t]. \tag{6.137}$$

Since the failure rate increases with decreased capacity, we have $\lambda(c_0) < \lambda(c_d)$. If we now use the definition of the time-dependent failure rate given in Eq. 4.15, we obtain, after evaluating the derivative,

$$\lambda(t) = \lambda_0 \left\{ \frac{1 + \dfrac{p_d}{1 - p_d} \dfrac{\lambda_d}{\lambda_0} \exp[-(\lambda_d - \lambda_0)t]}{1 + \dfrac{p_d}{1 - p_d} \exp[-(\lambda_d - \lambda_0)t]} \right\}, \tag{6.138}$$

where for brevity we have written $\lambda_0 = \lambda(c_0)$ and $\lambda_d = \lambda(c_d)$.

The wearin effect may be seen more explicitly by considering a system whose probability of defective construction is small, $p_d \ll 1$, but the defect greatly increased the failure rate, $\lambda_d \gg \lambda_0$. In this case the equation for $\lambda(t)$ reduces to

$$\lambda(t) \approx \lambda_0 \left(1 + \frac{p_d \lambda_d}{\lambda_0} e^{-\lambda_d t} \right). \tag{6.139}$$

Thus the failure rate decreases from an initial value of $\lambda_0 + p_d\lambda_d$ at zero time to the value λ_0 of the unflawed system after the defective units have failed.

EXAMPLE 6.8

A servomechanism is designed to have a constant failure rate and a design-life reliability of 0.99, in the absence of defects. A common manufacturing defect, however, is known to cause the failure rate to increase by a factor of 100. The purchaser requires the design-life reliability to be at least 0.975. (*a*) What fraction of the delivered servomechanisms may contain the defect if the reliability criterion is to be met? (*b*) If 10% of the servomechanisms contain the defect, how long must they be worn in before delivery to the purchaser?

Solution (*a*) Without the defect, the failure rate $\lambda_0 \equiv \lambda(c_0)$ may be found in terms of the design life T by $R_0(T) = e^{-\lambda_0 T}$; then

$$\lambda_0 T = \ln\left[\frac{1}{R(T)}\right] = \ln\left(\frac{1}{0.99}\right) = 0.01005.$$

To determine p, the acceptable fraction of units with defects, solve Eq. 6.137 with $t = T$ for p_d:

$$p_d = \frac{1 - R(T) \exp[+\lambda_0 T]}{1 - \exp[-(\lambda_d - \lambda_0)T]}.$$

With $\lambda_d \equiv \lambda(c_d) = 100\ \lambda_0$, $R(T) = 0.975$, and $\lambda_0 T = 0.01005$,

$$p_d = \frac{1 - 0.975e^{+0.01005}}{1 - e^{-99\times 0.01005}} = 0.024. \blacktriangleleft$$

(*b*) Recall the definition for reliability with wearin from Eq. 4.38. Combining Eq. 6.137 with this expression, we have, for a wearin period T_w;

$$R(T|T_w) = \frac{(1 - p_d)\exp[-\lambda_0(T + T_w)] + p_d \exp[-\lambda_d(T + T_w)]}{(1 - p_d)\exp(-\lambda_0 T_w) + p_d \exp(-\lambda_d T_w)}.$$

Solve for T_w:

$$T_w = \frac{1}{\lambda_d - \lambda_0}\ln\left[\frac{p_d}{1 - p_d}\,\frac{R(T|T_w) - \exp(-\lambda_d T)}{\exp(-\lambda_0 t) - R(T|T_w)}\right].$$

With $R(T|T_w) = 0.975$, $p_d = 0.1$, $\lambda_0 T = 0.01005$, and $\lambda_d T = 1.005$,

$$T_w = \frac{T}{99}\ln\left(\frac{0.1}{1 - 0.1}\,\frac{0.975 - e^{-100\times 0.01005}}{e^{-0.01005} - 0.975}\right)$$

$$= 0.015T \quad \text{or } 1\tfrac{1}{2}\% \text{ of the design life.} \blacktriangleleft$$

Wearout

As we have indicated, the decreasing failure rates of wearin are due inherently to the variance of capacity of a system. If the capacity of a system is known exactly, there is no wearin. In principle, wearout from aging may be viewed as a deterministic phenomenon that would be present even if both load and capacity were known exactly. To illustrate, suppose that a system has a capacity that is a known function of time, $c_0 = c_0(t)$, and that at any time there is no uncertainty in its value. If there is a constant load l, such as in Fig. 6.1*c* or a cyclic load of constant magnitude as in Fig. 6.1*d*, the system will fail at the time t_f for which

$$c_0(t_f) = l_o. \tag{6.140}$$

as illustrated in Fig. 6.15. The reliability for this deterministic system is then

$$R(t) = \begin{cases} 1, & t < t_f, \\ 0, & t > t_f. \end{cases} \tag{6.141}$$

Generally, neither load nor capacity is known exactly, and the probability density functions $f_c(c)$ and $f_l(l)$ give rise to a PDF of times to failure $f(t)$. The corresponding reliability $R(t)$ is then characterized by a failure rate that increases with time, provided only that the capacity is a decreasing function of time. We illustrate this effect with two simplified models. In the first the capacity is taken to be a random variable that is, however, a known function of time; in the second the capacity is assumed to be a known function of time in which there is no variability, whereas the load is treated as a random variable.

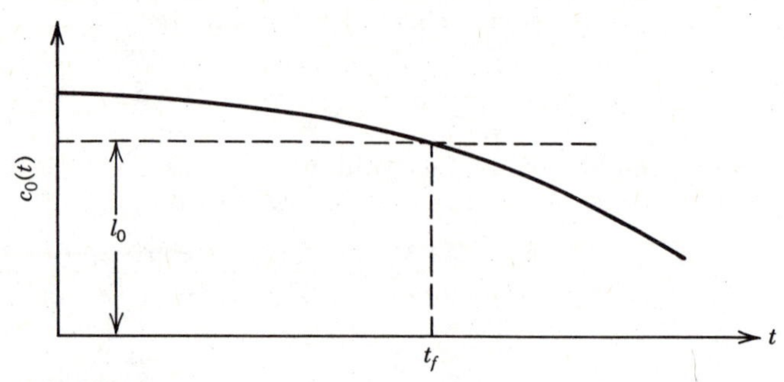

FIGURE 6.15 Capacity versus time for a system under constant loading.

Fixed-Load Magnitude

Consider the case of a fixed load, as in Figs. 6.1*c* and *d*. The oscillating load of known magnitude in Fig. 6.1*d*, in particular, is typical of fatigue-testing loads, as well as of many fatigue loads encountered in practice. Suppose that we let the load be l_0. Now suppose that the distribution of capacity at a time t is given by a lognormal distribution,

$$f_c(c) = \frac{1}{\sqrt{2\pi}\, s_c c} \exp\left(- \frac{1}{2s_c^2} \left\{ \ln\left[\frac{c}{c_0(t)}\right]\right\}^2 \right), \tag{6.142}$$

where the most probable value is given as the decreasing function of time

$$c_0(t) = c_1 t^{-\alpha}, \qquad \alpha > 1, \tag{6.143}$$

where c_1 and α are empirical constants.

In fatigue, for example, this decreasing capacity is due to the increase in crack lengths with time. The dispersion in the capacity c characterized by s_c is due to the probability distribution for the maximum crack size in the system at $t = 0$.

For a constant load l_0 the reliability $r(l_0)$ is given by Eq. 6.17. Thus combining Eqs. 6.142 with 6.17, we have

$$R(t) \equiv r(l_0) = \int_{l_0}^{\infty} \frac{1}{\sqrt{2\pi}\, s_c c} \exp\left(- \frac{1}{2s_c^2} \left\{ \ln\left[\frac{c}{c_0(t)}\right]\right\}^2 \right) dc. \tag{6.144}$$

Changing variables

$$x = \frac{1}{s_c} \ln\left[\frac{c}{c_0(t)}\right], \tag{6.145}$$

we may write the integral as a normal distribution

$$R(t) = \frac{1}{\sqrt{2\pi}} \int_{(1/s_c)\ln[l_0/c_0(t)]}^{\infty} e^{-\frac{1}{2}x^2}\, dx, \tag{6.146}$$

in which time appears only in the lower limit.

To demonstrate that the reliability given by Eq. 6.146 is characterized by the increasing failure rate of wear, we first obtain the PDF for failure, which from Eq. 4.10 is seen to be given by

$$f(t) = -\frac{dR(t)}{dt}. \tag{6.147}$$

Therefore, from Eqs. 6.146 and 6.147, we obtain

$$f(t) = -\frac{1}{s_c c_0(t)} \frac{dc_0(t)}{dt} \frac{1}{\sqrt{2\pi}} \exp\left(-\frac{1}{2s_c^2}\left\{\ln\left[\frac{l_0}{c_0(t)}\right]\right\}^2\right). \tag{6.148}$$

Finally, substitution of Eq. 6.143 into this expression yields

$$f(t) = \frac{1}{\sqrt{2\pi}st} \exp\left\{-\frac{1}{2s^2}\left[\ln\left(\frac{t}{t_0}\right)\right]^2\right\}, \tag{6.149}$$

where we have introduced the variables

$$s = s_c/\alpha, \tag{6.150}$$

and

$$t_0 = (c_1/l_0)^{1/\alpha}. \tag{6.151}$$

Here t_0 is the time to failure for the deterministic case in which $s_c = 0$.

It is seen that the PDF for time to failure is just the lognormal distribution. In Chapter 4 we demonstrated that this distribution has an increasing failure rate, provided that s/t_0 is sufficiently small. This is invariably the case in fatigue testing, making the lognormal distribution one of the most frequently used for the fitting of failure data. Other capacity distributions may be used. For example, it may be shown that a normal distribution $f_c(c)$ leads to a normal distribution in the time to failure; this is also shown in Chapter 4 to have an increasing failure rate.

EXAMPLE 6.9

The most probable strength of a steel lifting arm on a robot is given by $130N^{-0.1}$ kips, where N is the number of load cycles. If the loading is cyclic at the rate of 16/min, (*a*) what should the load be if the most probable life is designed to be one year? (*b*) If the initial strength is known with 90% confidence to within 20%, what fraction of the lifting arms are likely to fail during the first month, given the loading calculated in part *a*.

Solution (*a*) With $t = N/16$ min,

$$c_0(N) = 130N^{-0.1}, \qquad c_0(t) = 130(16t)^{-0.1},$$

$$c_0(t) = 98.5t^{-0.1}, \qquad \alpha = 0.1, \qquad c_1 = 98.5.$$

From Eq. 6.151, $l_0 = c_1 t_0^{-\alpha}$. For one year

$$t_0 = 365 \times 24 \times 60 \text{ min} = 525{,}600 \text{ min},$$

Hence

$$l_0 = 98.5 \times (525{,}600)^{-0.1} = 26.4 \text{ kips.} \blacktriangleleft$$

(*b*) From Eq. 3.105,

$$s_c = \frac{1}{1.645}\ln(1.2), \qquad s_c = 0.1108.$$

Then

$$s = \frac{s_c}{\alpha} = \frac{0.1108}{0.1} = 1.108.$$

For the lognormal distribution,

$$P\{\mathbf{t} < t\} = \Phi\left[\frac{1}{s}\ln\left(\frac{t}{t_0}\right)\right].$$

Since one month is $t_0/12$,

$$P\left\{\mathbf{t} < \frac{t_0}{12}\right\} = \Phi\left[\frac{1}{1.108}\ln\left(\frac{1}{12}\right)\right] = \Phi(-2.24)$$

From Appendix C,

$$\Phi(-2.24) = 0.0167. \blacktriangleleft$$

Therefore, fewer than 2% of the arms should fail in the first month.

Known Capacity

We now consider a system whose capacity is known with certainty, but its load is repetitive and of random magnitude, as in Fig. 6.14*a*.

Suppose that we let c_n be the capacity at the time of the nth loading. Then the probability of surviving the nth loading is just $r(c_n)$, given by Eq. 6.10. Since the magnitudes of the successive loads are independent of one another, we may write the probability of surviving the first n loads as

$$R_n = r(c_1)r(c_2)\cdots r(c_n). \tag{6.152}$$

Then, taking the exponential of $\ln R_n$, we obtain

$$R_n = \exp\left[+\sum_{n'=1}^{n}\ln r(c_{n'})\right] \tag{6.153}$$

Assuming that the probability of failure for any one loading is small, $p(c_n) = 1 - r(c_n) \ll 1$, we then obtain

$$R_n \simeq \exp\left[-\sum_{n'=1}^{n} p(c_{n'})\right]. \tag{6.154}$$

To illustrate that the failure rate increases with time, we must assume a specific model for the deterioration in c. Suppose that the c_n decreases with n so that $p(c_n)$ increases linearly with the load application according to

$$p(c_n) \approx p_0(1 + \epsilon n), \qquad \epsilon \ll 1. \tag{6.155}$$

We then obtain

$$\sum_{n'=1}^{n} p(c_{n'}) = n + \frac{\epsilon n^2}{2}. \tag{6.156}$$

Finally, if we assume that the loadings appear with a mean time between load application of τ, we may change variables to write the result in terms of time:

$$t = n\tau. \tag{6.157}$$

Therefore, Eqs. 6.154 through 6.157 yield

$$R(t) = \exp\left[-\frac{p_0}{\tau}\left(1 + \frac{\epsilon t}{2\tau}\right)t \right]. \tag{6.158}$$

Using the definition for the failure rate, Eq. 4.15, we see that for this simple model the failure rate increases:

$$\lambda(t) = \frac{p_0}{\Delta t}\left(1 + \frac{\epsilon t}{\tau}\right). \tag{6.159}$$

Bibliography

Ang, A. H-S., and W. H. Tang, *Probability Concepts in Engineering Planning and Design*, Vol. 1, Wiley, New York, 1975.

Augusti, G., A. Baratta, and F. Casciati, *Probabilistic Methods in Structural Engineering*, Chapman and Hall, London, England, 1984.

Freudenthal, A. M., J. M. Garrelts, and M. Shinozuka, "The Analysis of Structural Safety," *Journal of the Structural Division ASCE ST 1*, 267–325 (1966).

Gumbel, E. J., *Statistics of Extremes*, Columbia University Press, New York, 1958.

Haugen, E. B., *Probabilistic Mechanical Design*, Wiley, New York, 1980.

Haviland, R. D., *Engineering Reliability and Long Life Design*, Van Nostrand, New York, 1964.

Kapur, K. C., and L. R. Lamberson, *Reliability in Engineering Design*, Wiley, New York, 1977.

Thoft-Chirstensen, P., and M. J. Baker, *Structural Reliability Theory and Its Application*, Springer-Verlag, Berlin, 1982.

EXERCISES

6.1 A design engineer knows that one-half of the lightning loads on a surge protection system are greater than 500 V. Based on previous experience, such loads are known to follow the PDF:

$$f(v) = \gamma e^{-\gamma v}, \quad 0 \leq v < \infty.$$

(*a*) Estimate γ per volt.
(*b*) What is the mean load?
(*c*) For what voltage should the system be designed if the failure probability is not to exceed 5%?

6.2 The loading on industrial fasteners of fixed capacity is known to follow an exponential distribution. Thirty percent of the fasteners fail. If the fasteners are redesigned to double their capacity, what fraction will be expected to fail?

6.3 Suppose that the CDF for loading on a cable is

$$F_l(l) = 1 - \exp\left[-\left(\frac{l}{500}\right)^3\right],$$

where l is in pounds. To what capacity should the cable be designed if the probability of failure is to be no more than 0.5%?

6.4 Suppose that the PDFs for load and capacities are

$$f_l(l) = \gamma e^{-\gamma l}, \qquad 0 \leqslant l \leqslant \infty,$$

$$f_c(c) = \begin{cases} 0, & 0 \leqslant c < a. \\ 1/a, & a \leqslant c \leqslant 2a, \\ 0, & 2a < c \leqslant \infty. \end{cases}$$

Determine the reliability; evaluate all integrals.

6.5 The impact loading on a railroad coupling is expressed as an exponential distribution:

$$f_l(l) = \beta e^{-\beta l}.$$

The coupling is designed to have a capacity $\mathbf{c} = c_m$. However, because of material flaws, the PDF for the capacity is more accurately expressed as

$$f_c(c) = \begin{cases} \dfrac{\alpha e^{\alpha c}}{\exp(\alpha c_m) - 1}, & 0 \leqslant c \leqslant c_m, \\ 0, & c > c_m. \end{cases}$$

(*a*) Determine the reliability for a single loading, assuming that the flaws can be neglected.

(*b*) Recalculate part *a* using the capacity distribution with the flaws included.
(*c*) Show that the result of part *b* reduces to that of part *a* as $\alpha \to \infty$.
(*d*) Show that for $\alpha = 0$, the reliability is

$$r = 1 - \frac{1}{\beta c_m}[1 - e^{-\beta c_m}].$$

6.6 Consider a pressure vessel for which the capacity is defined as p, the maximum internal pressure that the vessel can withstand without bursting. This pressure is given by $p = \tau_0\sigma_m/2R$, where τ_0 is the unflawed thickness, σ_m is the stress at which failure occurs, and R is the radius. Suppose that the vessel thickness is $\tau(\geqslant \tau_0)$, but the distribution crack depths are the same as those given in Exercise 3.6.
(*a*) Show that the PDF for capacity is

$$f_{\mathrm{p}}(p) \equiv \begin{cases} \dfrac{2R}{\gamma\sigma_m}\dfrac{1}{e^{\tau/\gamma} - 1}\exp\left(\dfrac{2R}{\gamma\sigma_m}p\right), & 0 \leqslant p \leqslant \dfrac{\tau\sigma_m}{2R}, \\ 0, & p > \dfrac{\tau\sigma_m}{2R}. \end{cases}$$

(*b*) Normalize to $\tau\sigma_m/2R = 1$, then plot $f_{\mathrm{p}}(p)$ for $\gamma = \tau$, 0.5τ, and 0.1τ.
(*c*) Physically interpret the results of your plots.

6.7 In Exercise 6.6, suppose that the vessel is proof-tested at a pressure of $p = \tau\sigma_m/4R$. What is the probability of failure if
(*a*) $\gamma = 0.5\tau$?
(*b*) $\gamma = 0.1\tau$?

6.8 The twist strength of a standard bolt is 23 N · m with a standard deviation of 1.3 N · m. The wrenches used to tighten such bolts have an uncertainty of $\sigma = 2.0$ N · m in their torsion settings. If no more than 1 bolt in 1000 may fail from excessive tightening, what should the setting be on the wrenches? (Assume normal distributions.)

6.9 Steel cable strands have a normally distributed strength with a mean of 5000 lb and a standard deviation of 150 lb. The strands are incorporated into a crane cable that is proof-tested at 50,000 lb. It is specified that no more than 2% of the cables may fail the proof test. How many strands should be incorporated into the cable, assuming that the cable strength is the sum of the strand strengths?

6.10 Substitute the normal distributions for load and capacity, Eqs. 6.29 and 6.30, into the reliability expression, Eq. 6.20. Show that the resulting integral reduces to Eqs. 6.41 and 6.43.

6.11 It is estimated that the capacity of a newly designed structure is $\bar{c} = 10{,}000$ kips, $\sigma_c = 6000$ kips, normally distributed. The anticipated load on the structure will be $\bar{l}$ = 5000 kips, with an uncertainty of σ_l = 1500 kips, also normally distributed. Find the *un*reliability of the structure.

6.12 Suppose that steel wire has a mean tensile strength of 1200 lb. A cable is to be constructed with a capacity of 10,000 lb. How many wires are required for a reliability of 0.999 (*a*) if the wires have a 2% coefficient of variation? (*b*) If the wires have a 5% coefficient of variation? (*Note:* Assume that the strengths are normally distributed and that the cable strength is the sum of the wire strengths.)

6.13 Assume that the column in Exercise 3.20 is to be built with a safety factor of 1.6. If the strength of the column is normally distributed with a 20% coefficient of variation, what is the probability of failure?

6.14 Suppose that both load and capacity are known to a factor of two with 90% confidence. Assuming lognormal distributions, determine the safety factor c_0/l_0 necessary to obtain a reliability of 0.995.

6.15 The distribution of detectable flaw sizes in tubing is given by Eq. 6.61 with α = 17/cm. There are an average of three detectable flaws per centimeter of tubing.

(*a*) What fraction of the flaws will have a size larger than 0.8 cm?
(*b*) What is the probability of finding a flaw larger than 0.8 cm in a 100-m length of tubing?
(*c*) In 1000 meters of tubing?

6.16 Verify Eq. 6.80 using the fact that Euler's constant is the value of the integral

$$-\int_0^\infty e^{-x} \ln x \, dx = \gamma = 0.5772157.$$

6.17 The impact load on a landing gear is known to follow an extreme-value distribution of type I with a mean value of 2500 and a variance of 25×10^4. The capacity is approximated by a normal distribution with a mean value of 15,000 and a coefficient of variation of 0.05. Find the probability of failure per landing.

6.18 Consider a chain consisting of N links that is subjected to M loads. The capacity of a single link is described by the PDF $f_c(c)$. The PDF for any one of the loads is described by $f_l(l)$. Derive an expression in terms of $f_c(c)$ and $f_l(l)$ for the probability that the chain will fail from the M loadings.

6.19 Show in detail that Eq. 6.109 follows from Eq. 6.108.

6.20 Suppose that the design criteria for a structure is that the probability of an earthquake severe enough to do structural damage must be no more than 1.0% over the 40-year design life of the building.

(*a*) What is the probability of one or more earthquakes of this magnitude or greater occurring during any one year?
(*b*) What is the probability of the structure being subjected to more than one damaging earthquake over its design life?

6.21 A manufacturer of telephone switchboards was using switching circuits from a single supplier. The circuits were known to have a failure rate of 0.06/year. In its new board, however, 40% of the switching circuits came from a new supplier. Reliability testing indicates that the switchboards have a composite failure rate that is initially 80% higher than it was with circuits from the single supplier. The failure rate, however, appears to be decreasing with time.
(*a*) Estimate the failure rate of the circuits from the new supplier.
(*b*) What will the failure rate per circuit be for long periods of time?
(*c*) How long should the switchboards be worn in if the average failure rate of circuits should be no more than 0.1/year?

6.22 Suppose that a system has a time-independent failure rate that is a linear function of the system capacity c,

$$\lambda(c) = \lambda_0[1 + b(c_m - c)], \qquad b > 0,$$

where c_m is the design capacity of the system. Suppose that the presence of flaws causes the PDF or capacity of the system to be given by $f_c(c)$ in Exercise 6.5.
(*a*) Find the system failure rate.
(*b*) Show that it decreases with time.

6.23 The most probable strength of a steel beam is given by $24N^{-0.05}$ kips. This value is known to within 25% with 90% confidence.
(*a*) How many cycles will elapse before the beam loses 20% of its strength?
(*b*) Suppose that the cyclic load on the beam is 10 kips. How many cycles can be applied before the probability of failure reaches 10%?

6.24 A dam is built with a capacity to withstand a flood with a return period of 100 years. What is the probability that the capacity of the dam will be exceeded during its 40-year design life?

6.25 Suppose that the capacity of a system is given by

$$f_c(c) = \frac{1}{\sqrt{2\pi}\,\sigma_c} \exp\left\{-\frac{1}{2\sigma_c^2}[c - \bar{c}(t)]^2\right\},$$

where

$$\bar{c}(t) = c_0(1 - \alpha t).$$

If the system is placed under a constant load l,
(*a*) Find $f(t)$, the PDF for time to failure.
(*b*) Put $f(t)$ into a standard normal form and find σ_t and the MTTF.

CHAPTER 7

Redundancy

7.1 INTRODUCTION

It is a fundamental tenet of reliability engineering that as the complexity of a system increases, the reliability will decrease, unless compensatory measures are taken. Since a frequently used measure of complexity is the number of components in a system, the decrease in reliability may then be expressed in terms of the product rule derived in Chapter 2. To recapitulate, if the component failures are mutually independent, the reliability of a system with N nonredundant components is

$$R = R_1 R_2 \ldots R_n \ldots R_N, \tag{7.1}$$

where R_n is the reliability of the nth component.

The dramatic deterioration of system reliability that takes place with increasing numbers of components is illustrated graphically by considering systems with components of identical reliabilities. In Fig. 7.1 system reliability versus component reliability is plotted, each curve representing a system with a different number of components. It is seen, for example, that as the number of components is increased from 10 to 50, the component reliability must be increased from 0.978 to 0.996 to maintain a system reliability of 0.80.

An alternative to the requirements for components of increased reliability is to provide redundance in part or all of a system. In what follows, we examine a number of different redundant configurations and calculate the effect on system reliability and failure rates. We also discuss specifically several of the trade-offs between different redundant configurations as well as the increased problem of common-mode failures in highly redundant systems.

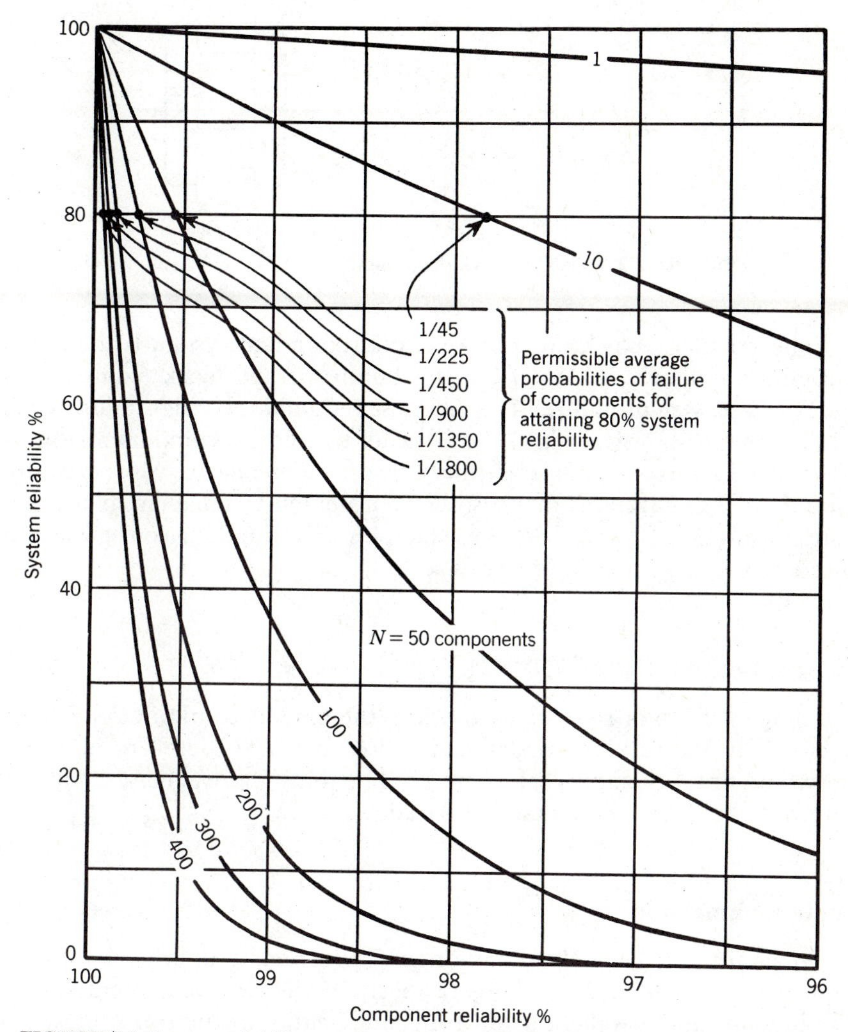

FIGURE 7.1 System reliability as a function of number and reliability of components. (From Norman H. Roberts, *Mathematical Methods of Reliability Engineering*, p. 112, McGraw-Hill, New York, 1964. Reprinted by permission.)

The graphical presentation of systems provided by reliability block diagrams adds clarity to the discussion of redundancy. In these diagrams, which have their origin in electric circuitry, a signal enters from the left and passes through the system and exits on the right. Each component is represented as a block in the system; when enough blocks fail so that all the paths by which the signal may pass from left (input) to right (output) are cut, the system is said to fail.

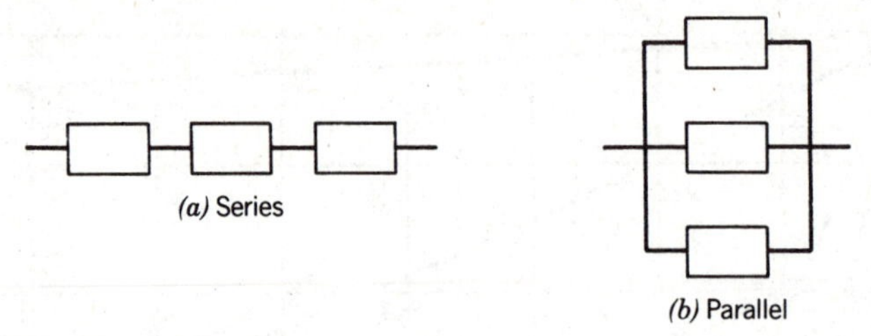

FIGURE 7.2 Reliability block diagrams.

The reliability block diagram of a nonredundant system is the series configuration shown in Fig. 7.2*a*; the failure of any block (component) clearly causes system failure. The simplest redundant configuration is the parallel system shown in Fig. 7.2*b*. Here all the blocks (components) must fail to cut the signal path and thus cause system failure. More general redundant configurations may also be represented as reliability block diagrams. Figures 7.10 and 7.12 are examples of redundant configurations considered in the following sections.

7.2 PARALLEL COMPONENTS

The simplest form of redundant system is the parallel configuration shown in Fig. 7.2*b*. We consider first the situation in which there are only two subsystems or components, before proceeding to the more general situations with three or more components in parallel.

Single Redundancy

Let X_1 signify the event that component 1 functions properly, and X_2 the event that component 2 functions properly. If the two-component system is redundant (i.e., parallel), it will function if either of the two subsystems function. Thus, if X is the event that the system functions, we have

$$X = X_1 \cup X_2, \tag{7.2}$$

and the system reliability is

$$R = P\{X_1 \cup X_2\}. \tag{7.3}$$

Therefore, using Eq. 2.10, we have

$$R = P\{X_1\} + P\{X_2\} - P\{X_1 \cap X_2\}. \tag{7.4}$$

We denote the component reliabilities by R_1 and R_2 and assume that the failures are independent. This assumption permits the last term in Eq.

7.4 to be replaced by R_1R_2, allowing us to write

$$R = R_1 + R_2 - R_1R_2. \tag{7.5}$$

If we assume constant component failure rates, the time dependence of the reliability is

$$R(t) = e^{-\lambda_1 t} + e^{-\lambda_2 t} - e^{-(\lambda_1+\lambda_2)t}. \tag{7.6}$$

The mean time to failure for the system can be calculated from Eq. 4.22:

$$\text{MTTF} = \frac{1}{\lambda_1} + \frac{1}{\lambda_2} - \frac{1}{\lambda_1 + \lambda_2}. \tag{7.7}$$

These results, and those obtained hereafter, are applicable only to active parallel systems in which all the components are in operation from $t = 0$. In standby systems, in which the second unit is not turned on until the first fails, the failures are not independent and therefore Eq. 7.5 does not apply. Standby systems are treated in Chapter 9.

EXAMPLE 7.1

The MTTF of a system with a constant failure rate has been determined. An engineer is to set the design life so that the end-of-life reliability is 0.9.
(*a*) Determine the design life in terms of the MTTF.
(*b*) If two of the systems are placed in active parallel, to what value may the design life be increased without causing a decrease in the end-of-life reliability?

Solution Let the failure rate be $\lambda \equiv 1/\text{MTTF}$.
(*a*) $R = e^{-\lambda T}$. Therefore, $T = (1/\lambda)\ln(1/R)$.

$$T = \ln\left(\frac{1}{R}\right) \times \text{MTTF} = \ln\left(\frac{1}{0.9}\right) \text{MTTF} = 0.105\ \text{MTTF}. \blacktriangleleft$$

(*b*) From Eq. 7.6, $R = 2e^{-\lambda T} - e^{-2\lambda T}$. Let $x \equiv e^{-\lambda T}$. Therefore, $x^2 - 2x + R = 0$. Solve the quadratic equation:

$$x = \frac{+2 \pm \sqrt{4 - 4R}}{2} = 1 - \sqrt{1 - R}.$$

The "+" solution is eliminated, since x cannot be greater than one. Since $x = e^{-\lambda T} = 1 - \sqrt{1 - R}$, then with $\lambda = 1/\text{MTTF}$,

$$T = \ln\left[\frac{1}{(1 - \sqrt{1 - R})}\right] \times \text{MTTF},$$

$$= \ln\left[\frac{1}{(1 - \sqrt{1 - 0.9})}\right] \times \text{MTTF} = 0.380\ \text{MTTF}. \blacktriangleleft$$

Thus the redundant system may have nearly four times the design life of the single system, even though it may be shown from Eq. 7.7 that the MTTF of the redundant system is only 50% longer.

Additional insight may be gained by considering the situation in which both components are identical. In this situation we let $\lambda_2 = \lambda_1$, and Eq. 7.6 reduces to

$$R(t) = 2e^{-\lambda_1 t} - e^{-2\lambda_1 t}. \tag{7.8}$$

Suppose that we now calculate the system failure rate, using Eq. 4.15. We obtain

$$\lambda(t) = -\frac{1}{R}\frac{dR}{dt} = \lambda_1 \left(\frac{1 - e^{-\lambda_1 t}}{1 - 0.5e^{-\lambda_1 t}} \right). \tag{7.9}$$

In Fig. 7.3 are plotted both the reliability and the failure rate for the parallel system, along with the results for a system consisting of a single component. The results for the failure rate are instructive. For even though the components have constant failure rates, the failure rate of the redundant system as a whole is a function of time. Characteristic of systems with complete redundancy, it has a zero failure rate at $t = 0$. The failure rate then increases to an asymptotic value equal to the failure rate of one component. Notice that when compared with the shapes of failure rates discussed in Chapter 4, this one resembles the wear situation in that it increases with time. The important difference, however, is that for redundant systems the rate is concave downward with an asymptotic upper limit equal to the component failure rate.

Multiple Redundancy

The reliability of a system can be further increased by putting increased numbers of components in parallel. For multiply redundant configurations we use a slightly different procedure for obtaining expressions for the system reliability in order to streamline the derivation.

Suppose that we have N components in parallel; if any one of the components functions, the system will function successfully. Thus, in order for the system to fail, all the components must fail. This may be written as follows. Let $\tilde{X}_i$ denote the event of the ith component failure and $\tilde{X}$ the system failure. Thus, for a system of N parallel components, we have

$$\tilde{X} = \tilde{X}_1 \cap \tilde{X}_2 \cap \ldots \cap \tilde{X}_N, \tag{7.10}$$

and the system reliability is

$$R = 1 - P\{\tilde{X}_1 \cap \tilde{X}_2 \cap \ldots \tilde{X}_N\}. \tag{7.11}$$

Moreover, if the failures of the components are mutually independent, we

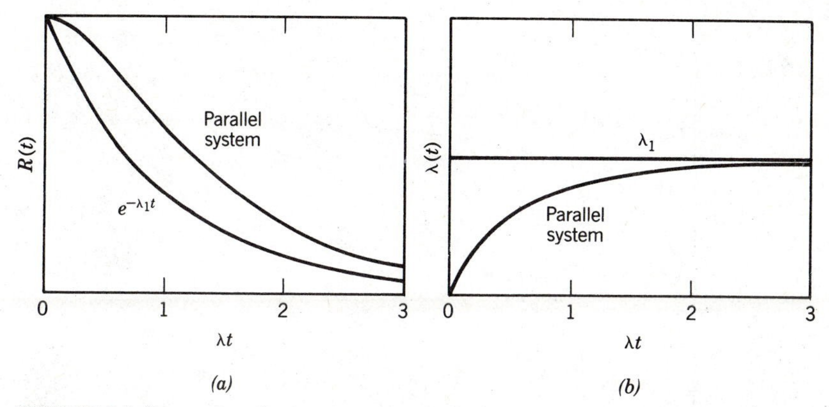

FIGURE 7.3 Properties of a two-component parallel system: (*a*) reliability, (*b*) failure rate.

may use the definition of independence, Eq. 2.8, to write

$$R = 1 - P\{\tilde{X}_1\}\,P\{\tilde{X}_2\}\,\ldots\,P\{\tilde{X}_N\}. \tag{7.12}$$

The $P\{\tilde{X}_i\}$ are the component failure probabilities; therefore, they are related to the reliabilities by

$$P\{\tilde{X}_i\} = 1 - R_i. \tag{7.13}$$

Consequently, we have

$$R = 1 - \prod_i (1 - R_i). \tag{7.14}$$

If the components are identical, simplifications may be made in the foregoing formulas. Suppose that all the R_i have the same value, $R_i = R_1$. Equation 7.14 then reduces to

$$R = 1 - (1 - R_1)^N. \tag{7.15}$$

The degree of improvement in system reliability brought about by multiple redundancy is indicated in Fig. 7.4, where system reliability is plotted versus component reliability for different numbers of parallel components.

We may use the binomial expansion, introduced in Chapter 2, to express the reliability in a form that is more convenient for evaluating the mean time to failure. The binomial coefficients allow us to write in general

$$(p + q)^N = \sum_{n=0}^{N} C_n^N p^{N-n} q^n, \tag{7.16}$$

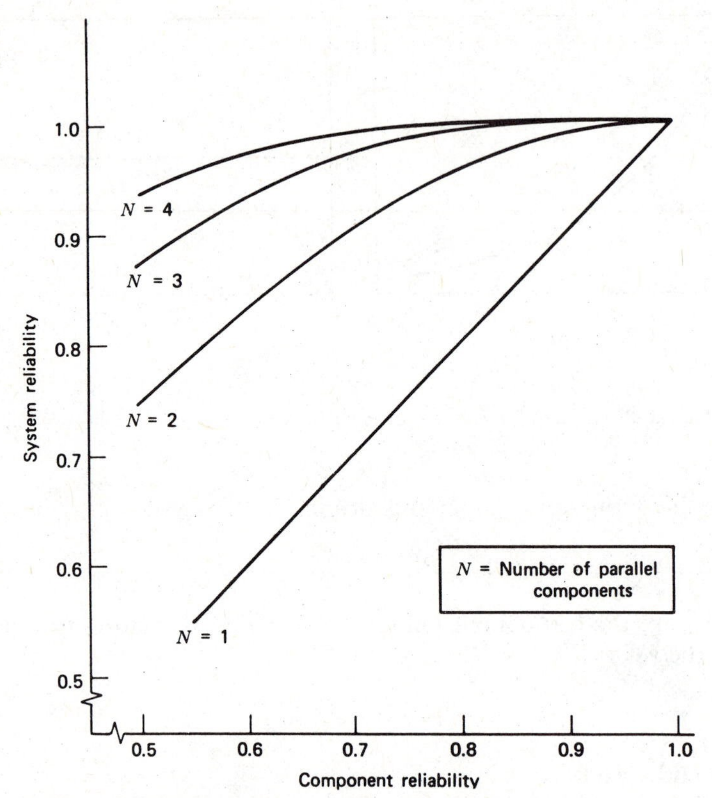

FIGURE 7.4 Reliability improvement by N parallel components. (From K. C. Kapur and L. R. Lamberson, *Reliability in Engineering Design*. Copyright © 1977, by John Wiley and Sons. Reprinted by permission.)

where the C_n^N are binomial coefficients given by Eq. 2.72. If we take $p = 1$ and $q = -R_1$, we obtain

$$(1 - R_1)^N = \sum_{n=0}^{N} C_n^N (-1)^n R_1^n. \tag{7.17}$$

Therefore, since $C_0^N = 1$, we may write Eq. 7.17 as

$$R = \sum_{n=1}^{N} (-1)^{n-1} C_n^N R_1^n. \tag{7.18}$$

To evaluate the MTTF, we assume a constant failure rate for each component. Then the preceding equation becomes

$$R(t) = \sum_{n=1}^{N} (-1)^{n-1} C_n^N e^{-n\lambda_1 t}. \tag{7.19}$$

Then, applying Eq. 4.22, to express the MTTF in terms of $R(t)$, we obtain

$$\text{MTTF} = \sum_{n=1}^{N} (-1)^{n-1} \frac{C_n^N}{n\lambda_1}. \tag{7.20}$$

EXAMPLE 7.2

How many identical components must be put in parallel to at least double the MTTF?

Solution Let MTTF_N signify the mean time to failure for N components in parallel. Therefore, from Eq. 7.20, we must have

$$\frac{\text{MTTF}_N}{\text{MTTF}_1} = \sum_{n=1}^{N} (-1)^{n-1} \frac{N!}{n(N-n)!n!} \geq 2$$

$$\frac{\text{MTTF}_2}{\text{MTTF}_1} = \frac{2!}{1(2-1)!1!} - \frac{2!}{2(2-2)!2!} = 1\frac{1}{2}$$

$$\frac{\text{MTTF}_3}{\text{MTTF}_1} = \frac{3!}{1(3-1)!1!} - \frac{3!}{2(3-2)!2!} + \frac{3!}{3(3-3)!3!} = 1\frac{5}{6}$$

$$\frac{\text{MTTF}_4}{\text{MTTF}_1} = \frac{4!}{1(4-1)!1!} - \frac{4!}{2(4-2)!2!} + \frac{4!}{3(4-3)!3!} - \frac{4!}{4(4-4)!4!} = 2\frac{1}{12}$$

$N = 4$. ◀

Independent Failure Modes

In Chapter 4 the concept of failure mode is introduced to describe the behavior of nonredundant systems in terms of component failures. If the component failures are independent, the system reliability may be represented as in Eq. 7.1, and the failure rates may be added as discussed in Chapter 4. Moreover, if a component may fail in more than one way (e.g., an electric component might fail either in short circuit or in open circuit), the component failure rate may be further divided into component failure mode contributions $\lambda_i = \lambda_{i1} + \lambda_{i2}$, provided that the component failure modes are independent. Such addition of failure rates also may be carried out for independent modes even though the failure rates are time-dependent.

We must now generalize the failure mode concept to include systems with redundance. Provided that the component failures are independent and each component has only one failure mode, a system of two components in parallel has only one failure mode; it requires both components to fail, and its failure rate, when the two components are identical, is given by Eq. 7.9. The same argument may be extended to three or more components in parallel: If the component failures are independent, there is only one failure mode consisting of all the parallel component failing. Similarly, a composite failure rate can be calculated; but, as in Eq. 7.9, the failure rate for a redundant set of components will be time-dependent, even though the component failure rates are taken to be constants.

Suppose now that the parallel set of components is placed within a larger system, such as that shown in Fig. 7.5, where for simplicity we assume that components 1 and 2, which appear in parallel, are identical. We may consider components 1 and 2 to constitute a composite element c with reliability given by Eq. 7.8,

$$R_c' = 2e^{-\lambda_1 t} - e^{-2\lambda_1 t} \tag{7.21}$$

and thus represent Fig. 7.5 as a series system, with reliability

$$R = R_a R_b R_c' R_d R_e. \tag{7.22}$$

Similarly, if each component has a constant failure rate, we may represent the system reliability as

$$R(t) = \exp\left[-\int_0^t \lambda(t')\,dt'\right], \tag{7.23}$$

where

$$\lambda(t) = \lambda_a + \lambda_b + \lambda_c(t) + \lambda_d + \lambda_e, \tag{7.24}$$

with $\lambda_c(t)$ the failure rate for the failure mode of the parallel subsystem, as given in Eq. 7.9.

When more complex parallel subsystems are considered, such as those represented in Fig. 7.10, more than one failure mode may appear. Moreover, even though the component failures may be independent, failure modes that involve more than one component may not be, for two or more failure modes sometimes require the same component failure. We defer the treatment of this question to Chapter 10, where it will be taken up as the analysis of systems with complex redundant structures. A second type of complication occurs when the failures of two or more components are not independent, but rather may be simultaneously brought about by the same cause. We treat such common-cause or common-mode failures next.

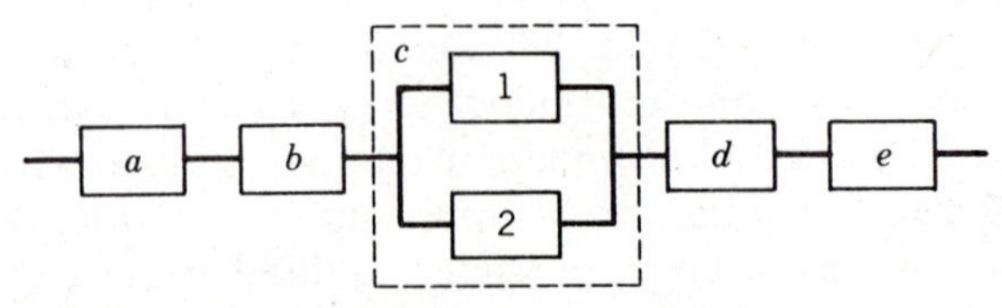

FIGURE 7.5 Reliability block diagram with a singly redundant component.

Common-Mode Failures

One must exercise extreme care in the use of multiply redundant components or subsystems. For with the very small system failure probabilities that may result, any phenomena that create dependence between the components failures can negate much of the benefit gained by the redundance. Such common-mode failures may be caused by common electric connections, shared environmental dust, humidity or vibration, common maintenance problems, and a host of other factors. We shall first illustrate the effect with a simple example.

Suppose that we consider an alarm system consisting of four independent alarms. We might calculate the reliability of the system from the reliabilities of the individual alarms using Eq. 7.15 with $N = 4$. However, if the signals from all four alarms pass through the same cable, we must also consider the possibility of cable failure. The cable then appears in series with the redundant alarms as shown schematically in Fig. 7.6. The system reliability is then calculated by placing the cable reliability, R', in series with the reliability for the parallel alarms:

$$R = R' [1 - (1 - R_1)^N]. \tag{7.25}$$

Clearly, unless the cable failure probability is made very small compared to that of the alarms, the reliability of the system will be limited primarily by the cable reliability.

Common-mode failures are also likely to originate within the redundant components, from common environmental effects of heat, vibration, or moisture; and from systematic maintenance errors carried out on redundant units or for other more subtle reasons. The most common technique for modeling such phenomena is through the use of the following failure rate model.* Suppose that λ is the total failure rate of a component. We divide this into two contributions

$$\lambda = \lambda_I + \lambda_c, \tag{7.26}$$

where λ_I is the rate of independent failures and λ_c is the common-mode failure rate. These partial failure rates may be used to express common-mode failure rates in redundant systems as follows. Define the factor β as the ratio

$$\beta = \lambda_c/\lambda. \tag{7.27}$$

Each of the parallel units then has an independent failure mode reliability,

* K. L. Flemming and P. H. Raabe, "A Comparison of Three Methods for the Quantitative Analysis of Common Cause Failures," General Atomic Report GA-A14568, 1978; see also *Fault Tree Handbook,* U.S. Nuclear Regulatory Commission, NUREG-0492, 1981.

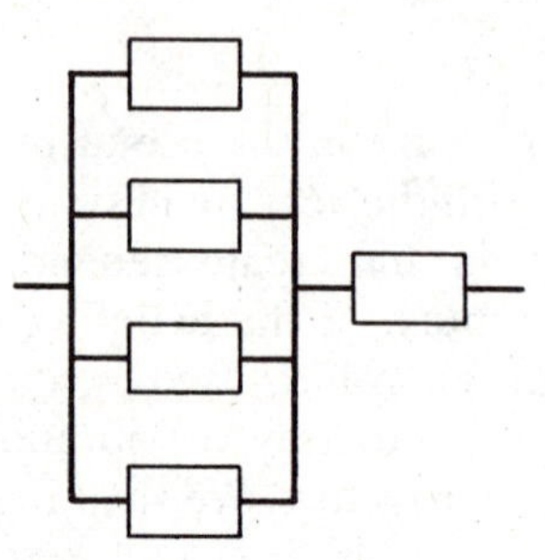

FIGURE 7.6 Reliability block diagram with a multiply redundant somponent.

accounting only for independent failure, of

$$R_I = e^{-\lambda_I t}, \tag{7.28}$$

and therefore the system reliability for independent failure is determined by using λ_I in Eq. 7.8. We multiply this system reliability by $e^{-\lambda_c t}$ to account for common-mode failures. Thus, for two units in parallel,

$$R(t) = (2e^{-\lambda_I t} - e^{-2\lambda_I t})e^{-\lambda_c t}, \tag{7.29}$$

or using $\lambda_c = \beta\lambda$ and $\lambda_I = (1 - \beta)\lambda$, we may write

$$R(t) = [2 - e^{-(1-\beta)\lambda t}]e^{-\lambda t}. \tag{7.30}$$

The β factor may also be applied to N identical units. We have, instead of Eq. 7.29,

$$R(t) = [1 - (1 - e^{-\lambda_I t})^N]e^{-\lambda_c t}. \tag{7.31}$$

EXAMPLE 7.3

A temperature sensor is to have a design-life reliability of no less than 0.98. Since a single sensor is known to have a reliability of only 0.90, the design engineer decides to put two of them in parallel. From Eq. 7.8 the reliability should then be 0.99, meeting the criterion. Upon reliability testing, however, the reliability is estimated to be only 0.97. The engineer first deduces that the degradation is due to common-mode failures and then considers two options: (1) putting a third sensor in parallel, and (2) reducing the probability of common-mode failures.

(*a*) Assuming that the sensors have constant failure rates, find the value of β that characterizes the common-mode failures.

(*b*) Will adding a third sensor in parallel meet the reliability criterion if nothing is done about common-mode failures?

(*c*) By how much must β be reduced if the two sensors in parallel are to meet the criterion?

Solution If the design-life reliability of a sensor is $R_1 = e^{-\lambda T} = 0.9$, then $\lambda T = \ln(1/R_1) = \ln(1/0.9) = 0.10536$.

(*a*) Let $R_2 = 0.97$ be the system reliability for two sensors in parallel. Then β is found in terms of R_2 from Eq. 7.30 to be

$$\beta = 1 + \frac{1}{\lambda T}\ln(2 - R_2\, e^{\lambda T}) = 1 + \frac{1}{0.10536}\ln\left(\frac{2 - 0.97}{0.9}\right),$$

$$= 0.2315. \blacktriangleleft$$

(*b*) The reliability for three sensors in parallel is given by Eq. 7.31 with $N = 3$. Using $\lambda_I = (1 - \beta)\lambda$ and $\lambda_c = \beta\lambda$, we may expand the bracketed term to obtain

$$R_3 = [3 - 3e^{-(1-\beta)\lambda T} + e^{-2(1-\beta)\lambda T}]\, e^{-\lambda T}.$$

From part *a* we have $(1 - \beta)\lambda T = (1 - 0.2315) \times 0.10536 = 0.08097$, and thus $e^{-(1-\beta)\lambda T} = 0.92222$. Thus the reliability is

$$R_3 = [3 - 3 \times 0.92222 + (0.92222)^2] \times 0.9 = 0.975$$

Therefore, the criterion is not met ◀ by putting a third sensor in parallel.

(*c*) To meet the criterion with two sensors in parallel, we must reduce β enough so that the equation in part *a* is satisfied with $R_2 = 0.98$. Thus

$$\beta = 1 + \frac{1}{0.10536}\ln\left(\frac{2 - 0.98}{0.9}\right) = 0.1165. \blacktriangleleft$$

Therefore, β must be reduced by at least

$$1 - \frac{0.1165}{0.2315} \approx 50\%. \blacktriangleleft$$

The limitations of the β-factor technique must be realized, particularly when it is applied to more than two units in parallel. When there are N units, for example, either only one unit fails at a time (independent failures) or all the units fail simultaneously (common-mode failures). In reality common modes that result from external shocks to the system, such as mechanical impact, electric surges, and so on, may cause one, two, or more simultaneous failures in a parallel system. Such situations may be treated using more subtle binomial failure models described elsewhere,* but only at the expense of having more than the single β parameter to be determined. Similarly, where the failure of one unit causes an increase in failure rate of the others, the Markov methods discussed in Chapter 9 must be employed.

The β-factor model for common-mode failures may also be applied to series components, but with quite different results. Suppose that components 1 and 2 are in series, each with a failure rate λ_1 that can be divided into independent and common-mode failures:

$$\lambda_1 = \lambda_I + \lambda_c. \tag{7.32}$$

The reliability of the series system is then

$$R(t) = e^{-2\lambda_I t}\, e^{-\lambda_c t}, \tag{7.33}$$

* *PRA Procedures Guide,* Vol. 1, U.S. Nuclear Regulatory Commission, NUREG/CR-2300, 1983.

or simply $e^{-\lambda t}$, provided that we define the system failure rate as

$$\lambda = 2\lambda_I + \lambda_c. \tag{7.34}$$

Since for this problem $\beta \equiv \lambda_c/\lambda_1$, we may write

$$\lambda = (2 - \beta)\lambda_1. \tag{7.35}$$

Thus we see that for a nonredundant system, the common-mode failure decreases the failure rate over what would occur were all component failures independent. Ordinarily, this is of little consequence, for if $\beta << 1$, only a small negative change in the failure rate will occur. Moreover, since neglecting common-mode failures in nonredundant systems is pessimistic (i.e., it underestimates the reliability), the correction can usually be neglected. In contrast, even small values of β may greatly decrease the reliability of redundant systems, and therefore common-mode failures are a very important consideration.

With the β-factor model for common-mode failures, we may incorporate the redundant system into reliability expressions for larger systems. Suppose once again that we consider the system in Fig. 7.5, but now we include the common-mode failure of the parallel subsystem by using Eq. 7.30 to represent the subsystem reliability R_c'. The system reliability is then given by Eq. 7.23 with a system failure rate of

$$\lambda(t) = \lambda_a + \lambda_b + \lambda_c'(t) + \beta\lambda_1 + \lambda_d + \lambda_e. \tag{7.36}$$

The parallel subsystem now makes two contributions to the system's failure rate: (1) $\lambda_c'(t)$, for independent mode failures, is represented once again by Eq. 7.9 but with λ_1 replaced by the independent failure rate $(1 - \beta)\lambda_1$ and (2) $\beta\lambda_1$ now represents the contribution from the common-mode failure.

Rare-Event Approximations

Redundant components are frequently applied in systems that are required to achieve a high degree of reliability, that is, systems for which R must be close to one. This is synonymous with the requirement that the system failure rate must be so small that $\int_0^t \lambda(t')\, dt' << 1$, for all values of t within the design life, or assuming hereafter a constant failure rate λ, that $\lambda t << 1$. In this situation the reliability $R = e^{-\lambda t}$ may be expanded as

$$R = 1 - \lambda t + \tfrac{1}{2}(\lambda t)^2 - \tfrac{1}{6}(\lambda t)^3 + \ldots, \tag{7.37}$$

and a reasonable approximation obtained by dropping all but the first two terms

$$R \simeq 1 - \lambda t. \tag{7.38}$$

The approximation is said to be conservative in that it may be shown to underestimate the reliability. It is called the rare-event approximation because it is applicable only when failures are rare events.

When we use rare-event approximations, for which the reliability is close to one, it often makes more sense to estimate the small value of the unreliability. The unreliability being defined by

$$\tilde{R} = 1 - R, \tag{7.39}$$

we obtain in the rare-event approximation

$$\tilde{R} \simeq \lambda t. \tag{7.40}$$

If t is a fixed time—say the design life—we may refer to unreliability and failure probability interchangeably.

We may use rare-event approximations to demonstrate more transparently the effects of redundancy and of common-mode failure. We consider first the case of a nonredundant or series system. Here the product rule, Eq. 7.1, yields for the reliability

$$R(t) \simeq \exp\left(-\sum_i \lambda_i t\right). \tag{7.41}$$

For the rare-event approximation to be valid, we require

$$\sum_i \lambda_i t << 1, \tag{7.42}$$

yielding

$$R(t) \simeq 1 - \sum_i \lambda_i t. \tag{7.43}$$

Thus, for two identical units in series with component failure rates λ_1,

$$R(t) \simeq 1 - 2\lambda_1 t. \tag{7.44}$$

For the reliability of the parallel system with two identical units, we expand Eq. 7.8 for $\lambda_1 t << 1$,

$$R(t) = 2[1 - \lambda_1 t + \tfrac{1}{2}(\lambda_1 t)^2 \ldots] - [1 - 2\lambda_1 t + \tfrac{1}{2}(2\lambda_1 t)^2 \ldots], \tag{7.45}$$

from which we obtain

$$R(t) \simeq 1 - (\lambda_1 t)^2. \tag{7.46}$$

This is indicative of redundant systems. The unreliability, $\tilde{R} = 1 - R$, is proportional to the square of $\lambda_1 t$; for N units in parallel,

$$R(t) \simeq 1 - (\lambda_1 t)^N. \tag{7.47}$$

Since $\lambda_1 t$ is a small number, the failure probability decreases as more units are added in parallel.

The deleterious effects of common-mode failures are also seen more clearly in the rare-event approximation. We obtain the reliability by writing Eq. 7.30 as

$$R(t) = 2e^{-\lambda t} - e^{-2\lambda t} e^{\beta\lambda t} \tag{7.48}$$

and by expanding:

$$R(t) = 2[1 - \lambda t + \tfrac{1}{2}(\lambda t)^2 \ldots] - [1 - 2\lambda t + \tfrac{1}{2}(2\lambda t)^2 \ldots] \times [1 + \beta\lambda t + \tfrac{1}{2}(\beta\lambda t)^2]. \tag{7.49}$$

Retaining only the leading terms, we have

$$R(t) \simeq 1 - \beta\lambda t - \left(1 - 2\beta + \frac{\beta^2}{2}\right)(\lambda t)^2 + \ldots. \tag{7.50}$$

Note that there is now a linear term in λt leading to a much larger failure probability unless β, the fraction of failures brought about by the common causes, is very small.

EXAMPLE 7.4

Suppose that a unit has a design-life reliability of 0.95. (*a*) Estimate the reliability if two of these units are put in parallel and there are no common-mode failures. (*b*) Estimate the maximum fraction β of common failures that is acceptable if the parallel units in part *a* are to retain a system reliability of at least 0.99.

Solution From Eq. 7.38 take $\lambda t = 0.05$.
(*a*) $R \approx 1 - (\lambda T)^2$, $R = 0.9975$. ◀
(*b*) From Eq. 7.50,

$$\tilde{R} = 1 - R = 0.01 \approx \beta\lambda T + \left(1 - 2\beta + \frac{\beta^2}{2}\right)(\lambda T)^2.$$

Thus, with $\lambda T \approx 0.05$, we have

$$0.00125\beta^2 + 0.045\beta - 0.0075 = 0.$$

Therefore,

$$\beta = \frac{-0.045 \pm (2.0625 \times 10^{-3})^{1/2}}{0.0025}.$$

For β to be positive, we must take the positive root. Therefore, $\beta \leq 0.166$. ◀

7.3 REDUNDANCY ALLOCATION

High reliability can be achieved in a variety of ways; the choice will depend on the nature of the equipment, its cost, and its mission. If we were to provide an emergency power supply for a hospital, an air traffic control system, or a nuclear power plant, for example, the most cost-effective solution might well be to use commercially available diesel generators as the components in a redundant configuration. On the other hand, the use of redundancy may not be the optimal solution in systems in which the minimum size and weight are overriding considerations: for example, in satellites or other space applications, in well-logging equipment, and in pacemakers and similar biomedical applications. In such applications space or weight limitations may dictate an increase in component reliability rather than redundancy. Then more emphasis must be placed on manufacturing quality control or on controlling the operating environment.

Once a decision is made that redundancy is required, a number of design trade-offs must be examined to determine how redundancy is to be employed. If the entire system is not to be duplicated, it must be decided which components should be duplicated. Consider, for example, the simple two-component system shown in Fig. 7.7*a*. If the reliability $R_a = R_1R_2$ is not large enough, which component should be made redundant? Depending on the choice, the system *b* or *c* will result. It immediately follows that

$$R_b = (2 - R_1)R_1R_2, \tag{7.51}$$

$$R_c = R_1(2 - R_2)R_2. \tag{7.52}$$

Or taking the differences of the results, we have

$$R_b - R_c = R_1R_2(R_2 - R_1). \tag{7.53}$$

Not surprisingly, we see from this expression that the greatest reliability is achieved in the redundant configuration if we duplicate the component with the least reliability; if $R_2 > R_1$, then system R_b is preferable, and conversely. This rule of thumb can be generalized to systems with any number of nonredundant components; the largest gains are to be achieved by making the least reliable components redundant. In reality, the relative costs of the components also must be considered, and if one component is much more expensive than a second, duplicating the second may increase reliability more per dollar spent than duplicating the first. Generally, however, the greater danger is wasting resources in making highly reliable components redundant, when the system reliability is limited most severely by a less reliable component which is not made redundant. Since component costs are normally available, the greatest impediment to making an informed choice is lack of reliability data for the components involved.

EXAMPLE 7.5

Suppose that in the system shown in Fig. 7.7*a* the two components have the same cost, and $R_1 = 0.7$, $R_2 = 0.95$. If it is permissible to add two components to the

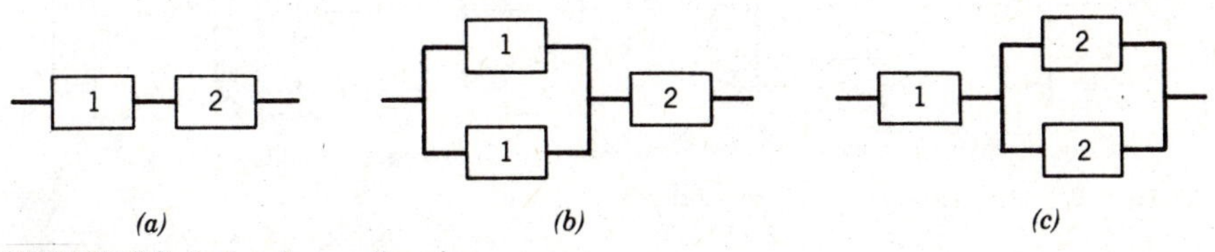

FIGURE 7.7 Redundancy allocation.

system, would it be preferable to replace component 1 by three components in parallel or to replace components 1 and 2 each by simple parallel systems?

Solution If component 1 is replaced by three components in parallel, then from Eq. 7.15

$$R_a = [1 - (1 - R_1)^3]R_2 = 0.973 \times 0.95 = 0.92435.$$

If each of the two components is replaced by a simple parallel system,

$$R_b = [1 - (1 - R_1)^2][1 - (1 - R_2)^2] = 0.91 \times 0.9975 = 0.9077.$$

In this problem the reliability R_1 is so low that even the reliability of a simple parallel system, $2R_1 - R_1^2$, is smaller than that of R_2. Thus replacing component 1 by three parallel components ◀ yields the higher reliability.

Trade-offs in the allocation of redundancy often involve considerations other than simply cost effectiveness. This is illustrated in the following discussions of high- and low-level redundancy, of *m/N* systems, and of fail-safe and fail-to-danger considerations.

High- and Low-Level Redundancy

After a decision has been made to provide redundance within a system, a number of questions must be answered. One of the most fundamental of these concerns the level at which redundance is to be provided. For example, consider the system consisting of three subsystems, as shown in Fig. 7.8. In high-level redundancy, the entire system is duplicated, as indicated in Fig. 7.8*a,* whereas in low-level redundancy the duplication takes place at the subsystem or component level indicated in Fig. 7.8*b.*

Indeed, the concept of the level at which redundance is applied can be further generalized to lower and lower levels. If each of the blocks in the diagram is a subsystem, each consisting of components, we might place the redundance at a still lower component level. For example, computer redundance might be provided at the highest level by having redundant computers, at an intermediate level by having redundant circuit boards within

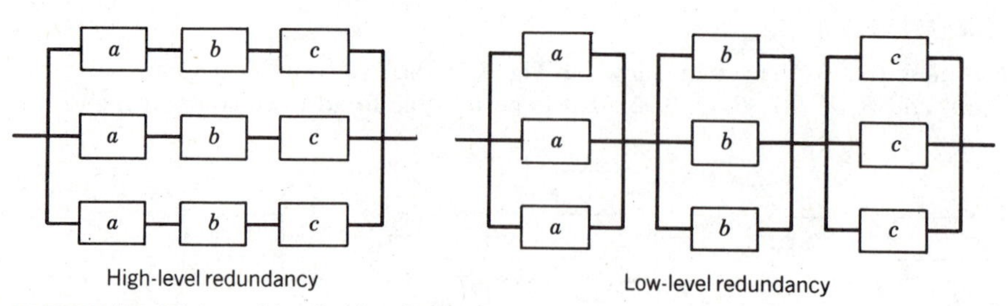

FIGURE 7.8 High- and low-level redundancy.

a single computer, or at the lowest level by having redundant chips on the circuit boards.

Suppose that we determine the reliability of each of the systems in Fig. 7.8 with the component failures assumed to be mutually independent. The reliability of the system without redundance is then

$$R = R_a R_b R_c. \tag{7.54}$$

The reliability of the two redundant configurations may be determined by considering them as composites of series and parallel configurations.

For the high-level redundance shown in Fig. 7.8*a*, we simply take the parallel combination of the two series systems. Since the reliability of each series subsystem is given by Eq. 7.54, R_{HL} of the high-level redundant system is given by

$$R_{\mathrm{HL}} = 2R - R^2, \tag{7.55}$$

or equivalently,

$$R_{\mathrm{HL}} = 2R_a R_b R_c - R_a^2 R_b^2 R_c^2. \tag{7.56}$$

Conversely, to calculate R_{LL}, the reliability of the low-level redundant system, we first consider the parallel combinations of component types *a*, *b*, and *c* separately. Thus the two components of type *a* in parallel yield

$$R_A = 2R_a - R_a^2, \tag{7.57}$$

and similarly,

$$R_B = 2R_b - R_b^2; \qquad R_C = 2R_c - R_c^2. \tag{7.58}$$

The low-level redundant system then consists of a series combination of the three redundant subsystems. Hence

$$R_{\mathrm{LL}} = R_A R_B R_C, \tag{7.59}$$

or, inserting Eqs. 7.57 and 7.58 into this expression, we have

$$R_{\mathrm{LL}} = (2R_a - R_a^2)(2R_b - R_b^2)(2R_c - R_c^2). \tag{7.60}$$

Both the high- and the low-level redundant systems have the same number of components. They do not result, however, in the same reliability. This may be demonstrated by calculating the quantity $R_{\mathrm{LL}} - R_{\mathrm{HL}}$. For simplicity we do this for systems in which all the components have the same reliability, R^*, and thus

$$R_{\mathrm{HL}} = 2R_*^3 - R_*^6 \tag{7.61}$$

and

$$R_{\mathrm{LL}} = (2R^* - R_*^2)^3. \tag{7.62}$$

After some algebra we have

$$R_{LL} - R_{HL} = 6R_*^3(1 - R_*)^2. \tag{7.63}$$

Consequently,

$$R_{LL} > R_{HL}. \tag{7.64}$$

In general, regardless of how many components the original system has in series, and regardless of whether two or more components are put in parallel, low-level redundancy yields higher reliability, but only if two very important conditions are met.

First, the reliabilities of the components cannot depend on the configuration in which they are located. Second, and more important, the failures must be truly independent in both configurations. In reality, the common-mode failures that significantly reduce the gains of redundancy are more likely to occur with low-level than with high-level redundancy. In high-level redundancy the components of like character are likely to be more isolated physically and therefore less susceptible to common local stresses. For example, a faulty connector may cause a circuit board to overheat and then the two redundant chips on that board to fail. But if the redundant chips are on different circuit boards in a high-level redundant system, this common-mode failure mechanism will not exist. Physical isolation is, in general, easier to achieve in high-level than in low-level redundancy; this in turn may eliminate many causes of common-mode failures, such as local flooding and overheating.

Some quantitative insight into the problem of common-mode failures with low-level redundance can be gained as follows. Suppose that we consider the same high- and low-level redundant systems for which the results are given by Eqs. 7.61 and 7.62, and let the component reliability be represented by $R_* = e^{-\lambda t}$. Suppose further that because components in the high-level system are physically isolated, there are no significant common-mode failures. Then we may write simply

$$R_{HL} = e^{-3\lambda t}(2 - e^{-3\lambda t}). \tag{7.65}$$

However, in the low-level system, we specify that some fraction β of the failure rate λ is due to common-mode failures. In this case the quantities R_a, R_b, and R_c will not reduce to Eq. 7.62 yielding

$$R_{LL} = (2e^{-\lambda t} - e^{-2\lambda t})^3, \tag{7.66}$$

where there are no common-mode failures. Rather, the β-factor model replaces Eqs. 7.57 and 7.58 by Eq. 7.30 to yield

$$R_A = R_B = R_C = 2e^{-\lambda t} - e^{-2\lambda t}e^{\beta\lambda t}. \tag{7.67}$$

Then, using Eq. 7.59, we obtain for the low-level redundant system the smaller value

$$R_{LL} = (2e^{-\lambda t} - e^{-2\lambda t}e^{\beta\lambda t})^3, \tag{7.68}$$

This must be compared to Eq. 7.65 to determine which provides the greater reliability.

EXAMPLE 7.6

Suppose that the design-life reliability of each of the components in the high- and low-level redundant systems pictures in Fig. 7.8 is 0.99. What fraction of the failure rate in the low-level system may be due to common-mode failures, without the advantage of low-level redundancy being lost?

Solution Set $R_{HL} = R_{LL}$, using Eqs. 7.65 and 7.68 at the end of the design life:

$$e^{-3\lambda T}(2 - e^{-3\lambda T}) = (2e^{-\lambda T} - e^{-2\lambda T + \beta\lambda T})^3.$$

Solving for β yields

$$\beta = \frac{1}{\lambda T}\ln[2 - (2 - e^{-3\lambda T})^{1/3}] + 1.$$

Since $e^{-\lambda T} = 0.99$, $\lambda T = 0.01005$. Thus

$$\beta = \frac{1}{0.01005}\ln[2 - (2 - 0.99^3)^{1/3}] + 1 = 0.0197. \blacktriangleleft$$

m/N Parallel Systems

In the parallel systems just discussed, when any one of the two or more subsystems functions, the system operates successfully. Such systems are often referred to as $1/N$ systems, where N is the number of units in parallel. In what follows, we discuss the m/N system where m (instead of 1) is the minimum number of the N units that must function for the system to operate successfully. Such systems are popular for relief valves, pumps, motors, and other equipment which must have a minimum capacity to meet design criteria. With such systems it is often possible to increase the reliability of the system without increasing the cost by a commensurate amount. This is true particularly if the large capacity of the system requires more than one unit of a commercially available size to function at normal capacity. As we shall see in the next subsection, m/N configurations are also popular in instrumentation and control systems to prevent spurious fail-safe operation of a unit from causing undesirable consequences.

An m/N system may be represented in a reliability block diagram, as shown for a 2/3 system in Fig. 7.9. The block representing each component, however, must be repeated in the diagram. For the three components shown, the reliabilities of the three parallel paths are R_1R_2, R_2R_3, and R_1R_3. Putting

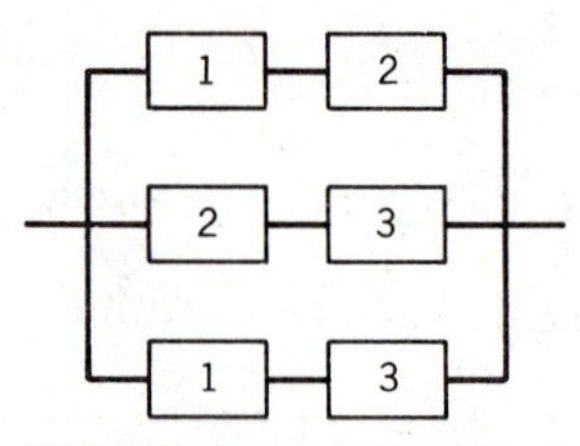

FIGURE 7.9 Reliability block diagram for a $\frac{2}{3}$ system.

these paths in parallel, we obtain

$$R = 1 - (1 - R_1R_2)(1 - R_2R_3)(1 - R_1R_3), \tag{7.69}$$

or if all the components have reliability R_*,

$$R = 1 - (1 - R_*^2)^3. \tag{7.70}$$

If the components are identical, the reliability of an m/N system may be determined more conveniently using the binomial distribution introduced in Chapter 2. Suppose that p is the probability of failure over some period of time for one unit. That is,

$$p = 1 - R_*, \tag{7.71}$$

where R_* is the component reliability. From the binomial distribution the probability that n units will fail is just

$$P\{\mathrm{n} = n\} = C_n^N p^n (1 - p)^{N-n}. \tag{7.72}$$

In order for an m/N system to function, there must be no more than $N - m$ failures:

$$P\{\mathrm{n} \leq N - m\} = \sum_{n=0}^{N-m} C_n^N p^n (1 - p)^{N-n}. \tag{7.73}$$

The probability that no more than $N - m$ will fail, however, is just the probability that the system will function; it is the reliability. Thus, combining Eqs. 7.71 and 7.73, we have

$$R = \sum_{n=0}^{N-m} C_n^N (1 - R_*)^n R_*^{N-n}. \tag{7.74}$$

Alternately, since

$$P\{\mathrm{n} > N - m\} = \sum_{n=N-m+1}^{N} C_n^N p^n (1 - p)^{N-n} \tag{7.75}$$

is the probability that the system will fail, we may also write the system reliability as

$$R = 1 - \sum_{n=N-m+1}^{N} C_n^N (1 - R_*)^n R_*^{N-n}. \tag{7.76}$$

Equations 7.74 and 7.76 are identical in value. Depending on the ratio of m to N, one may be more convenient than the other to evaluate. For example, in a $1/N$ system Eq. 7.76 is simpler to evaluate, since the sum on the right-hand side has only one term, $n = N$, yielding

$$R = 1 - (1 - R_*)^N. \tag{7.77}$$

In dealing with redundant configurations, whether of the $1/N$ or m/N variety, we can simplify the calculations substantially with little loss of accuracy if the component failure probabilities are small (i.e., when the component reliabilities approach one). In these situations a reasonable approximation is often to include only the single most dominant term in Eq. 7.76. To illustrate, suppose that R_* is very close to one; we may replace it by one in the R_*^{N-n} term to yield

$$R \approx 1 - \sum_{n=N-m+1}^{N} C_n^N (1 - R_*)^n. \tag{7.78}$$

We note, however, that the terms in the $(1 - R_*)^n$ series decrease very rapidly in magnitude as the exponent is increased. Consequently, we need to include only the term taken to the lowest power of $1 - R_*$ in the series. Thus we have an approximate value for the reliability:

$$R \approx 1 - C_{N-m+1}^N (1 - R_*)^{N-m+1}. \tag{7.79}$$

The unreliability corresponding to this expression is, of course,

$$\tilde{R} \approx C_{N-m+1}^N \tilde{R}_*^{N-m+1}, \tag{7.80}$$

where $\tilde{R}_* = 1 - R_*$.

If Eqs. 7.79 and 7.80 are to be valid, $\tilde{R}_*$, the component unreliability, must be very small. If a constant failure rate model is employed, then according to Eq. 7.40 we have $\tilde{R}_* = \lambda t$. Thus we may write

$$\tilde{R} \simeq C_{N-m+1}^N (\lambda t)^{N-m+1}. \tag{7.81}$$

EXAMPLE 7.7

A pressure vessel is equipped with six relief valves. Pressure transients can be controlled successfully by any three of these valves. If the probability that any one of these valves will fail to operate on demand is 0.04, what is the probability on demand that the relief valve system will fail to control a pressure transient? Assume that the failures are independent.

Solution In this situation, the foregoing equations are valid if unreliability is defined as demand failure probability. Using the rare-event approximation, we have from Eq. 7.80, with $N = 6$ and $m = 3$,

$$\tilde{R} \approx C_4^6(0.04)^4 = \frac{6!}{2!4!}(0.04)^4 = 15 \times 256 \times 10^{-8}$$

$$\tilde{R} \approx 0.38 \times 10^{-4}. \blacktriangleleft$$

Fail-Safe and Fail-to-Danger

Thus far we have lumped all failures together. When we are determining the reliability, however, there are situations in which different modes of failure can have quite different effects. Consider an alarm system, or for that matter, safety-related systems discussed more generally in Chapter 10. The alarm may fail in one of two ways. It may fail to function even though the danger is present or it may give a spurious or false alarm even though no danger is present. The first of these is referred to as fail-to-danger and the second as fail-safe. Generally, the probability of fail-to-danger is made much smaller than the fail-safe probability. Even then, small fail-safe probabilities are also required. If too many spurious alarms are sounded, they will tend to be ignored. Then, when the real danger is present, the alarm is also likely to be ignored. This difficulty can be circumvented by automating the safety actions, but then each spurious alarm may lead to a significant economic loss. This would certainly be the case were a chemical plant, a nuclear reactor, or any other industrial installation shut down frequently by the spurious operation of safety systems.

The distinction between fail-safe and fail-to-danger has at least two important implications for reliability engineering. First, many design alterations that may be made to decrease the fail-to-danger probability are likely to increase the fail-safe probability. An obvious example is that of power supply failures, which are often a primary cause of failure of crudely designed safety systems. Often, the system can be redesigned so that if the power supply fails, the system will fail-safe instead of to-danger. Specifically, instead of leaving the system unprotected following the failure, the power supply failure will cause the system to function spuriously. Of course, if no change is made in the probability of power supply failure, the reduction in the probability for system fail-to-danger will be compensated for by the increased number of spurious operations.

A second implication for reliability engineering is that the more redundancy is used to reduce the probability of fail-to-danger, the more fail-safe incidents are likely to occur. To demonstrate this, consider a $1/N$ parallel system with which are associated two failure probabilities p_d and p_s, for fail-to-danger and fail-safe, respectively. The fail-to-danger unreliability for the system is found by noting that all units must fail. Hence

$$\tilde{R}_d = p_d^N. \tag{7.82}$$

However, the system fail-safe unreliability is calculated by noting that any one-unit failure with probability p_s will cause the system to fail-safe. Thus

$$\tilde{R}_s = 1 - (1 - p_s)^N. \tag{7.83}$$

Using the approximation $p_s << 1$, we see that the fail-safe probability grows linearly with the number of units in parallel,

$$\tilde{R}_s \approx Np_s. \tag{7.84}$$

The m/N configuration has been extensively used in electronic and other protection systems to limit the number of spurious operations at the same time that the redundancy provides high reliability. In such systems the fail-to-danger unreliability is given by Eq. 7.75:

$$\tilde{R}_d = \sum_{n=N-m+1}^{N} C_n^N p_d^n (1 - p_d)^{N-n}. \tag{7.85}$$

With the rare-event approximation this reduces to a form analogous to Eq. 7.80:

$$\tilde{R}_d \simeq C_{N-m+1}^N p_d^{N-m+1}. \tag{7.86}$$

Conversely, at least m spurious signals must be generated for the system to fail-safe. Assuming independent failures with probability p_s, we have

$$\tilde{R}_s = P\{n \geq m\} = \sum_{n=m}^{N} C_n^N p_s^n (1 - p_s)^{N-n}. \tag{7.87}$$

Again using the rare-event approximation that $p_s << 1$, we may approximate this expression by

$$\tilde{R}_s \simeq C_m^N p_s^m. \tag{7.88}$$

From Eqs. 7.86 and 7.88 the trade-off between fail-to-danger and spurious operation is seen. The fail-safe unreliability is decreased by increasing m, and the fail-to-danger unreliability is decreased by increasing $N - m$. Of course, as N becomes too large, common-mode failures may severely limit further improvement.

EXAMPLE 7.8

You are to design an m/N detection system. The number of components, N, must be as small as possible to minimize cost. The fail-to-danger and the fail-safe probabilities for the identical components are

$$p_d = 10^{-2}, \qquad p_s = 10^{-2}.$$

Your design must meet the following criteria:

1. Probability of system fail-to-danger $< 10^{-4}$.
2. Probability of system fail-safe $< 10^{-2}$.

What values of m and N should be used?

Solution Make a table of unreliabilities (i.e., the failure probabilities) for fail-safe and fail-to-danger using the rare-event approximations given by Eqs. 7.88 and 7.84:

m/N	$\tilde{R}_s$ Eq. 7.88	$\tilde{R}_d$ Eq. 7.86
1/1	$p_s = 10^{-2}$	$p_d = 10^{-2}$
1/2	$2p_s = 2 \times 10^{-2}$	$p_d^2 = 10^{-4}$
2/2	$p_s^2 = 10^{-4}$	$2p_d = 2 \times 10^{-2}$
1/3	$3p_s = 3 \times 10^{-2}$	$p_d^3 = 10^{-6}$
2/3	$3p_s^2 = 3 \times 10^{-4}$	$3p_d^2 = 3 \times 10^{-4}$
3/3	$p_s^3 = 10^{-6}$	$3p_d = 3 \times 10^{-2}$
1/4	$4p_s = 4 \times 10^{-2}$	$p_d^4 = 10^{-8}$
2/4	$6p_s^2 = 6 \times 10^{-4}$ ◀	$4p_d^3 = 4 \times 10^{-6}$ ◀
3/4	$4p_s^3 = 4 \times 10^{-6}$	$6p_d^2 = 6 \times 10^{-6}$
4/4	$p_s^4 = 10^{-8}$	$4p_d = 4 \times 10^{-2}$

At least four components are required to meet both criteria. They are met by a 2/4 ◀ system.

7.4 REDUNDANCY IN COMPLEX CONFIGURATIONS

Thus far we have discussed simple redundance, the level at which redundance is applied, and redundancy in m/N systems. Systems, of course, may take on more complex configurations than those shown in the preceding figures, and within such configurations redundant components or subsystems may be employed. Then it must be possible to evaluate the system reliability in terms of that of the components. In what follows we examine the analysis of redundancy in two classes of systems: those that may be analyzed in terms of series and parallel configurations, and those in which the components are linked in such a way that they cannot.

Series-Parallel Configurations

As long as a system can be decomposed into series and parallel subsystem configurations, the techniques of the preceding sections can be employed repeatedly to derive expressions for system reliability. As an example con-

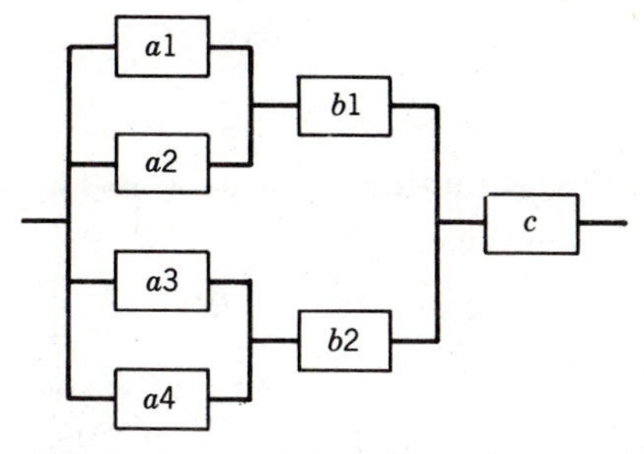

FIGURE 7.10 Reliability block diagram of a series–parallel configuration.

sider the reliability block diagram shown for a system in Fig. 7.10. Components $a1$ through $a4$ have reliability R_a, and components $b1$ and $b2$ have reliability R_b. For the following analysis to be valid, the failures of the components must be independent of one another.

We begin by noting that there are two sets of subsystems of type a components, consisting of a simple parallel configuration as shown in Fig. 7.11*a*. Thus we define the reliability of these configurations as

$$R_A = 2R_a - R_a^2. \tag{7.89}$$

The system configuration then appears as the reduced block diagram shown in Fig. 7.11*b*. We next note that each newly defined subsystem A is in series

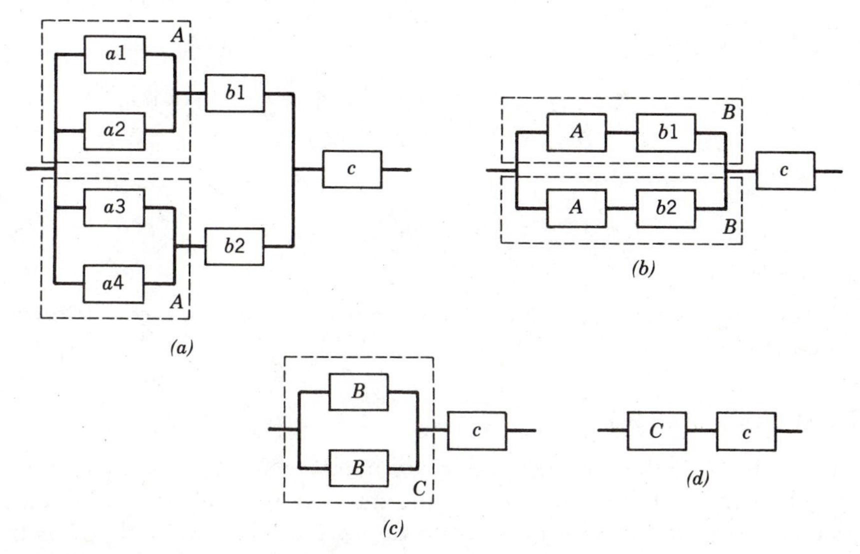

FIGURE 7.11 Decomposition of the system in Fig. 7.10.

with a component of type b. We may therefore define a subsystem B by

$$R_B = R_A R_b, \tag{7.90}$$

and the reduced block diagram then appears as in Fig. 7.11c. Since the two subsystems B are in parallel, we may write

$$R_C = 2R_B - R_B^2 \tag{7.91}$$

to yield the simplified configuration shown in Fig. 7.11d. Finally, the total system consists of the series of subsystems C and component c. Thus

$$R = R_C R_c. \tag{7.92}$$

Having derived an expression for the system reliability, we may combine Eqs. 7.89 through 7.92 to obtain the system reliability in terms of that of R_a, R_b, and R_c:

$$R = (2R_a - R_a^2)R_b[2 - (2R_a - R_a^2)R_b]R_c. \tag{7.93}$$

EXAMPLE 7.9

Suppose that in Fig. 7.10, $R_a = R_b = e^{-\lambda t} \equiv R_*$ and $R_c = 1$. Find R in the rare-event approximation.

Solution We simplify Eq. 7.93,

$$R = R_*^2(2 - R_*)[2 - (2 - R_*)R_*^2]$$

and write it as a polynomial in R_*:

$$R = 4R_*^2 - 2R_*^3 - 4R_*^4 + 4R_*^5 - R_*^6$$

Then we expand $R_*^N = e^{-N\lambda t} \simeq 1 - N\lambda t + \frac{1}{2}N^2(\lambda t)^2 - \cdots$ to obtain for small λt

$$R \simeq 4[1 - 2\lambda t + 2(\lambda t)^2] - 2[1 - 3\lambda t + \tfrac{9}{2}(\lambda t)^2] - 4[1 - 4\lambda t + 8(\lambda t)^2]$$
$$+ 4[1 - 5\lambda t + \tfrac{1}{2}25(\lambda t)^2] - 1 + 6\lambda t - 18(\lambda t)^2$$
$$R \simeq (4 - 2 - 4 + 4 - 1) - (8 - 6 - 16 + 20 - 6)(\lambda t)$$
$$- (-8 + 9 + 32 - 50 + 18)(\lambda t)^2 + \cdots$$
$$R \simeq 1 - (\lambda t)^2. \blacktriangleleft$$

Had the coefficient of the $(\lambda t)^2$ term also been zero, we would have needed to carry terms in $(\lambda t)^3$.

Linked Configurations

In some situations the linkage of the components or subsystems is such that the foregoing technique of decomposing into parallel and series configurations cannot be applied. Such is the case for the system configuration shown in Fig. 7.12, consisting of subsystem types 1, 2, and 3, with reliabilities R_1, R_2, and R_3.

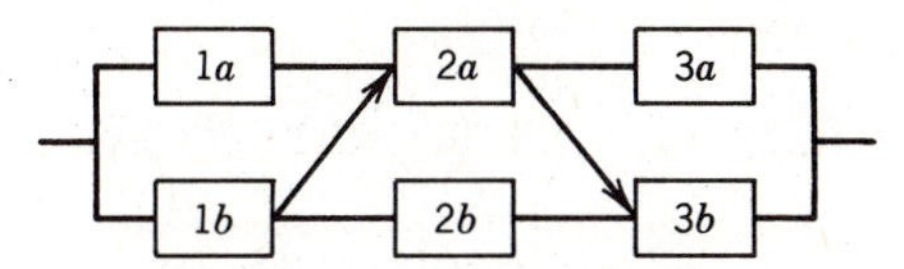

FIGURE 7.12 Reliability block diagram of a cross-linked system.

To analyze this system, we consider two situations, when subsystem 2*a* is failed, and when subsystem 2*a* is operating. Since R, the system reliability, is just the probability of success, we may write it as the sum of two mutually exclusive states, component 2*a* failed and component 2*a* operational. To do this, we first define R^-, the reliability, given the failure of 2*a*, and R^+, given the successful operation of 2*a*. Then, since $1 - R_2$ is just the probability that 2*a* will fail and R_2 the probability that it will not, we may write the system reliability as

$$R = R^-(1 - R_2) + R^+R_2. \tag{7.94}$$

We must now evaluate the conditional reliabilities R^+ and R^-. For R^- in which 2*a* has failed, we disconnect all the paths leading through 2*a* in Fig. 7.12; the result appears in Fig. 7.13*a*. Conversely, for R^+ in which 2*a* is functioning, we pass a path through 2*a*, thereby bypassing 2*b* with the result shown in Fig. 7.13*b*.

We see that when 2*a* is failed, the reduced system consists of a series of three subsystems, 1*b*, 2*b*, and 3*b*; subsystems 1*a* and 3*a* no longer make any contribution to the value of R^-. We obtain

$$R^- = R_1R_2R_3. \tag{7.95}$$

When 2*a* is operating, we have a series combination of two parallel configurations, 1*a* and 1*b* in the first and 3*a* and 3*b* in the second; since component 2*b* is always bypassed, it has no effect on R^+. Therefore, we have

$$R^+ = (2R_1 - R_1^2)(2R_3 - R_3^2). \tag{7.96}$$

Finally, substituting these expressions into Eq. 7.94, we find the system reliability to be

$$R = R_1R_2R_3(1 - R_2) + (2R_1 - R_1^2)(2R_3 - R_3^2)R_2^2. \tag{7.97}$$

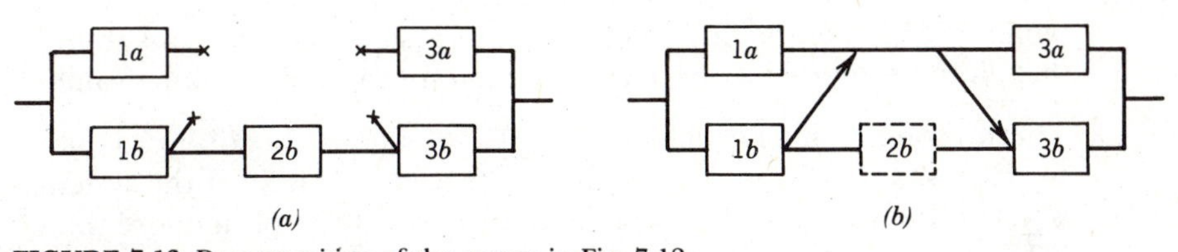

FIGURE 7.13 Decomposition of the system in Fig. 7.12.

EXAMPLE 7.10

Evaluate Eq. 7.97 in the rare-event approximation when $R_n = e^{-\lambda t}$ for all n.

Solution Let $R_* \equiv R_n$. Then Eq. 7.97 becomes

$$R = R_*^3(1 - R_*) + (2 - R_*)^2 R_*^4.$$

Writing this expression as a polynomial in R_*, we have

$$R = R_*^3 + 3R_*^4 - 4R_*^5 + R_*^6.$$

Now we expand $R_* \approx 1 - \lambda t$, and therefore $R^N \approx 1 - N\lambda t$; we have

$$R \approx 1 - 3\lambda t + 3(1 - 4\lambda t) - 4(1 - 5\lambda t) + 1 - 6\lambda t,$$
$$R \approx 1 - \lambda t. \blacktriangleleft$$

Since the coefficient in the linear term in λt is nonzero, we are finished; had it been zero, we would need to carry the $(\lambda t)^2$ terms in all the expansions.

Bibliography

Arsenault, J. E., and J. A. Roberts (eds.), *Reliability and Maintainability of Electronic Systems,* Computer Science Press, Potomac, MD, 1980.

Barlow, R. E., and F. Proschan, *Mathematical Theory of Reliability,* Wiley, New York, 1965.

Fault Tree Handbook, U.S. Nuclear Regulatory Commission, NUREG-0492, 1981.

Henley, E. J., and H. Kumanoto, *Reliability Engineering and Risk Assessment,* Prentice-Hall, Englewood Cliffs, NJ, 1981.

Ireson, W. G. (ed.), *Reliability Handbook,* McGraw-Hill, New York, 1966.

Kapur, K. C., and L. R. Lamberson, *Reliability in Engineering Design,* Wiley, New York, 1977.

Roberts, N. H., *Mathematical Methods in Reliability Engineering,* McGraw-Hill, New York, 1964.

Sandler, G. H. *System Reliability Engineering,* Prentice-Hall, Englewood Cliffs, NJ, 1963.

EXERCISES

7.1 A nonredundant system with 100 components has a design-life reliability of 0.90. The system is redesigned so that it has only 70 components. Estimate the design life of the redesigned systems, assuming that all the components have constant failure rates of the same value.

7.2 Thermocouples of a particular design have a failure rate of $\lambda = 0.008$/hr. How many thermocouples must be placed in parallel if the system is to run for 100 hrs with a system failure probability of no more than 0.05? Assume that all failures are independent.

7.3 A disk drive has a constant failure rate and an MTTF of 5000 hr.
(*a*) What will the probability of failure be for one year of operation?
(*b*) What will the probability of failure be for one year of operation if two of the drives are placed in parallel and the failures are independent?
(*c*) What will the probability of failure be for one year of operation if the common-mode errors are characterized by $\beta = 0.2$?

7.4 Consider a 1/3 system in active parallel, each unit of which has a constant failure rate λ.
(*a*) Plot the system failure rate $\lambda(t)$ in units of λ versus λt from $\lambda t = 0$, to large enough λt to approach an asymptotic system failure rate.
(*b*) What is the asymptotic value $\lambda(\infty)$?
(*c*) At what interval should the system be shut down and failed components replaced if there is a criterion that $\lambda(t)$ should not exceed 1/3 of the asymptotic value?

7.5 Find the variance in the time to failure, assuming a constant failure rate λ:
(*a*) For two units in series.
(*b*) For two units in parallel.
(*c*) Which is larger?

7.6 Consider two components with the same MTTF. One has an exponential distribution, the other a Rayleigh distribution (see Exercise 7.14). If they are placed in parallel, find the system MTTF in terms of the component MTTF.

7.7 Verify Eq. 7.47.

7.8 Consider a system with three identical components with failure rate λ_1. Find the system failure rate:
(*a*) For all three components in series.
(*b*) For all three components in parallel.
(*c*) For two components in parallel and the third in series.
(*d*) Plot the results for parts *a, b,* and *c* on the same scale for $0 \leq t \leq 5/\lambda$.

7.9 The design criterion for the ac power system for a reactor is that its failure probability be less than 2×10^{-5}/year. Off-site power failures may be expected to occur about once in 5 years. If the on-site ac power system consists of two independent diesel generators, each of which is capable of meeting the ac power requirements, what is the maximum failure probability per year that each diesel generator can have if the design criterion is to be met? If three independent diesel generators are used in parallel, what is the value of the maximum failure probability? (Neglect common-mode failures.)

7.10 Suppose that in Exercise 7.9 one-fourth of the diesel generator failures are caused by common-mode effects and therefore incapacitate all the parallel systems. Under these conditions what is the maximum failure probability (i.e., random and common-mode) that is allowable if two diesel generators are used? If three diesel generators are used?

7.11 Suppose that two identical units are placed in parallel. Each has a Weibull distribution with known θ and $m > 1$.
(*a*) Determine the system reliability.
(*b*) Find a rare-event approximation for part *a*.

7.12 Suppose that the units in Exercise 7.11 each have a Weibull distribution with $m = 2$. By how much is the MTTF increased by putting them in parallel?

7.13 A component has a one-year design-life reliability of 0.9; two such components are placed in parallel. What is the one-year reliability of the resulting system:
(*a*) In the absence of common-mode failures?
(*b*) If 20% of the failures are common-mode failures?

7.14 Suppose that a system consists of two subsystems in parallel. The reliability of each subsystem is given by the Rayleigh distribution

$$R(t) = e^{-(t/\theta)^2}.$$

Assuming that common-mode failures may be neglected, determine the system MTTF.

7.15 An engineer designs a system consisting of two subsystems in series. The reliabilities are $R_1 = 0.98$ and $R_2 = 0.94$. The cost of the two subsystems is about equal. The engineer decides to add two redundant components. Which of the following would it be better to do?
(*a*) Duplicate subsystems 1 and 2 in high-level redundance.
(*b*) Duplicate subsystems 1 and 2 in low-level redundance.
(*c*) Replace the second subsystem with 1/3 redundance.
Justify your answer.

7.16 For a system with two units in parallel, each with a failure rate λ and common-mode failures specified by β, find the MTTF in terms of λ and β.

7.17 Two identical components, each with a constant failure rate, are in series. To improve the reliability two configurations are considered: (*a*) for high-level redundancy, (*b*) for low-level redundancy. Calculate the system MTTF in terms of MTTF_0, the system mean time to failure without redundance.

7.18 A radiation-monitoring system consists of a detector, an amplifier, and an annunciator. Their lifetime reliabilities and costs are, respectively, 0.83 (\$1200), 0.58 (\$2400), and 0.69 (\$1600).

(*a*) How would you allocate redundancy to achieve a system lifetime reliability of 0.995?

(*b*) What is the cost of the system?

7.19 Suppose that a system consists of two components, each with a failure rate λ, placed in series. A redundant system is built consisting of four components. Derive expressions for the system failure rates (*a*) for high-level redundancy, (*b*) for low-level redundancy. (*c*) Plot the results of parts *a* and *b* along with the failure rate of the nonredundant system for $0 \leq t \leq 2/\lambda$.

7.20 For constant failure rates evaluate R_{HL} and R_{LL} for high- and low-level redundancy in the rare-event approximation beginning with Eqs. 7.61 and 7.62.

7.21 A system consists of three components in series, each with a reliability of 0.96. A second set of three components is purchased and a redundant system is built. What is the reliability of the redundant system (*a*) with high-level redundancy, (*b*) with low-level redundancy?

7.22 The failure rate on a jet engine is $\lambda = 10^{-3}$/hr. What is the probability that more than two engines on a four-engine aircraft will fail during a 2-hr flight? Assume that the failures are independent.

7.23 The shutdown system on a nuclear reactor consists of four independent subsystems, each consisting of a control rod bank and its associated drives and actuators. Insertion of any three banks will shut down the reactor. The probability that a subsystem will fail is 0.2×10^{-4} per demand. What is the probability per demand that the shutdown system will fail, assuming that common-mode failures can be neglected?

7.24 The identical components of the system below have fail-to-danger probabilities of $p_d = 10^{-2}$ and fail-safe probabilities of $p_s = 10^{-1}$.

(*a*) What is the system fail-to-danger probability?

(*b*) What is the system fail-safe probability?

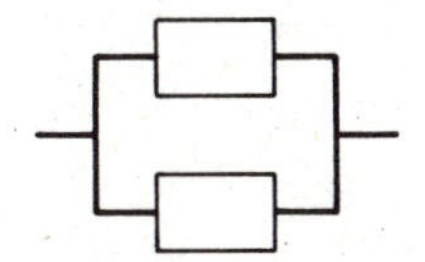

7.25 Calculate the reliabilities of the following systems:

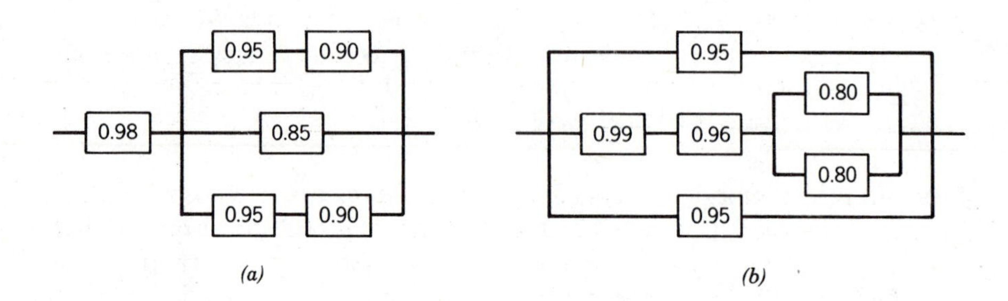

7.26 Calculate the reliability, $R(t)$, for the following systems, assuming that all the components have failure rate λ. Then use the rare-event approximation to simplify the result.

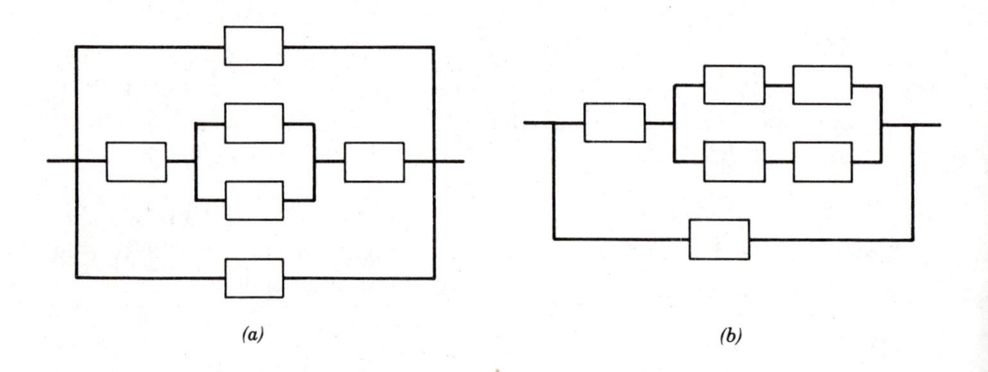

7.27 Calculate the reliability for the following system, assuming that all the component failure rates are equal. Then use the rare-event approximation to simplify your result.

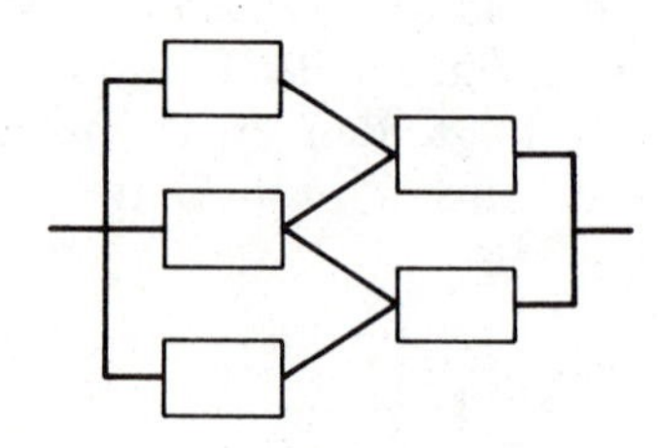

7.28 Given the following component reliabilities, calculate the reliability of the two systems.

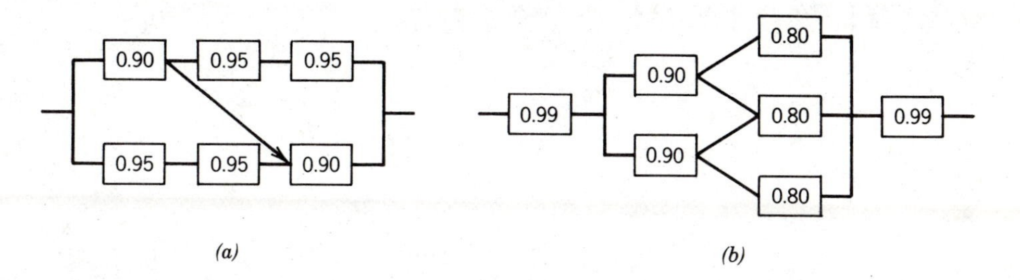

(a) *(b)*

7.29 Calculate the reliabilities of the following two systems, assuming that all the component reliabilities are equal. Then determine which system has the higher reliability.

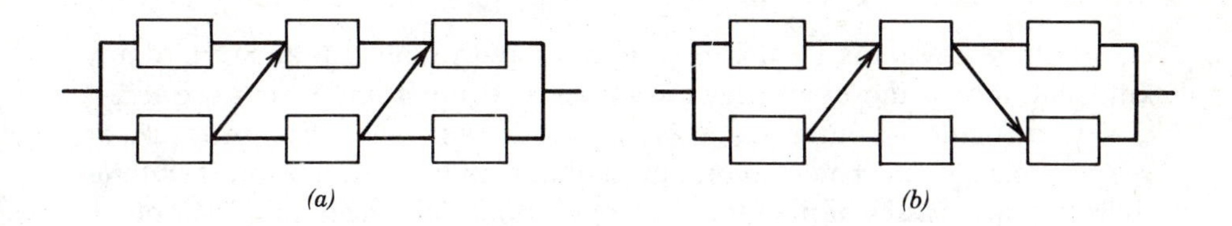

(a) *(b)*

CHAPTER 8

Maintained Systems

8.1 INTRODUCTION

Relatively few systems are designed to operate without maintenance of any kind, and for the most part they must operate in environments where access is very difficult, in outer space or high-radiation fields, for example, or where replacement is more economical than maintenance. For most systems there are two classes of maintenance, one or both of which may be applied. In preventive maintenance, parts are replaced, lubricants changed, or adjustments made before failure occurs. The objective is to increase the reliability of the system over the long term by staving off the aging effects of wear, corrosion, fatigue, and related phenomena. In contrast, repair or corrective maintenance is performed after failure has occurred in order to return the system to service as soon as possible. Although the primary criteria for judging preventive-maintenance procedures is the resulting increase in reliability, a different criterion is needed for judging the effectiveness of corrective maintenance. The criterion most often used is the system availability, which is defined roughly as the probability that the system will be operational when needed.

The amount and type of maintenance that is applied depend strongly on its costs as well as the cost and safety implications of system failure. Thus, for example, in determining the maintenance for an electric motor used in a manufacturing plant, we would weigh the costs of preventive maintenance against the money saved from the decreased number of failures. The failure costs would need to include, of course, both those incurred in repairing or replacing the motor, and those from the loss of production during the unscheduled downtime for repair. For an aircraft engine the trade-off would be much different; the potentially disastrous consequences of engine failure would eliminate repair maintenance as a primary consideration. Concern would be with how much preventive maintenance can be afforded and with the possibility of failures induced by faculty maintenance.

In both preventive and corrective maintenance, human factors play a very strong role. It is for this reason that laboratory data are often not representative of field data. In field service the quality of preventive maintenance is not likely to be as high. Moreover, repairs carried out in the field are likely to take longer and to be less than perfect. The measurement of maintenance quantities thus depends strongly on human reliability so that there is great difficulty in obtaining reproducible data. The numbers depend not only on the physical state of the hardware, but also on the training, vigilance, and judgment of the maintenance personnel. These quantities in turn depend on many social and psychological factors that vary to such an extent that the probabilities of maintenance failures and repair times are generally more variable than the failure rates of the hardware.

In this chapter we first examine preventive maintenance. Then we define and discuss availability and other quantities needed to treat corrective maintenance. Subsequently, we examine the repair of two types of failure: those that are revealed (i.e., immediately obvious) and those that are unrevealed (i.e., are unknown until tests are run to detect them). Finally, we examine the relation of a system to its components from the point of view of corrective maintenance.

8.2 PREVENTIVE MAINTENANCE

In this section we examine the effects of preventive maintenance on the reliability of a system or component. We first consider ideal maintenance in which the system is restored to an as-good-as-new condition each time maintenance is applied. We then examine more realistic situations in which the improvement in reliability brought about by maintenance must be weighed against the possibility that faulty maintenance will lead to system failure. Finally, the effects of preventive maintenance on redundant systems are examined.

Idealized Maintenance

Suppose that we denote the reliability of a system without maintenance as $R(t)$, where t is the operation time of the system; it includes only the intervals when the system is actually operating, and not the time intervals during which it is shut down. If we perform maintenance on the system at time intervals T, then, as indicated in Fig. 8.1, for $t < T$ maintenance will have no effect on reliability. That is, if $R_M(t)$ is the reliability of the maintained system,

$$R_M(t) = R(t), \qquad 0 \leq t < T. \tag{8.1}$$

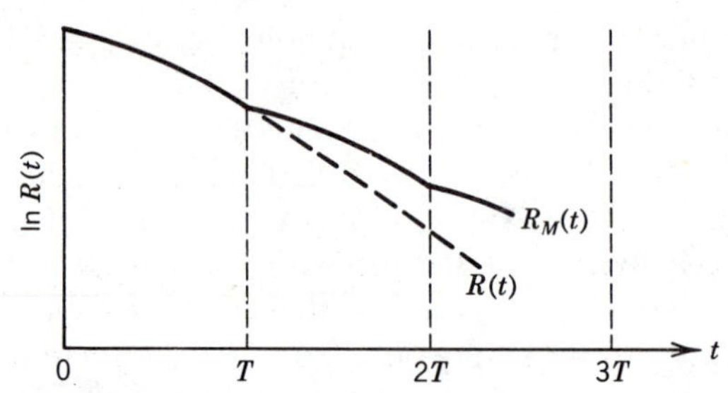

FIGURE 8.1 The effect of preventive maintenance on reliability.

Now suppose that we perform maintenance at T, restoring the system to an as-good-as-new condition. This implies that the maintained system at $t > T$ has no memory of accumulated wear effects for times before T. Thus, in the interval $T < t \leq 2T$, the reliability is the product of the probability $R(T)$ that the system survived to T, and the probability $R(t - T)$ that a system as good as new at T will survive for a time $t - T$ without failure:

$$R_M(t) = R(T)R(t - T), \qquad T \leq t < 2T. \tag{8.2}$$

Similarly, the probability that the system will survive to time t, $2T \leq t < 3T$, is just the reliability $R_M(2T)$ multiplied by the probability that the newly restored system will survive for a time $t - 2T$:

$$R_M(t) = R(T)^2 R(t - 2T), \qquad 2T \leq t < 3T. \tag{8.3}$$

The same argument may be used repeatedly to obtain the general expression

$$R_M(t) = R(T)^N R(t - NT), \qquad NT \leq t < (N + 1)T, \quad N = 0, 1, 2, \ldots. \tag{8.4}$$

The MTTF for a system with preventive maintenance can be determined by replacing $R(t)$ by $R_M(t)$ in Eq. 4.22:

$$\text{MTTF} = \int_0^\infty R_M(t)\, dt. \tag{8.5}$$

To evaluate this expression, we first divide the integral into time intervals of length T:

$$\text{MTTF} = \sum_{N=0}^{\infty} \int_{NT}^{(N+1)T} R_M(t)\, dt. \tag{8.6}$$

Then, inserting Eq. 8.4, we have

$$\text{MTTF} = \sum_{N=0}^{\infty} \int_{NT}^{(N+1)T} R(T)^N R(t - NT)\, dt. \tag{8.7}$$

Setting $t' = t - NT$ then yields

$$\text{MTTF} = \sum_{N=0}^{\infty} R(T)^N \int_0^T R(t')\,dt'. \tag{8.8}$$

Then, evaluating the infinite series,

$$\sum_{n=0}^{\infty} R(T)^N = \frac{1}{1 - R(T)}, \tag{8.9}$$

we have

$$\text{MTTF} = \frac{\int_0^T R(t)\,dt}{1 - R(T)}. \tag{8.10}$$

We would now like to estimate how much improvement, if any, in reliability we derive from the preventive maintenance. The first point to be made is that in random or chance failures (i.e., those represented by a constant failure rate λ), idealized maintenance has no effect. This is easily proved by putting $R(t) = e^{-\lambda t}$ on the right-hand side of Eq. 8.4. We obtain

$$R_M(t) = (e^{-\lambda t})^N e^{-\lambda(t-NT)} = e^{-N\lambda t} e^{-\lambda(t-NT)} = e^{-\lambda t} \tag{8.11}$$

or simply

$$R_M(t) = R(t), \qquad 0 \le t \le \infty. \tag{8.12}$$

Preventive maintenance has a quite definite effect, however, when aging or wear causes the failure rate to become time-dependent. To illustrate this effect, suppose that the reliability can be represented by the two-parameter Weibull distribution described in Chapter 4. For the system without maintenance we have

$$R(t) = \exp\left[-\left(\frac{t}{\theta}\right)^m\right]. \tag{8.13}$$

Equation 8.4 then yields for the maintained system

$$R_M(t) = \exp\left[-N\left(\frac{T}{\theta}\right)^m\right] \exp\left[-\left(\frac{t - NT}{\theta}\right)^m\right], \tag{8.14}$$

$$NT \le t < (N + 1)T,$$

$$N = 0, 1, 2, \ldots .$$

To examine the effect of maintenance, we calculate the ratio $R_M(t)/R(t)$. The relationship is simplified if we calculate this ratio at the time of maintenance $t = NT$:

$$\frac{R_M(NT)}{R(NT)} = \exp\left[-N\left(\frac{T}{\theta}\right)^m + \left(\frac{NT}{\theta}\right)^m\right]. \tag{8.15}$$

Thus there will be a gain in reliability from maintenance only if the argument of the exponential is positive, that is, if $(NT/\theta)^m > N(T/\theta)^m$. This reduces to the condition

$$N^{m-1} - 1 > 0. \tag{8.16}$$

This states simply that m must be greater than one for maintenance to have a positive effect on reliability; it corresponds to a failure rate that is increasing with time through aging. Conversely, for $m < 1$, preventive maintenance decreases reliability. This corresponds to a failure rate that is decreasing with time through early failure. Specifically, if new defective parts are introduced into a system that has already been "worn in," increased rates of failure may be expected. These effects on reliability are illustrated in Fig. 8.2 where Eq. 8.14 is plotted for both increasing ($m > 1$) and decreasing ($m < 1$) failure rates, along with random failures ($m = 1$).

Naturally, a system may have several modes of failure corresponding to increasing and decreasing failure rates. For example, in Chapter 4 we note that the bathtub curve for a device may be expressed as the sum of Weibull distributions

$$\int_0^t \lambda(t')\, dt' = \left(\frac{t}{\theta_1}\right)^{m_1} + \left(\frac{t}{\theta_2}\right)^{m_2} + \left(\frac{t}{\theta_3}\right)^{m_3}, \tag{8.17}$$

where $m_1 < 1$ and $m_3 > 1$ correspond to wearin and wearout, and $m_2 = 1$ is due to the time-independent contribution to the failure rate. For this

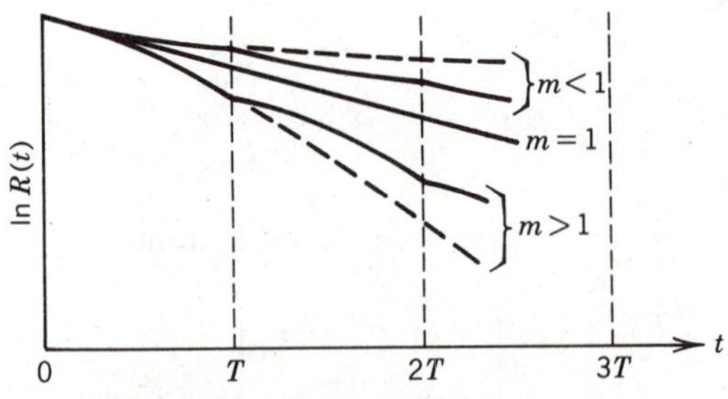

FIGURE 8.2 The effect of preventive maintenance on reliability: $m > 1$, increasing failure rate; $m < 1$, decreasing failure rate; $m = 1$, constant failure rate.

system we must choose the maintenance interval for which the positive effect on wearout time is greater than the negative effect on wearin time. In practice, the terms in Eq. 8.17 may be due to different components of the system. Thus we would perform preventive maintenance only on the components for which the wearout effect dominate. For example, we may replace worn spark plugs in an engine without even considering replacing a fuel injection system with a new one, which might itself be defective.

EXAMPLE 8.1

A compressor is designed for 5 years of operation. There are two significant contributions to the failure rate. The first is due to wear of the thrust bearing and is described by a Weibull distribution with $\theta = 7.5$ year and $m = 2.5$. The second, which includes all other causes, is a constant failure rate of $\lambda_0 = 0.013$/year.
(*a*) What is the reliability if no preventive maintenance is performed over the 5-year design life?
(*b*) If the reliability of the 5 year design life is to be increased to at least 0.9 by periodically replacing the thrust bearing, how frequently must it be replaced?

Solution Let $T_d = 5$ be the design life.
(*a*) The system reliability may be written as

$$R(T_d) = R_0(T_d)R_M(T_d),$$

where

$$R_0(T_d) = e^{-\lambda_0 T_d} = e^{-0.013\times 5} = 0.9371,$$

is the reliability if only the constant failure rate is considered. Similarly,

$$R_M(T_d) = e^{-(T_d/\theta)^m} = e^{-(5/7.5)^{2.5}} = 0.6957$$

is the reliability if only the thrust bearing wear is considered. Thus,

$$R(T_d) = 0.9371 \times 0.6957 = 0.6519. \blacktriangleleft$$

(*b*) Suppose that we divide the design life into N equal intervals; the time interval, T, at which maintenance is carried out is then $T = T_d/N$. Correspondingly, $T_d = NT$. For bearing replacement at time interval T, we have from Eq. 8.14,

$$R_M(T_d) = \exp\left[-N\left(\frac{T_d}{N\theta}\right)^m\right] = \exp\left[-N^{1-m}\left(\frac{T_d}{\theta}\right)^m\right].$$

For the criterion to be met, we must have

$$R_M(T_d) = \frac{R(T_d)}{R_0(T_d)} \geq \frac{0.9}{0.9371}, \qquad R_M(T_d) \geq 0.9604.$$

With $(T_d/\theta)^m = (5/7.5)^{2.5} = 0.36289$, we calculate

$$R_M(T_d) = \exp(-0.36289N^{-1.5}).$$

N	1	2	3	4	5
$R_M(T_d)$	0.696	0.880	0.933	0.956	0.968

Thus the criterion is met for $N = 5$, ◀ and the time interval for bearing replacement is $T = T_d/N = \frac{5}{5} = 1$ year. ◀

In Chapter 4 we state that even when wear is present, a constant failure rate model may be a reasonable approximation, provided that preventive maintenance is carried out, with timely replacement of wearing parts. Although this may be intuitively clear, it is worthwhile to demonstrate it with our present model. Suppose that we have a system for which wearin effects can be neglected, allowing us to ignore the first term in Eq. 8.17 and write

$$R(t) = \exp\left[-\frac{t}{\theta_2} - \left(\frac{t}{\theta_3}\right)^{m_3}\right]. \tag{8.18}$$

The corresponding expression for the maintained system given by Eq. 8.4 becomes

$$R_M(t) = \exp\left[-N\left(\frac{T}{\theta_3}\right)^{m_3}\right]\exp\left[-\frac{t}{\theta_2} - \left(\frac{t - NT}{\theta_3}\right)^{m_3}\right],$$

$$NT \le t \le (N+1)T. \tag{8.19}$$

For a maintained system the failure rate may be calculated by replacing R by R_M in Eq. 4.15:

$$\lambda_M(t) = -\frac{1}{R_M(t)}\frac{d}{dt}R_M(t). \tag{8.20}$$

Thus, taking the derivative, we obtain

$$\lambda_M(t) = \frac{1}{\theta_2} + \frac{m_3}{\theta_3}\left(\frac{t - NT}{\theta_3}\right)^{m_3 - 1}, \qquad NT \le t < (N+1)T. \tag{8.21}$$

Provided that the second term, the wear term, is never allowed to become substantial compared to the first, the random-failure term, the overall failure rate may be approximated as a constant by averaging over the interval T. This is illustrated for a typical set of parameters in Fig. 8.3.

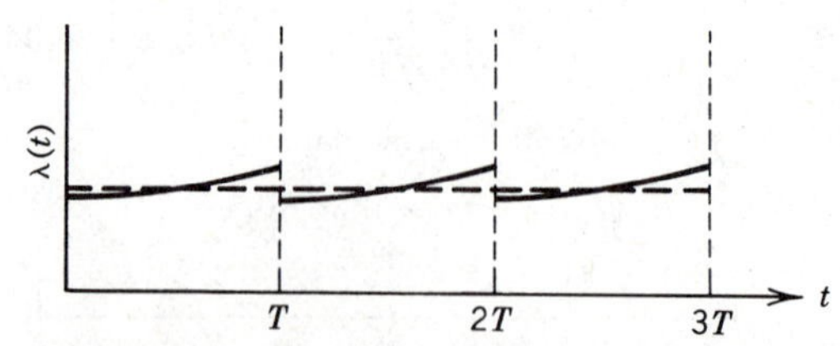

FIGURE 8.3 Failure rate for a system with preventive maintenance.

Imperfect Maintenance

Next consider the effect of a less-than-perfect human reliability on the overall reliability of a maintained system. This enters through a finite probability p that the maintenance is carried out unsatisfactorily, in such a way that the faulty maintenance causes a system failure immediately thereafter. To take this into account in a simple way, we multiply the reliability by the maintenance nonfailure probability, $1 - p$, each time that maintenance is performed. Thus Eq. 8.4 is replaced by

$$R_M(t) = R(T)^N(1 - p)^N R(t - NT), \qquad NT < t < (N + 1)T, \quad N = 0, 1, 2, \ldots . \tag{8.22}$$

The trade-off between the improved reliability from the replacement of wearing parts and the degradation that can come about because of maintenance error may now be considered. Since random failures are not affected by preventive maintenance, we consider the system in which only aging is present, by using Eq. 8.13 with $m > 1$. Once again the ratio R_M/R after the Nth preventive maintenance is a useful indication of performance. Note that for $p << 1$, we may approximate

$$(1 - p)^N \approx e^{-Np} \tag{8.23}$$

to obtain

$$\frac{R_M(NT)}{R(NT)} = \exp\left[-N\left(\frac{T}{\theta}\right)^m - Np + \left(\frac{NT}{\theta}\right)^m\right]. \tag{8.24}$$

For there to be an improvement from the imperfect maintenance, the argument of the exponential in this expression must be positive. This reduces to the condition

$$p < (N^{m-1} - 1)\left(\frac{T}{\theta}\right)^m. \tag{8.25}$$

Consequently, the benefits from imperfect maintenance are not seen for a long time when either N or T is large. This is plausible because after a long time wear effects degrade the reliability enough that the positive effect of maintenance compensates for the probability of maintenance failure. This is illustrated in Fig. 8.4.

EXAMPLE 8.2

Suppose that in Example 8.1 the probability of faulty bearing replacement causing failure of the compressor is $p = 0.02$. What will the design-life reliability be with the annual replacement program?

Solution At the end of the design life ($T_d = 5$ years) maintenance will have been performed four times. From the preceding problem we take the perfect maintenance result to be

$$R(T_d) = R_0 R_M = 0.937 \times 0.968 = 0.907.$$

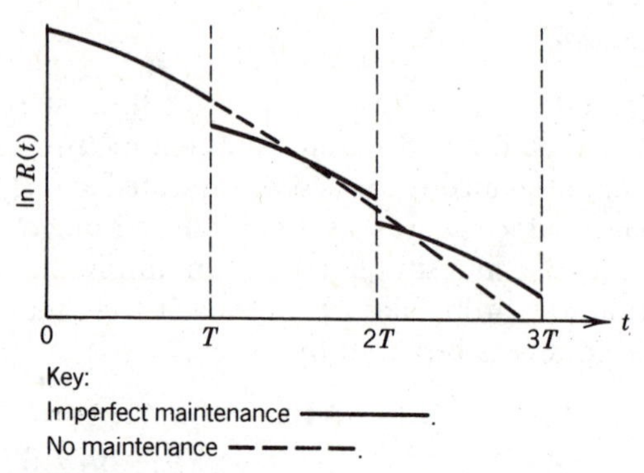

FIGURE 8.4 The effect of imperfect preventive maintenance on reliability.

With imperfect maintenance,

$$R(T_d) = R_0 R_M (1 - p)^4 = 0.907 \times 0.98^4 = 0.907 \times 0.922 = 0.836. \blacktriangleleft$$

In evaluating the trade-off between maintenance and aging, we must examine the failure mode very closely. Suppose, for example, that we consider the maintenance of an engine. If after maintenance the engine fails to start, but no damage is done, the failure may be corrected by redoing the maintenance. In this case p may be set equal to zero in the model just given, with the understanding that preventive maintenance includes a checkout and repairing maintenance errors.

The situation is potentially more serious if the maintenance failure damages the system or is delayed because it is an induced early-failure. We consider each of these problems separately. Suppose first that after maintenance the engine is started and is irreparably damaged by the maintenance error. Whether maintenance is desirable in these circumstances strongly depends on the failure mode that the maintenance is meant to prevent. If the engine's normal mode of failure is simply to stop running because a component is worn, with no damage to the remainder of the engine, it is unlikely that even the increased reliability provided by the preventive maintenance is economically worthwhile. Provided that there are no safety issues at stake, it may be more expedient to wait for failure, and then repair, rather than to chance damage to the system through faulty maintenance. If we are concerned about servicing an aircraft engine, however, the situation is entirely different. Damaging or destroying an occasional engine on the ground following faulty maintenance may be entirely justified in order to decrease the probability that wear will cause an engine to fail in flight.

Consider, finally, the situation in which the maintenance does not cause immediate failure but adds a wearin failure rate. This may be due to the replacement of worn components with defective new ones. However, it is equally likely to be due to improper installation or reassembly of the system, thereby placing excessive stress on one or more of the components. After the first repair, we then have a failure rate described by a bathtub curve, as in Eq. 8.17, with the first term stemming at least in part from imperfect maintenance. The reliability is then determined by inserting Eq. 8.17 into Eq. 8.4. If we assume that the early failure term is due to faulty maintenance, it may be shown by again calculating $R_M(NT)/R(NT)$ that the reliability is improved only if

$$(1 - N^{m_1-1})\left(\frac{T}{\theta_1}\right)^{m_1} < (N^{m_3-1} - 1)\left(\frac{T}{\theta_3}\right)^{m_3}, \qquad m_1 < 1, \quad m_3 > 1. \tag{8.26}$$

Whether or not an increase in overall reliability is the only criterion to be used once again depends on whether the failure modes are comparable in the system damage that is done. If no safety questions are involved, it is primarily a question of weighing the costs of repairing the failures caused by aging against those induced by maintenance errors. This might be the case, for example, with an automobile engine. With an aircraft engine, however, prevention of failure in flight must be the overriding criterion; the cost of repairing the engine following failure, of course, is not relevant if the plane crashes. In this, and similar situations, the more important consideration is often the effect of maintenance errors on redundant systems because maintenance is one of the primary causes of common-mode failures. We examine these next.

Redundant Components

The foregoing expressions for $R_M(t)$ may be used in calculating the reliability of redundant systems as in Chapter 7, but only if the maintenance failures on different components are independent of one another. This stipulation is frequently difficult to justify. Although some maintenance failures are independent, such as the random neglect to tighten a bolt, they are more likely to be systematic; if the wrong lubricant is put in one engine, it is likely to be put in a second one also.

The common-mode failure model introduced in Chpater 7 may be applied with some modification to treat such dependent maintenance failures. As an example we consider a parallel system consisting of two identical components. If the maintenance is imperfect but independent, we may insert Eq. 8.22 into Eq. 7.5 to obtain

$$R_I(t) = 2R(T)^N(1 - p)^N R(t - NT) - R(T)^{2N}(1 - p)^{2N} R(t - NT)^2, \qquad NT \leq t < (N + 1)T, \quad N = 0, 1, 2, \ldots. \tag{8.27}$$

Suppose that a maintenance failure on one component implies that the same failure occurs simultaneously in the other. We account for this by separating out the maintenance failures into a series component, much as we did with the common-mode failure rate λ_c in Chapter 7. Thus the system failure is modeled by taking the reliability for perfect maintenance (i.e., $p = 0$) and multiplying by $1 - p$ for each time that maintenance is performed. Thus, for dependent maintenance failures,

$$R_D(t) = \{2R(T)^N R(t - NT) - R(T)^{2N} R(t - NT)^2\}(1 - p)^N, \quad \begin{array}{l} NT \le t < (N+1)T \\ N = 0, 1, 2, \ldots . \end{array} \tag{8.28}$$

The degradation from maintenance induced common-mode failures is indicated by the ratio of Eqs. 8.28 to 8.27. We find

$$\frac{R_D(NT)}{R_I(NT)} = \frac{1 - \frac{1}{2}R(T)^N}{1 - \frac{1}{2}(1 - p)^N R(T)^N}. \tag{8.29}$$

The value of this ratio is less than one, and it decreases each time imperfect preventive maintenance is performed.

8.3 REPLACEMENT POLICY

For systems subject to wear—and therefore to an increasing failure rate—preventive maintenance can substantially increase reliability, provide that the maintenance itself does not introduce a significant number of defects. Even in the idealized situation in which maintenance returns the system to an as-good-as-new condition, however, the increased reliability must be weighed against the maintenance costs.

Maintenance costs must be considered when determining the frequency and conditions under which to replace parts or components of a system in order to minimize the overall cost of operations. The situation varies widely with the nature of the components. This may be understood by considering some typical parts of an automobile. At one extreme preventive maintenance has no effect on totally random phenomena. Thus, for example, we do not replace bumpers, for they are usually damaged by random impacts and not wear. For other components the wear effects may provide unambiguous signs that the time of failure is approaching. These outwardly visible indications signal when the component is nearly worn out and replacement is required; the squeaking of brakes, for example, is often an indication that the linings should be replaced. For other components wear may be the primary mode of failure, but no outward signs may herald approaching failure. Brush wear in an automobile's alternator is one of many component failures that give no sign of what is happening.

One of the challenges of design is to incorporate monitoring systems that give clear indications of wear. Although the capital cost of the equipment may be increased, it will no longer be necessary to replace parts at fixed intervals, some of which may be capable of much longer operation. Wear monitoring is the most essential in life support systems in which component failure could have catastrophic results. There are, however, many situations in which failures caused by aging cannot be anticipated through monitoring. Consideration then must be given to how often parts should be replaced. It is these situations with which we next deal.

Age Replacement

Consider the following part replacement policy: The part is replaced after it has been in operation for time T. Note that with this so-called age replacement policy, if a part fails at some time t_f, it is replaced and the next replacement does not take place until $t_f + T$ or at the time of the next failure, whichever comes first.

To determine the cost of age replacement, we denote the cost of replacing the unfailed part during preventive maintenance as C_p. The cost of part failure and consequent replacement is C_f. In all situations $C_f > C_p$, for the failure itself will have significant economic consequences beyond the cost of part replacement. The costs of wornout brushes in an automobile's alternator might include towing, lost work time, temporary lodging, and so on. The operational cost of the component over a span of time is then

$$C = N_f C_f + N_m C_c, \tag{8.30}$$

where N_f and N_m are, respectively, the expected number of failures and the number of replacements of unfailed parts during t. To evaluate N_f and N_m, we first evaluate the total number of replacements,

$$N = N_f + N_m, \tag{8.31}$$

approximately as follows.

Suppose that the total time of operation is very long compared to the mean time between replacements. Thus we assume that N is large. We may then modify Eq. 4.106 to estimate N:

$$N \approx \frac{t}{\text{MTBR}}. \tag{8.32}$$

Note, however, that the mean time between replacement (MTBR) is not the same as MTBF, for replacement occurs at failure or at age T, whichever comes first.

To calculate MTBR, we define $R'(t)$ as the probability that a part will operate for a time greater than t without replacement. Since the part is replaced automatically at time T, we have

$$R'(t) = \begin{cases} R(t), & t \le T, \\ 0, & t > T, \end{cases} \tag{8.33}$$

where $R(t)$ is the part reliability. With this definition we may determine the MTBR from Eq. 4.22 but with $R(t)$ replaced by $R'(t)$:

$$\text{MTBR} = \int_0^\infty R'(t)\, dt = \int_0^T R(t)\, dt. \tag{8.34}$$

Therefore, the number of replacements over a long time span is

$$N = \frac{t}{\displaystyle\int_0^T R(t)\, dt}. \tag{8.35}$$

Since only a fraction $R(T)$ of the parts will survive to the next preventive maintenance, the number of unfailed parts replaced will be

$$N_m = R(T)N = \frac{tR(T)}{\displaystyle\int_0^T R(t)\, dt}. \tag{8.36}$$

Moreover, since a fraction $1 - R(T)$ will fail, we have

$$N_f = [1 - R(T)]N = \frac{t[1 - R(T)]}{\displaystyle\int_0^T R(t)\, dt}. \tag{8.37}$$

With these two quantities determined, we can use Eq. 8.30 to express the cost as

$$C(T) = \frac{t[1 - R(T)]}{\displaystyle\int_0^T R(t)\, dt} C_f + \frac{tR(T)}{\displaystyle\int_0^T R(t)\, dt} C_c. \tag{8.38}$$

We may now face the question of choosing an age replacement interval T that will minimize the cost. From elementary calculus this is seen to be identical to finding the T that satisfies the condition

$$\frac{d}{dT} C(T) = 0. \tag{8.39}$$

If we perform this operation on Eq. 8.38 and use the definition of the failure rate $\lambda(t)$, we obtain, after some algebra,

$$R(T) + \lambda(T) \int_0^T R(t)\, dt - 1 = \frac{C_m}{C_f - C_m}. \tag{8.40}$$

For many models of the failure rate this is a transcendental equation that must be solved numerically if the optimum T is to be estimated. Here we consider only the widely used two-parameter Weibull distribution, for with it we can obtain some relatively elementary results. From Eqs. 4.55 and 4.58 we have

$$R(t) = e^{-(t/\theta)^m}; \qquad \lambda(t) = \frac{m}{\theta}\left(\frac{t}{\theta}\right)^{m-1}, \tag{8.41}$$

where for increasing failure rates we must have $m > 1$. Therefore, Eq. 8.40 becomes

$$e^{-(T/\theta)^m} + \frac{m}{\theta}\left(\frac{T}{\theta}\right)^{m-1} \int_0^T e^{-(t/\theta)^m}\, dt - 1 = \frac{C_m}{C_f - C_m}. \tag{8.42}$$

Suppose that we consider the case $C_f >> C_m$, where the cost of failure is much greater than that of preventive replacement. As we might expect, frequent preventive maintenance can then be shown to be justified. This leads to the condition $T << \theta$, thus allowing us to expand the reliability as

$$R(t) = 1 - \left(\frac{T}{\theta}\right)^m + \cdots \tag{8.43}$$

on the left-hand side of Eq. 8.42. We obtain

$$1 - \left(\frac{T}{\theta}\right)^m + \cdots + \frac{m}{\theta}\left(\frac{T}{\theta}\right)^{m-1}\left[T - \frac{1}{m+1}\left(\frac{T}{\theta}\right)^{m+1} + \cdots\right] - 1 \approx \frac{C_m}{C_f}. \tag{8.44}$$

Retaining only terms of the order T^m, we have

$$(m - 1)\left(\frac{T}{\theta}\right)^m \approx \frac{C_m}{C_f} \tag{8.45}$$

or

$$T \approx \theta\left[\frac{1}{m-1}\frac{C_m}{C_f}\right]^{1/m}. \tag{8.46}$$

To illustrate, suppose that we take $C_f = 50\, C_m$; then we obtain for different values of m:

m	1.5	2.0	2.5	3.0	3.5	4.0
T/θ	0.117	0.141	0.177	0.215	0.252	0.286
$R(T/\theta)$	0.961	0.980	0.987	0.990	0.992	0.993

As indicated by the reliability, this corresponds to a very small fraction of the parts that fail before replacement.

Note that for $m \leq 1$ there is no optimal replacement age. This is to be expected, since $m = 1$ corresponds to random failures, and $m < 1$ to decreasing failure rates. To repeat, replacement of parts when the rate of failure is decreasing through early failures will then increase the number of failures.

Batch Replacement

There are alternatives to age replacement. One of the most common of these is batch replacement. With batch replacement the component is replaced at times T, $2T$, $3T$, . . ., *and* whenever a failure occurs. Batch replacement differs from age replacement in that there is no need to keep track of the times of the preceding replacements. Thus, when we are faced with maintaining large numbers of relatively inexpensive parts, it is likely to be the preferred alternative. But for fewer more valuable parts the record keeping of age replacement may be justified. For example, bulbs in a large number of streetlights are more likely to be replaced in batches; turbine blades on a large turbogenerator will be replaced when they have acquired a specified age.

There are other differences between age and batch replacement. They become obvious if we compare the two methods applied to the same part with the same replacement interval T. Generally, batch replacement will remove more unfailed parts than age replacement. After a part has failed and been replaced, the new part will itself be removed at the next batch replacement, before it has acquired an age T. Conversely, there will be fewer failures with batch replacement, since some of the parts will be replaced before time T.

The cost of batch replacement over a long period of time can again be calculated from Eq. 8.30, where we have $N_m = t/T$. The number of failures is given by $N_f = \overline{N}(T)/T$, where $\overline{N}(T)$ is the expected number of failures between batch replacements. This can be calculated from the Poisson distribution derived in Chapter 4 for constant failure rates. For the time-dependent failure rates the calculation of $\overline{N}(T)$ is a more prodigious task, requiring concepts from renewal theory* that are beyond the scope of this book. This being the case, the estimate of the optimal replacement interval T may involve extra numerical computations even for the Weibull or other standard time-to-failure distributions. Suffice it to state that since for batch replacements N_f will be smaller and N_m larger than for age replacement, the optimum for $C_f > C_m$ will occur for somewhat larger values of T.

8.4 CORRECTIVE MAINTENANCE

With or without preventive maintenance, the definition of reliability has been central to all our deliberations. This is no longer the case, however, when we consider the many classes of systems in which corrective mainte-

* R. E. Barlow, and F. Proschan, *Mathematical Theory of Reliability,* Wiley, New York, 1965.

nance plays a substantial role. Now we are interested not only in the probability of failure, but also in the number of failures and, in particular, in the times required to make repairs. For such considerations two new reliability parameters become the focus of attention. Availability is the probability that a system is available for use at a given time. Roughly, it may be viewed as a fraction of time that a system is in an operational state. Maintainability is a measure of how fast a system may be repaired following failure. Both availability and maintainability, however, require more formal definitions if they are to serve as a quantitative basis for the analysis of repairable systems.

Availability

For repairable systems a fundamental quantity of interest is the availability. It is defined as follows:

$$A(t) = \text{probability that a system is performing satisfactorily at time } t. \tag{8.47}$$

This is referred to as the point availability. Often it is necessary to determine the interval or mission availability. The interval availability is defined by

$$A^*(T) = \frac{1}{T}\int_0^T A(t)\,dt. \tag{8.48}$$

It is just the value of the point availability averaged over some interval of time, T. This interval may be the design life of the system or the time to accomplish some particular mission. Finally, it is often found that after some initial transient effects the point availability assumes a time-independent value. In these cases the steady-state or asymptotic availability is defined as

$$A^*(\infty) = \lim_{T\to\infty} \frac{1}{T}\int_0^T A(t)\,dt. \tag{8.49}$$

If a system or its components cannot be repaired, the point availability is just equal to the reliability. The probability that it is available at t is just equal to the probability that it has not failed between 0 and t:

$$A(t) = R(t). \tag{8.50}$$

Combining Eqs. 8.48 and 8.50, we obtain

$$A^*(T) = \frac{1}{T}\int_0^T R(t)\,dt. \tag{8.51}$$

Thus, as T goes to infinity, the numerator, according to Eq. 4.22, becomes the MTTF, a finite quantity. The denominator, T, however, becomes infinite. Thus the steady-state availability of a nonrepairable system is

$$A^*(\infty) = 0. \tag{8.52}$$

Since all systems eventually fail, and there is no repair, the availability averaged over an infinitely long time span is zero.

PROBLEM 8.3

A nonrepariable system has a known MTTF and is characterized by a constant failure rate. The system mission availability must be 0.95. Find the maximum design life that can be tolerated in terms of the MTTF.

Solution For a constant failure rate the reliability is $R = e^{-\lambda t}$. Insert this into Eq. 8.51 to obtain

$$A^*(T) = \frac{1}{\lambda T}(1 - e^{-\lambda T}).$$

Expanding the exponential then yields

$$A(T) = \frac{1}{\lambda t}(1 - 1 + \lambda T - \tfrac{1}{2}(\lambda T)^2 + \cdots).$$

Thus $A(T) \approx 1 - \frac{1}{2}\lambda T$, for $\lambda T << 1$ or $0.95 = 1 - \frac{1}{2}\lambda T$. Then $\lambda T = 0.1$, but MTTF $= 1/\lambda$. Therefore, $T = 0.1 \times$ MTTF. ◀

Maintainability

We may now proceed to the quantitative description of repair processes and the definition of maintainability. Suppose that we let $\mathbf{t}$ be the time required to repair a system, measured from the time of failure. If all repairs take the same length of time, $\mathbf{t}$ is just a number, say $\mathbf{t} = \tau$. In reality, repairs require different lengths of time, and even the time to perform a given repair is uncertain because circumstances, skill level, and a host of other factors vary. Therefore $\mathbf{t}$ is normally not a constant but rather a random variable. This variable can be considered in terms of distribution functions as follows.

Suppose that we define the PDF for repair as

$$m(t)\,\Delta t = P\{t \le \mathbf{t} \le t + \Delta t\}. \tag{8.53}$$

That is, $m(t)\,\Delta t$ is the probability that repair will require a time between t and $t + \Delta t$. The CDF corresponding to Eq. 8.53 is defined as the maintainability

$$M(t) = \int_0^t m(t')\,dt', \tag{8.54}$$

and the mean time to repair or MTTR is then

$$\text{MTTR} = \int_0^\infty t m(t)\,dt. \tag{8.55}$$

Analogous to the derivations of the failure rate given in Chapter 4, we may define the instantaneous repair rate as

$$\nu(t)\,\Delta t = \frac{P\{t \le \mathbf{t} \le t + \Delta t\}}{P\{\mathbf{t} > t\}}; \tag{8.56}$$

$\nu(t)\,\Delta t$ is the conditional probability that the system will be repaired between t and $t + \Delta t$, given that it is failed at t. Noting that

$$M(t) = P\{\mathbf{t} \le t\} = 1 - P\{\mathbf{t} \ge t\}, \tag{8.57}$$

we then have

$$\nu(t) = \frac{m(t)}{1 - M(t)}. \tag{8.58}$$

Equations 8.54 and 8.58 may be used to express the maintainability and the PDF in terms of the repair rate. To do this, we differentiate Eq. 8.54 to obtain

$$m(t) = \frac{d}{dt} M(t), \tag{8.59}$$

and combine this result with Eq. 8.58 to yield

$$\nu(t) = [1 - M(t)]^{-1} \frac{d}{dt} M(t). \tag{8.60}$$

Moving dt to the left and integrating between 0 and t, we obtain

$$\int_0^t \nu(t')\, dt' = \int_0^{M(t)} \frac{dM}{1 - M}. \tag{8.61}$$

Evaluating the integral on the right-hand side and solving for the maintainability, we have

$$M(t) = 1 - \exp\left[-\int_0^t \nu(t')\, dt' \right]. \tag{8.62}$$

Finally, we may use Eq. 8.59 to express the PDF for repair times as

$$m(t) = \nu(t) \exp\left[-\int_0^t \nu(t')\, dt' \right]. \tag{8.63}$$

A great many factors go into determining both the mean time to repair and the PDF, $m(t)$, by which the uncertainties in repair time are characterized. These factors range from the ability to diagnose the cause of failure, on the one hand, to the availability of equipment and skilled personnel to carry out the repair procedures on the other. The determining factors in estimating repair time vary greatly with the type of system that is under consideration. This may be illustrated with the following comparison.

In many mechanical systems the causes of the failure are likely to be quite obvious. If a pipe ruptures, a valve fails to open, or a pump stops running, the diagnoses of the component in which the mechanical failure has occurred may be straightforward. The primary time entailed in the repair is then determined by how much time is required to extract the component from the system and install the new component, for each of these processes may involve a good deal of metal cutting, welding, or other time-consuming procedures.

In contrast, if a computer fails, maintenance personnel may spend most of the repair procedure time in diagnosing the problem, for it may take considerable effort to understand the nature of the failure well enough to be able to locate the circuit board, chip, or other component that is the cause. Conversely, it may be a rather straightforward procedure to replace the faulty component once it has been located.

In both of these examples we have assumed that the necessary repair parts are available at the time they are needed and that it is obvious how much of the system should be replaced to eliminate the fault. In fact, both the availability of parts and the level of repair involve subtle economic trade-offs between the cost of inventory, personnel, and system downtime.

For example, suppose that the pump fails because bearings have burned out. We must decide whether it is faster to remove the pump from the line and replace it with a new unit or to tear it down and replace only the bearings. If the entire pump is to be replaced, on-site inventories of spare pumps will probably be necessary, but the level of skill needed by repair personnel to install the new unit may not be great. Conversely, if most of the pump failures are caused by bearing failures, it may make sense to stock only bearings on site and to repack the bearings. But here repair personnel will need different and perhaps greater training and skill. Such trade-offs are typical of the many factors that must be considered in maintainability engineering, the discipline that optimizes $M(t)$ at a high level with as low a cost as possible.

8.5 REPAIR: REVEALED FAILURES

In this section we examine systems for which the failures are revealed, so that repairs can be immediately initiated. In these situations two quantities are of primary interest, the number of failures over a given span of time and the system availability. The number of failures is needed in order to calculate a variety of quantities including the cost of repair, the necessary repair parts inventory, and so on. Provided that the MTTR is much smaller than the MTTF, reasonable estimates for the number of failures can be obtained using the Poisson distribution as in Chapter 4, and neglecting the system downtime for repair. For availability calculations, repair time must be considered or else we would obtain simply $A(t) = 1$. Ordinarily, this is not an acceptable approximation, for even small values of the unavailability $\tilde{A}(t)$ are frequently important, whether they be due to the risk incurred through the unavailability of a critical safety system or to the production loss during the downtimes of an assembly line.

In what follows two models for repair are developed to estimate the availability of a system, constant repair rate and constant repair time. It will be clear from comparing these that most of the more important results depend primarily on the MTTR, not on the details of the repair distribution.

Constant Repair Rates

To calculate availability, we must take the repair rate into account, even though it may be large compared to the failure rate. We assume that the distribution of times to repair can be characterized by a constant repair rate

$$\nu(t) = \nu. \tag{8.64}$$

The PDF of times to repair is then exponential,

$$m(t) = \nu e^{-\nu t}, \tag{8.65}$$

and the mean time to repair is simply

$$\text{MTTR} = 1/\nu. \tag{8.66}$$

Although the exponential distribution may not reflect the details of the distribution very accurately, it provides a reasonable approximation for predicting availabilities, for these tend to depend more on the MTTR than on the details of the distribution. As we shall illustrate, even when the PDF of the repair is bunched about the MTTR rather than being exponentially distributed, the constant repair rate model correctly predicts the asymptotic availability.

Suppose that we consider a two-state system; it is either operational, state 1, or it is failed, state 2. Then $A(t)$ and $\tilde{A}(t)$, the availability and unavailability, are the probabilities that the state is operational or failed, respectively, at time t, where t is measured from the time at which the system operation commences. We therefore have the initial conditions $A(0) = 1$ and $\tilde{A}(0) = 0$, and of course,

$$A(t) + \tilde{A}(t) = 1. \tag{8.67}$$

A differential equation for the availability may be derived in a manner similar to that used for the Poisson distribution in Chapter 4. We consider the change in $A(t)$ between t and $t + \Delta t$. There are two contributions. Since $\lambda\,\Delta t$ is the conditional probability of failure during Δt, given that the system is available at t, the loss of availability during Δt is $\lambda\,\Delta t\,A(t)$. Similarly, the gain in availability is equal to $\nu\ \Delta t\ \tilde{A}(t)$, where $\nu\ \Delta t$ is the conditional probability that the system is repaired during Δt, given that it is unavailable at t. Hence it follows that

$$A(t + \Delta t) = A(t) - \lambda\ \Delta t\ A(t) + \nu\ \Delta t\ \tilde{A}(t). \tag{8.68}$$

Rearranging terms and eliminating $\tilde{A}(t)$ with Eq. 8.67, we obtain

$$\frac{A(t + \Delta t) - A(t)}{\Delta t} = -(\lambda + \nu)A(t) + \nu. \tag{8.69}$$

Since the expression on the left-hand side is just the derivative with respect to time, Eq. 8.69 may be written as the differential equation,

$$\frac{d}{dt} A(t) = -(\lambda + \nu)A(t) + \nu. \tag{8.70}$$

We now may use an integrating factor of $e^{\lambda+\nu}$, along with the initial condition $A(0) = 1$ to obtain

$$A(t) = \frac{\nu}{\lambda + \nu} + \frac{\lambda}{\lambda + \nu} e^{-(\lambda+\nu)t}. \tag{8.71}$$

Note that the availability begins at $A(0) = 1$ and decreases monotonically to an asymptotic value $1/(1 + \lambda/\nu)$, which depends only on the ratio of failure to repair rate. The interval availability may be obtained by inserting Eq. 8.71 into Eq 8.48 to yield

$$A^*(T) = \frac{\nu}{\lambda + \nu} + \frac{\lambda}{(\lambda + \nu)^2 T}[1 - e^{-(\lambda+\nu)T}], \tag{8.72}$$

and the asymptotic availability is obtained by letting T go to infinity. Thus

$$A^*(\infty) = \frac{\nu}{\lambda + \nu}. \tag{8.73}$$

Finally, note from Eqs. 8.71 and 8.73 that for constant repair rates

$$A^*(\infty) = A(\infty). \tag{8.74}$$

Since, in most instances, repair rates are much larger than failure rates, a frequently used approximation comes from expanding Eq. 8.73 and deleting higher terms in λ/ν. We obtain after some algebra

$$A^*(\infty) \simeq 1 - \lambda/\nu. \tag{8.75}$$

The ratio in Eq. 8.73 may be expressed in terms of the mean time between failures and the mean time to repair. Since MTTF = $1/\lambda$ and MTTR = $1/\nu$, we have

$$A(\infty) = \frac{\text{MTTF}}{\text{MTTF} + \text{MTTR}}. \tag{8.76}$$

This expression is sometimes used for the availability even though neither failure or repair is characterized well by the exponential distribution. This is often quite adequate, for, in general, when availability is averaged over a reasonable period T of time, it is insensitive to the details of the failure or repair distributions. This is indicated for constant repair times in the following section.

EXAMPLE 8.4

In the following table are times (in days) over a 6-month period at which failure of a production line occurred (t_f) and times (t_r) at which the plant was brought back on line following repair.

i	t_{fi}	t_{ri}	i	t_{fi}	t_{ri}
1	12.8	13.0	6	56.4	57.3
2	14.2	14.8	7	62.7	62.8
3	25.4	25.8	8	131.2	134.9
4	31.4	33.3	9	146.7	150.0
5	35.3	35.6	10	177.0	177.1

(*a*) Calculate the 6-month-interval availability from the plant data.
(*b*) Estimate MTTF and MTTR from the data.
(*c*) Estimate the interval availability using the results of part *b* and Eq. 8.67, and compare this result to that of part *a*.

Solution During the 6 months (182.5 days) there are 10 failures and repairs. (*a*) From the data $\tilde{A}(T)$ is just the fraction of that time for which the system is inoperable. Thus

$$\tilde{A}(T) = \frac{1}{T}\sum_{i=1}^{10}(t_{ri} - t_{fi})$$

$$= \frac{1}{182.5}(0.2 + 0.6 + 0.4 + 1.9 + 0.3 + 0.9 + 0.1 + 3.7 + 3.3 + 0.1)$$

$$\tilde{A}(T) = 0.0630$$

$$A(T) = 1 - 0.063 = 0.937. \blacktriangleleft$$

(*b*) Taking $t_{r0} = 0$, we first estimate the MTTF and MTTR from the data:

$$\text{MTTF} = \frac{1}{N}\sum_{i=1}^{10}(t_{fi} - t_{ri-1})$$

$$= \tfrac{1}{10}(12.8 + 1.2 + 10.6 + 5.6 + 2.0 + 20.8 + 5.4 + 68.4 + 11.8 + 27.0)$$

$$\text{MTTF} = \tfrac{1}{10}\,165.6 = 16.56. \blacktriangleleft$$

$$\text{MTTR} = \frac{1}{N}\sum_{i=1}^{10}(t_{ri} - t_{fi}) = \frac{T}{10}\frac{1}{T}\sum_{i=1}^{10}(t_{fi} - t_{ri}) = \frac{182.5}{10}\tilde{A}(T)$$

$$= 1.15 \text{ days.} \blacktriangleleft$$

(*c*) $A(T) = \dfrac{\nu}{\nu + \lambda} = \dfrac{1}{1 + \dfrac{\text{MTTR}}{\text{MTTF}}} = \dfrac{1}{1 + \dfrac{0.85}{16.5}} = 0.935.$ ◀

Constant Repair Times

In the foregoing availability model we have used a constant repair rate, as we shall also do throughout much of the remainder of this chapter. Before proceeding, however, we repeat the calculation of the system availability using a repair model that is quite different; all the repairs are assumed to require exactly the same time, τ. Thus the PDF for time to repair has the form

$$m(t) = \delta(t - \tau), \tag{8.77}$$

where δ is the Dirac delta function discussed in Chapter 3. Although the availability is more difficult to calculate with this model, the result is instructive. It will be seen that whereas the details of the time dependence of $A(t)$ differ, the general trends are the same, and the asymptotic value is still given by Eq. 8.76.

A differential equation may be obtained for the availability, with the initial condition $A(0) = 1$. Since all repairs require a time τ, there are no repairs for $t < \tau$. Thus instead of Eq. 8.68, we have only the failure term on the right-hand side,

$$A(t + \Delta t) = A(t) - \lambda\, \Delta t\, A(t), \qquad 0 \leq t \leq \tau, \tag{8.78}$$

which corresponds to the differential equation

$$\frac{d}{dt} A(t) = -\lambda A(t), \qquad 0 \leq t \leq \tau. \tag{8.79}$$

For times greater than τ, repairs are also made; the number of repairs made during Δt is just equal to the number of failures during Δt at a time τ earlier: $\lambda\, \Delta t\, A(t - \tau)$. Thus the change in availability during Δt is

$$A(t + \Delta t) = A(t) - \lambda\, \Delta t\, A(t) + \lambda\, \Delta t\, A(t - \tau), \qquad t > \tau, \tag{8.80}$$

which corresponds to the differential equation

$$\frac{d}{dt} A(t) = -\lambda A(t) + \lambda A(t - \tau), \qquad t > \tau. \tag{8.81}$$

Equations 8.80 and 8.81 are more difficult to solve than those for the constant repair rate. During the first interval, $0 \leq t \leq \tau$, we have simply

$$A(t) = e^{-\lambda t}, \qquad 0 \leq t \leq \tau. \tag{8.82}$$

For $t > \tau$, the solution in successive intervals depends on that of the preceding interval. To illustrate, consider the interval $N\tau \leq t \leq (N + 1)\tau$. Applying an integrating factor $e^{\lambda t}$ to Eq. 8.81, we may solve for $A(t)$ in terms of $A(t - \tau)$:

$$A(t) = A(N\tau)e^{-\lambda(t-N\tau)} + \int_{N\tau}^{t} dt'\, \lambda e^{-\lambda(t-t')}A(t' - \tau), \; N\tau \le t \le (N + 1)\tau. \tag{8.83}$$

For $N = 1$, we may insert Eq. 8.82 on the right-hand side to obtain

$$A(t) = e^{-\lambda t} + \lambda(t - \tau)e^{-\lambda(t-\tau)}, \qquad \tau \le t \le 2\tau. \tag{8.84}$$

For $N = 2$ there will be three terms on the right-hand side, and so on. The general solution for arbitrary N appears quite similar to the Poisson distribution:

$$A(t) = \sum_{n=0}^{N} \frac{[\lambda(t - n\tau)]^n}{n!} e^{-\lambda(t-n\tau)}, \qquad N\tau \le t \le (N + 1)\tau. \tag{8.85}$$

The solutions for the constant repair rate and the constant repair time models are plotted for the point availability $A(t)$ in Fig. 8.5 for $\tau = 1/\nu$. Note that the discrete repair time leads to breaks in the slope of the availability curve, whereas this is not the case with the constant failure rate model. However, both curves follow the same general trend downward and converge to the same asymptotic value. Thus, if we are interested only in the general characteristics of availability curves, which ordinarily is the case, the constant repair rate model is quite adequate, even though some of the structure carried by a more precise evaluation of the repair time PDF may be lost. Moreover, to an even greater extent than with failure rates, not enough data are available in most cases to say much about the spread of repair times about the MTTR. Therefore, the single-parameter exponential distribution may be all that can be justified, and Eq. 8.76 provides a reasonable estimate of the availability.

8.6 TESTING AND REPAIR: UNREVEALED FAILURES

As long as system failures are revealed immediately, the time to repair is the primary factor in determining the system availability. When a system is not in continuous operation, however, failures may occur but remain undiscovered. This problem is most pronounced in backup or other emergency equipment that is operated only rarely, or in stockpiles of repair parts or other materials that may deteriorate with time. The primary loss of availability then may be due to failures in the standby mode that are not detected until an attempt is made to use the system.

A primary weapon against these classes of failures is periodic testing. As we shall see, the more frequently testing is carried out, the more failures will be detected and repaired soon after they occur. However, this must be weighed against the expense of frequent testing, the loss of availability

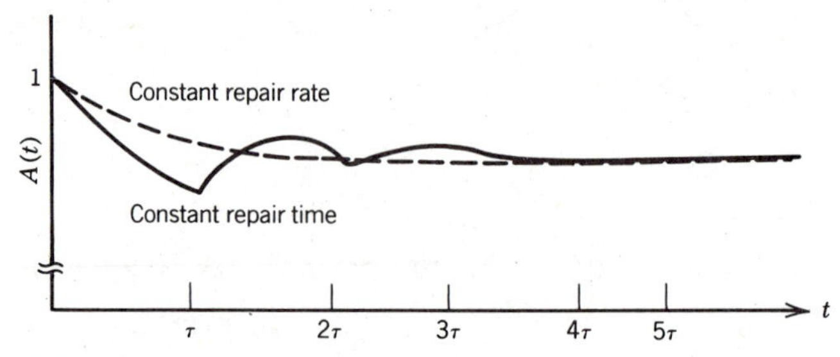

FIGURE 8.5 Availability for different repair models.

through downtime for testing, and the possibility of excessive component wear from too-frequent testing.

Idealized Periodic Tests

Suppose that we first consider the effect of a simple periodic test on a system whose reliability can be characterized by a constant failure rate:

$$R(t) = e^{-\lambda t}. \tag{8.86}$$

The first thing that should be clear is that system testing has no positive effect on reliability. For unlike preventive maintenance the test will only catch failures after they occur. Testing may at the same time increase wear on some systems. Therefore, if carried out too frequently, tests may cause deterioration of the reliability.

Testing, however, has a very definite positive effect on availability. To see this in the simplest case, suppose that we perform a system test at time interval T_0. In addition, we make the following three assumptions: (1) The time required to perform the test is negligible, (2) the time to perform repairs is negligible, and (3) the repairs are carried out perfectly and restore the system to an as-good-as-new condition. Later, we shall examine the effects of relaxing these assumptions.

Suppose that we test a system with reliability given by Eq. 8.86 at time interval T_0. As indicated, if there is no repair, the availability is equal to the reliability. Thus, before the first test,

$$A(t) = R(t), \qquad 0 \leq t < T_0. \tag{8.87}$$

Since the system is repaired perfectly and restored to an as-good-as-new state at $t = T_0$, we will have $R(T_0) = 1$. Then since there is no repair between T_0 and $2T_0$, the availability will again be equal to the reliability, but now the reliability is evaluated at $t - T_0$:

$$A(t) = R(t - T_0), \qquad T_0 \leq t < 2T_0. \tag{8.88}$$

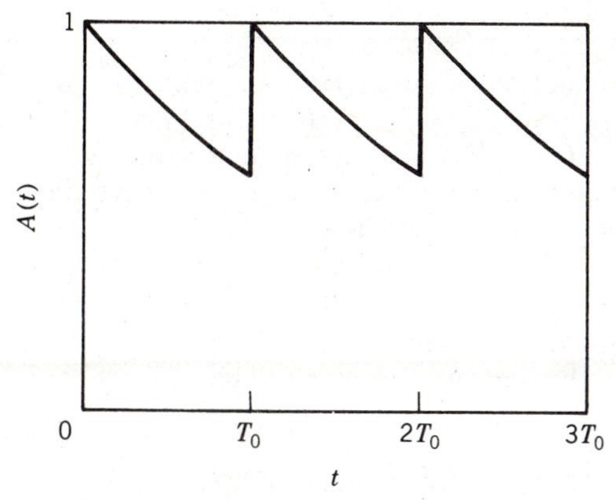

FIGURE 8.6 Availability with idealized periodic testing for unrevealed failures.

This pattern repeats itself as indicated in Fig. 8.6. The general expression is

$$A(t) = R(t - NT_0), \qquad NT_0 \leq t < (N + 1)T_0. \tag{8.89}$$

For the situation indicated in Fig. 8.6, the interval and the asymptotic availability have the same value, provided that the integral in Eq. 8.48 is taken over a multiple of T_0, say mT_0. We have

$$A^*(mT_0) = \frac{1}{mT_0}\int_0^{mT_0} A(t)\, dt = \frac{1}{T_0}\int_0^{T_0} A(t)\, dt. \tag{8.90}$$

Since the interval availability is independent of the number of intervals over which $A^*(T)$ is calculated, so will the asymptotic availability $A^*(\infty)$:

$$A^*(\infty) = \lim_{m\to\infty} \frac{1}{mT_0}\int_0^{mT_0} A(t)\, dt = \frac{1}{T_0}\int_0^{T_0} A(t)\, dt. \tag{8.91}$$

The effect of the testing interval on availability may be seen by combining Eqs. 8.86 and 8.91. We obtain

$$A^*(\infty) = \frac{1}{\lambda T_0}(1 - e^{-\lambda T_0}). \tag{8.92}$$

Ordinarily, the test interval would be small compared to the MTTF: $\lambda T_0 << 1$. Therefore, the exponential may be expanded, and only the leading terms are retained to make the approximation

$$A^*(\infty) \simeq 1 - \tfrac{1}{2}\lambda T_0. \tag{8.93}$$

EXAMPLE 8.5

Annual inspection and repair are carried out on a large group of smoke detectors of the same design in public buildings. It is found that 15% of the smoke detectors are not functional. If it is assumed that the failure rate is constant,
(*a*) In what fraction of fires will the detectors offer protection?
(*b*) If the smoke detectors are required to offer protection for at least 99% of fires, how frequently must inspection and repair be carried out?

Solution With inspection and repair at interval T_0, the fraction of detectors that are operational at the time of inspection will be

$$R = e^{-\lambda T_0} = 0.85.$$

Then $\lambda T_0 = -\ln(0.85) = 0.162$. Since $T_0 = 1$ year, $\lambda = 0.162/\text{year}$.
(*a*) If we assume that the fires are uniformly distributed in time, the fractional protection is just equal to the interval availability; from Eq. 8.92,

$$A^*(\infty) = \frac{1}{\lambda T_0}(1 - e^{-\lambda T_0}) = \frac{1}{0.162}(1 - 0.85) = 0.926. \blacktriangleleft$$

(*b*) For this high availability the rare-event approximation, Eq. 8.93, may be used:

$$0.99 = A^*(\infty) \approx 1 - \tfrac{1}{2}\lambda T_0.$$

Thus from Eq. 8.93,

$$T_0 = \frac{2[1 - A^*(\infty)]}{\lambda} = \frac{2(1 - 0.99)}{0.162} = 0.123 \text{ year}$$
$$= 0.123 \times 12 \text{ months} \approx 1\tfrac{1}{2} \text{ months.} \blacktriangleleft$$

Real Periodic Tests

Equation 8.93 indicates that we may achieve availabilities as close to one as desired merely by decreasing the test interval T_0. This is not the case, however, for as the test interval becomes smaller, a number of other factors—test time, repair time, and imperfect repairs—become more important in estimating availability.

When we examine these effects, it is useful to visualize them as modifications in the curve shown in Fig. 8.6. The interval or asymptotic availability may be pictured as proportional to the area under the curve within one test interval, divided by T. Thus we may view each of the factors listed earlier in terms of the increase or decrease that it causes in the area under the curve. In particular, with reasonable assumptions about the ratios of the various parameters involved, we may derive approximate expressions similar to Eq. 8.93 that are quite simple, but at the same time are not greatly in error.

Consider first the effect of a nonnegligible test time, t_t. During the test we assume that the system must be taken off line, and the system has an availability of zero during the test. The point availability will then appear as the solid line in Fig. 8.7. Provided that we again assume that $\lambda T_0 << 1$, so that Eq. 8.93 holds, and that $t_t << T_0$, the test time, is small compared to the test interval, we may approximate the contribution of the test to system downtime as t_t/T_0. The availability indicated in Eq. 8.93 is therefore decreased to

$$A^*(\infty) \simeq 1 - \tfrac{1}{2}\lambda T_0 - \frac{t_t}{T_0}. \tag{8.94}$$

We next consider the effect of a nonzero time to repair on the availability. The probability of finding a failed system at the time of testing is just one minus the point availability at the time the test is carried out. For small T_0 this probability may be shown to be approximately λT_0. Since $1/\nu$ is the mean time to repair, the contribution to the unavailability over the period T_0 is $\lambda T_0/\nu$, or dividing by the interval T_0, we find, as in Eq. 8.75, the loss of availability to be approximately λ/ν. We may therefore modify our availability by subtracting this term to yield

$$A^*(\infty) \simeq 1 - \tfrac{1}{2}\lambda T_0 - \frac{t_t}{T_0} - \frac{\lambda}{\nu}. \tag{8.95}$$

The effect of this contribution to the system unavailability is indicated by the dotted line in Fig. 8.7.

Examination of Eq. 8.95 is instructive. Clearly, decreases in failure rate and in test time t_t increase the availability, as do increases in the repair rate ν. It may also be shown that the more perfect the repair, the higher the availability. Decreasing the test interval, however, may either increase or decrease the availability, depending on the value of the other parameters. For, as indicated in Eq. 8.95, it appears in both the numerator and the denominator of terms.

Suppose that we differentiate Eq. 8.95 with respect to T_0 and set the

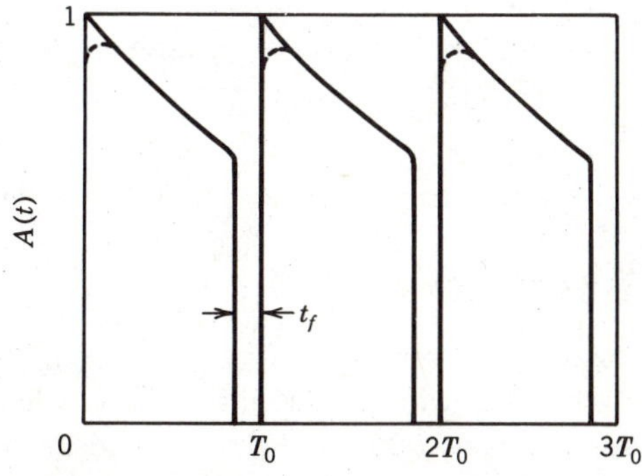

FIGURE 8.7 Availability with realistic periodic testing for unrevealed failures.

result equal to zero in order to determine the maximum availability:

$$\frac{d}{dT_0} A^*(\infty) = -\tfrac{1}{2}\lambda + \frac{t_t}{T_0^2} = 0. \tag{8.96}$$

The optimal test interval is then

$$T_0 = \left(\frac{2t_t}{\lambda}\right)^{1/2}. \tag{8.97}$$

Substitution of this expression back into Eq. 8.95 yields a maximum availability of

$$A^*(\infty) = 1 - (2\lambda t_t)^{1/2} - \frac{\lambda}{\nu}. \tag{8.98}$$

If the test interval is longer than Eq. 8.97, undetected failures will lower availability. However, if a shorter test interval is employed, the loss of availability during testing will not be fully compensated for by earlier detection of failures. As indicated in Eq. 8.97, the test interval should increase as the failure rate decreases, and decrease as the testing time can be decreased. Other trade-offs may need to be considered as well. For example, will hurrying to decrease the test time increase the probability that failures will be missed?

EXAMPLE 8.6

A sulfur dioxide scrubber is known to have a MTBF of 137 days. Testing the scrubber requires half a day, and the mean time to repair is 4 days. (*a*) Choose the test period to maximize the availability. (*b*) What is the maximum availability?

Solution (*a*) From Eq. 8.97, with MTBF $= 1/\lambda$,

$$T_0 = (2t_t \,\mathrm{MTBF})^{1/2} = (2 \times 0.5 \times 137)^{1/2} = 11.7 \text{ days.} \blacktriangleleft$$

(*b*) From Eq. 8.98,

$$A^*(\infty) = 1 - \left(\frac{2t_t}{\mathrm{MTBF}}\right)^{1/2} - \frac{\mathrm{MTTR}}{\mathrm{MTBF}},$$

$$A^*(\infty) = 1 - \left(\frac{2 \times 0.5}{137}\right)^{1/2} - \frac{4}{137} = 0.885. \blacktriangleleft$$

8.7 SYSTEM AVAILABILITY

Thus far we have examined only the effects on availability of the failure and repair of a system as a whole. But just as for reliability, it is often instructive to examine the availability of a system in terms of the component availabilities. Not only are data more likely to be available at the component level, but the analysis can provide insight into the gains made through redundant configurations, and through different testing and repair strategies.

Since availability, like reliability, is a probability, system availabilities can be determined from parallel and series combinations of component availabilities. In fact, the techniques developed in Chapter 7 for combining reliabilities are also applicable to point availabilities, but only provided that both the failure and repair rates for the components are independent of one another. If this is not the case, either the β-factor method described in Chapter 7 or the Markov methods discussed in the following chapter may be required to model the component dependencies. In this chapter we consider situations in which the component properties are independent of one another, deferring analysis of component dependencies to the following chapter.

In what follows we estimate point availabilities of systems in terms of components. The appropriate integral is then taken to obtain interval and asymptotic availabilities. When the component availabilities become time-independent after a long period of operation, steady-state availabilities may be calculated simply by letting $t \rightarrow \infty$ in the point availabilities. In testing or other situations in which there is a periodicity in the point availability the point availability must be averaged over a test period, even though the system has been in operation for a substantial length of time. Very often when repair rates are much higher than failure rates, simplifying approximations, in which λ/ν is assumed to be very small, are of sufficient accuracy and lead to additional physical insight in comparing systems.

For systems without redundancy the availability obeys the product law introduced in Chapter 7. Suppose that we let $\tilde{X}$ represent the failed state of the system, and X the unfailed or operational state of the system. Similarly, let $\tilde{X}_i$ represent the failed state of component i, and X_i the unfailed state of the same component. In a nonredundant system, all the components must be available for the system to be available:

$$X = X_1 \cap X_2 \cap \ldots \cap X_M. \tag{8.99}$$

Since the availability is defined as just the probability that the system is available, we have

$$A(t) = \prod_i A_i(t). \tag{8.100}$$

where the $A_i(t)$ are the independent component availabilities.

For redundant (i.e., parallel) systems, all the components must be unavailable if the system is to be unavailable. Thus, if $\tilde{X}$ signifies a failed system and $\tilde{X}_i$ the failed state of component i, we have

$$\tilde{X} = \tilde{X}_1 \cap \tilde{X}_2 \cap \tilde{X}_3 \cap \ldots \cap \tilde{X}_M. \tag{8.101}$$

Since the unavailability is one minus the availability, we have

$$1 - A(t) = [1 - A_1(t)][1 - A_2(t)] \ldots [1 - A_M(t)], \tag{8.102}$$

or more compactly,

$$A(t) = 1 - \prod_i [1 - A_i(t)]. \tag{8.103}$$

Comparing Eqs. 8.100 and 8.103 with Eqs. 7.1 and 7.14 indicates that the same relationships hold for point availabilities as for reliabilities. The other relationships derived in Chapter 7 also hold when the assumption that the components are mutually independent is made throughout.

Revealed Failures

Suppose that we now apply the constant repair rate model to each component. According to Eq. 8.71, the component availabilities are then

$$A_i(t) = \frac{\nu_i}{\nu_i + \lambda_i} + \frac{\lambda_i}{\nu_i + \lambda_i} e^{-(\lambda_i + \nu_i)t}. \tag{8.104}$$

This relationship may be applied in the foregoing equations to estimate system availability.

If we are interested only in asymptotic availability, we may delete the second term of Eq. 8.104 to obtain

$$A_i(\infty) = \frac{\nu_i}{\nu_i + \lambda_i}. \tag{8.105}$$

Combining this expression with Eq. 8.100, we have for a nonredundant system

$$A(\infty) = \prod_i \frac{\nu_i}{\nu_i + \lambda_i}. \tag{8.106}$$

If we further make the reasonable assumption that repair rates are large compared to failure rates, $\nu_i >> \lambda_i$, then

$$A_i(\infty) \simeq 1 - \frac{\lambda_i}{\nu_i}. \tag{8.107}$$

With this expression substituted into Eq. 8.100 to estimate the availability of a nonredundant system, we obtain

$$A(\infty) \simeq \prod_i \left(1 - \frac{\lambda_i}{\nu_i}\right). \tag{8.108}$$

But since we have already deleted higher-order terms in the ratios λ_i/ν_i, for consistency we also should eliminate them from this equation. This yields

$$A(\infty) \simeq 1 - \sum_i \frac{\lambda_i}{\nu_i}. \tag{8.109}$$

Thus the rapid deterioration of the availability with increased number of components is seen. If we further assume that all the repair rates can be replaced by an average value $\nu_i = \nu$, Eq. 8.109 becomes

$$A(\infty) \simeq 1 - \lambda/\nu, \tag{8.110}$$

where

$$\lambda = \sum_i \lambda_i. \tag{8.111}$$

Therefore, we obtain the same result as given for the system as a whole, provided that we sum the component failure rates as in Chapter 4.

The effect of redundancy may be seen by inserting Eq. 8.105 into Eq. 8.103, the availability of a parallel system. For N identical units with $\lambda_i = \lambda_1$ and $\nu_i = \nu$, we have

$$A(\infty) = 1 - \left(\frac{\lambda_1}{\lambda_1 + \nu}\right)^N. \tag{8.112}$$

If we consider the case where $\nu >> \lambda_1$, then

$$A(\infty) \simeq 1 - \left(\frac{\lambda_1}{\nu}\right)^N, \tag{8.113}$$

or correspondingly for the unavailability,

$$\tilde{A}(\infty) \simeq \left(\frac{\lambda_1}{\nu}\right)^N. \tag{8.114}$$

The analogy to Eq. 7.47 for the reliability of parallel systems is clear; both unreliability and unavailability are proportional to the Nth power of the failure rate. The foregoing relationships assume that there are no common-mode failures. If there are, the β-factor method of Chapter 7 may be adapted, putting a fictitious component in series with a failure and a repair rate for the common-mode failure. Once again the presence of common-mode failure limits the gains that can be made through the use of parallel configurations, although not as severely as for systems that cannot be repaired. Suppose we consider as an example N units in parallel, each having a failure rate λ divided into independent and common-mode failures as in Eqs. 7.25 and 7.31. We have

$$A(\infty) = \{1 - [1 - A_i(\infty)]^N\} A_c(\infty), \tag{8.115}$$

where A_i are the availabilities with only the independent failure rate λ_I taken into account, and A_c is the common-mode availability with failure rate λ_c. We assume that both common and independent failure modes have the same repair rate. Thus

$$A(\infty) = \left[1 - \left(\frac{\lambda_I}{\lambda_I + \nu}\right)^N\right] \frac{\nu}{\lambda_c + \nu}. \tag{8.116}$$

This may also be written in terms of β factors by recalling that $\lambda_I \equiv (1 - \beta)\lambda$ and $\lambda_c \equiv \beta\lambda$.

EXAMPLE 8.7

A system has a ratio of $\nu/\lambda = 100$. What will the asymptotic availability be (*a*) for the system, (*b*) for two of the systems in parallel with no common-mode failures, and (*c*) for two systems in parallel with $\beta = 0.2$?

Solution (*a*) $A(\infty) = \dfrac{100}{1 + 100} = 0.990.$ ◀

(*b*) $A(\infty) = 1 - \left(\dfrac{1}{1 + 100}\right)^2 = 0.99990.$ ◀

(*c*) $\dfrac{\lambda_I}{\nu} = (1 - \beta)\dfrac{\lambda}{\nu} = (1 - 0.2)\dfrac{1}{100} = 0.8 \times 10^{-2}$

$\dfrac{\lambda_c}{\nu} = \beta\dfrac{\lambda}{\nu} = 2 \times 10^{-3}.$

Therefore, from Eq. 8.116,

$$A(\infty) = \left[1 - \left(\frac{0.8 \times 10^{-2}}{1 + 0.8 \times 10^{-2}}\right)^2\right] \frac{1}{2 \times 10^{-3} + 1} = 0.9979. \blacktriangleleft$$

Unrevealed Failures

In the derivations just given it is assumed that component failures are detected immediately and that repair is initiated at once. Situations are also encountered in which the component failures go undetected until periodic testing takes place. The evaluation of availability then becomes more complex, for several testing strategies may be considered. Not only is the test interval T_0 subject to change, but the testing may be carried out on all the components simultaneously or in a staggered sequence. In either event the calculation of the system availability is now more subtle, for the point availabilities will have periodic structures, and they must be averaged over a test period in order to estimate the asymptotic availability.

To illustrate, consider the effects of simultaneous and staggered testing patterns on two simple component configurations: the nonredundant configuration consisting of two identical components in series, and the completely redundant configuration consisting of two identical components in parallel. For clarity we consider the idealized situation in which the testing time and the time to repair can be ignored. The failure rates are assumed to be constant.

We begin by letting $A_1(t)$ and $A_2(t)$ be the component point availabilities. Since the testing is carried out at intervals of T_0, we need only determine the system point availability $A(t)$ between $t = 0$ and $t = T_0$, for the asymptotic mission availability is then obtained by averaging $A(t)$ over the test period:

$$A^*(\infty) = A^*(T_0) = \frac{1}{T_0}\int_0^{T_0} A(t)\, dt. \tag{8.117}$$

Simultaneous Testing

When both components are tested at the same time, $t = 0, T_0, 2T_0, \ldots$, the point availabilities are given by

$$A_1(t) = e^{-\lambda t}, \qquad 0 \le t < T_0, \tag{8.118}$$

and

$$A_2(t) = e^{-\lambda t}, \qquad 0 \le t < T_0. \tag{8.119}$$

For the series system we have

$$A(t) = A_1(t)A_2(t), \tag{8.120}$$

or

$$A(t) = e^{-2\lambda t}, \qquad 0 \le t < T_0. \tag{8.121}$$

For the parallel system we obtain

$$A(t) = A_1(t) + A_2(t) - A_1(t)A_2(t), \tag{8.122}$$

or

$$A(t) = 2e^{-\lambda t} - e^{-2\lambda t}, \qquad 0 \le t < T_0. \tag{8.123}$$

The availabilities are plotted as solid lines in Fig. 8.8*a* and *b*, respectively. The asymptotic availability obtained from Eq. 8.117 for the series system is

$$A_s^*(T_0) = \frac{1}{2\lambda T_0}(1 - e^{-2\lambda T_0}) \tag{8.124}$$

whereas that of the parallel system is

$$A_p^*(T_0) = \frac{1}{2\lambda T_0}(3 - 4e^{-\lambda T_0} + e^{-2\lambda T_0}). \tag{8.125}$$

Staggered Testing

We now consider the testing of components at staggered intervals of $T_0/2$. We assume that component 1 is tested at $0, T_0, 2T_0, \ldots$, whereas component 2 is tested at the half-intervals $T_0/2, 3T_0/2, \ldots$. The point availabilities within any interval after the first one are given by

$$A_1(t) = e^{-\lambda t}, \qquad 0 \le t < T_0, \tag{8.126}$$

and

$$A_2(t) = \begin{cases} \exp\left[-\lambda\left(t + \dfrac{T_0}{2}\right)\right] & 0 \le t < \dfrac{T_0}{2} \\ \exp\left[-\lambda\left(t - \dfrac{T_0}{2}\right)\right], & \dfrac{T_0}{2} \le t < T_0. \end{cases} \tag{8.127}$$

To determine the point system availability, we combine these two equations with Eqs. 8.120 and 8.122, respectively, for the series and parallel configurations. The results are plotted as dotted lines in Figs. 8.8*a* and 8.8*b*.

To calculate the asymptotic availabilities for staggered testing, we first note from Fig. 8.8 that the system point availabilities for both series and parallel situations have a periodicity over the half-intervals $T_0/2$. Therefore, instead of averaging $A(t)$ over an entire interval as in Eq. 8.117, we need to average it over only the half-interval. Hence

$$A^*(T_0) = \frac{2}{T_0}\int_0^{T_0/2} A(t)\, dt. \tag{8.128}$$

For the series configuration we calculate $A_1(t)A_2(t)$ from Eqs. 8.126 and 8.127,

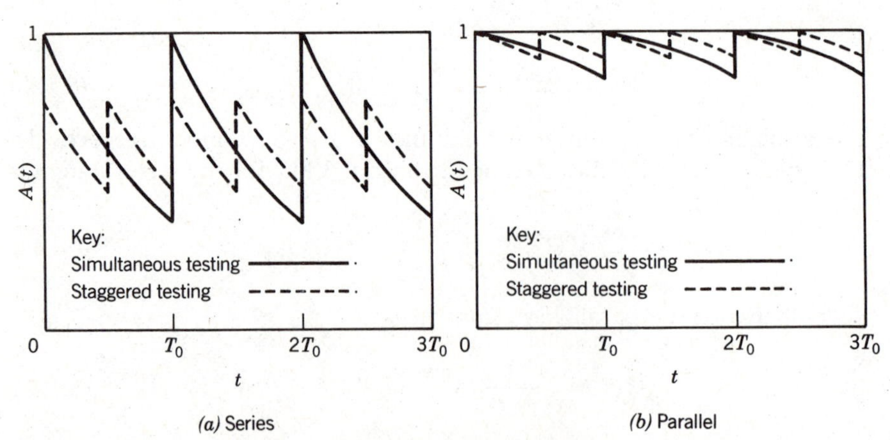

FIGURE 8.8 Availability for a two-component system with unrevealed failures.

substitute the result into Eq. 8.128, and carry out the integral to obtain

$$A_s^*(T_0) = \frac{1}{2\lambda T_0}(e^{-\lambda T_0/2} - e^{-3\lambda T_0/2}). \tag{8.129}$$

Similarly, for the parallel configuration we form $A(t)$ by substituting Eqs. 8.126 and 8.127 into Eq. 8.122, combine the result with Eq. 8.128, and perform the integral to obtain

$$A_p^*(T_0) = \frac{1}{\lambda T_0}(2 - 2e^{-\lambda T_0} - e^{-\lambda T_0/2} + e^{-3\lambda T_0/2}). \tag{8.130}$$

Although the point availabilities plotted as dotted lines in Fig. 8.8 are interesting in understanding the effects of staggering on the availability, the asymptotic values are often more useful, for they allow us to compare the strategies with a single number. Evaluation of the appropriate expressions indicates that in the nonredundant (series) configuration higher availability is obtained from simultaneous testing, whereas staggered testing yields the higher availability for redundant (parallel) configurations.

This behavior can be understood explicitly if the expressions for the asymptotic availability are expanded in powers of λT_0, since for small failure rates the lowest-order terms in λT_0 will dominate the expressions. The results of such expansions are presented in Table 8.1.

The effects of staggered testing become more pronounced when repair time, testing time, or both are not negligible. We can see, for example, that even for a zero failure rate, the testing time t_t will decrease the availability of the series system by t_t/T_0 if the systems are tested simultaneously. If the tests are staggered in the series system, the availability will decrease by $2t_t/T_0$. Conversely, in the parallel system simultaneous testing with no failures will decrease the availability by t_t/T_0, but if the tests are staggered so that they do not take both components out at the same time, the availability does not decrease.

TABLE 8.1 Av ilability $A^*(T_0)$ for Unrevealed Failures

Testing	Series System	Parallel System
Simultaneous	$1 - \lambda T_0 + \frac{2}{3}(\lambda T_0)^2$	$1 - \frac{1}{3}(\lambda T_0)^2$
Staggered	$1 - \lambda T_0 + \frac{13}{24}(\lambda T_0)^2$	$1 - \frac{5}{14}(\lambda T_0)^2$

EXAMPLE 8.8

A voltage monitor achieves an average availability of 0.84 when it is tested monthly; the repair time is negligible. Since the 0.84 availability is unacceptably low, two monitors are placed in parallel. What will the availability of this twin system be (*a*) if the monitors are tested monthly at the same time, (*b*) if they are tested monthly at staggered intervals?

Solution First we must find λT_0. Try Eq. 8.93, the rare-event approximation:

$$0.84 = 1 - \tfrac{1}{2}\lambda T_0; \qquad \lambda T_0 \approx 0.32.$$

This is too large for the exponential expansion to be used. Therefore, we use Eq. 8.92 instead. We obtain a transcendental equation

$$0.84 = \frac{1}{\lambda T_0}(1 - e^{-\lambda T_0}).$$

Solving iteratively, we find that

λT_0	0.320	0.340	.360	.380
$(1/0.84)(1 - e^{-\lambda T_0})$	0.326	0.343	.3599	.376

Therefore,

$$\lambda T_0 \approx .36;$$

(*a*) From Eq. 8.125 we find for simultaneous testing

$$A_p^*(T_0) = \frac{1}{2 \times 0.36}(3 - 4e^{-0.36} + e^{-2\times 0.36}) = 0.967. \blacktriangleleft$$

(*b*) From Eq. 8.130 we find for staggered testing

$$A_p^*(T_0) = \frac{1}{0.36}(2 - 2e^{-0.36} - e^{-0.36/2} + e^{-3\times 0.36/2}) = 0.978. \blacktriangleleft$$

These results can be generalized to combinations of series and parallel configurations. However, the evaluation of the integral in Eq. 8.117 over the test period may become tedious. Moreover, the evaluation of maintenance, testing, and repair policies become more complex in real systems that contain combinations of revealed and unrevealed failures, large numbers of components, and dependencies between components. Some of the more common types of dependencies are included in the following chapter.

Bibliography

Barlow, R. E., and F. Proschan, *Mathematical Theory of Reliability,* Wiley, New York, 1965.

Frankel, E. G., *Systems Reliability and Risk Analysis,* Martinius Nijhoft, The Hague, Holland, 1984.

Gertsbakh, I. B., *Models for Preventive Maintenance,* North-Holland Publishing Co., Amsterdam, Holland, 1977.

Green, A. E., *Safety Systems Analysis,* Wiley, New York, 1983.

Jardine, A. K. S., *Maintenance, Replacement, and Reliability,* Wiley, New York, 1973.

McCormick, N. J., *Reliability and Risk Analysis,* Academic Press, New York, 1981.

Pieruschla, E., *Principles of Reliability,* Prentice-Hall, Englewood Cliffs, NJ, 1963.
Sandler, G. H., *System Reliability Engineering,* Prentice-Hall, Englewood Cliffs, NJ, 1963.

EXERCISES

8.1 Without preventive maintenance the reliability of a condensate demineralizer is characterized by

$$\int_0^t \lambda(t')\, dt' = 1.2 \times 10^{-2}t + 1.1 \times 10^{-9}t^2$$

where t is in hours. The design life is 10,000 hr.
(*a*) What is the design-life reliability?
(*b*) Suppose that by overhaul the demineralizer is returned to as-good-as-new condition. How frequently should such overhauls be performed to achieve a design-life reliability of at least 0.95?
(*c*) Repeat part *b* for a target reliability of at least 0.975.

8.2 Repeat part *b* of Exercise 8.1 assuming that there is a 1% probability that faulty overhaul will cause the demineralizer to fail destructively immediately following start-up. Is it possible to achieve the 0.95 reliability? If so, how many overhauls are required?

8.3 Discuss under what conditions preventative maintenance can increase the reliability of a simple parallel system, even though the component failure rates are time-independent. Justify your results.

8.4 Derive Eq. 8.26.

8.5 Derive an equation analogous to Eqs. 8.27 and 8.28 that includes a probability p_I of independent maintenance failure and a probability p_c of common-mode maintenance failure.

8.6 Suppose that a device has a failure rate of

$$\lambda(t) = (0.015 + 0.02t)/\text{year},$$

where t is in years.
(*a*) Calculate the reliability for a 5-year design life assuming that no maintenance is performed.
(*b*) Calculate the reliability for a 5-year design life assuming that annual preventive maintenance restores the system to an as-good-as-new condition.
(*c*) Repeat part *b* assuming that there is a 5% chance that the preventive maintenance will cause immediate failure.

8.7 Show that preventive maintenance has no effect on the MTTF for a system with a constant failure rate.

8.8 Suppose that the times to failure of an unmaintained component may be given by a Weibull distribution with $m = 2$. Perfect preventive maintenance is performed at intervals $T = 0.25\theta$.
(*a*) Find the MTTF of the maintained system in terms of θ.
(*b*) Determine the percentage increase in the MTTF over that of the unmaintained system.

8.9 Solve Exercise 8.8 approximately for the situation in which $T << \theta$.

8.10 The reliability of a device is given by the Rayleigh distribution

$$R(t) = e^{-(t/\theta)^2}$$

The MTTF is considered to be unacceptably short. The design engineer has two alternatives: a second identical system may be set in parallel or (perfect) preventive maintenance may be performed at some interval T. At what interval T must the preventive maintenance be performed to obtain an increase in the MTTF equal to what would result from the parallel configuration without preventive maintenance? (*Note*: See the solution for Exercise 7.14.)

8.11 The following table gives a series of times to repair (man-hours) obtained for a diesel engine.

11.6	7.9	27.7	17.8	8.9	22.5
3.3	33.3	75.3	9.4	28.5	5.4
10.3	1.1	7.8	41.9	13.3	5.3

(*a*) Estimate the MTTR.
(*b*) Estimate the repair rate and its 90% confidence interval assuming that the data is exponentially distributed.

8.12 Find the asymptotic availability for the systems shown in Exercise 7.26, assuming that all the components are subject only to revealed failures and that the repair rate is ν. Then approximate your result for the case $\nu/\lambda >> 1$.

8.13 A computer has an MTTF = 34 hr and an MTTR = 2.5 hr.
(*a*) What is the availability?
(*b*) If the MTTR is reduced to 1.5 hr, what MTTF can be tolerated without decreasing the availability of the computer?

8.14 The following table gives the times at which a system failed (t_f) and the times at which the subsequent repairs were completed (t_r) over a 2000-hr period.

t_f	t_r	t_f	t_r
51	52	1127	1134
90	92	1236	1265
405	412	1297	1303
507	529	1372	1375
535	539	1424	1439
615	616	1531	1552
751	752	1639	1667
760	766	1789	1795
835	839	1796	1808
881	884	1859	1860
933	941	1975	1976
1072	1091		

(*a*) Calculate the average availability over the time interval $0 \leq t \leq t_{max}$ directly from the data.

(*b*) Assuming constant failure and repair rates, estimate λ and μ from the data.

(*c*) Use the values of λ and μ obtained in part *b* to estimate $A(t)$ and the time-averaged availability for the interval $0 \leq t \leq t_{max}$. Compare your results to part *a*.

8.15 A system consists of two subsystems in series, each with $\nu/\lambda = 10^2$ as its ratio of repair rate to failure rate. Assuming revealed failures, what is the availability of the system after an extended period of operation?

8.16 Reliability testing has indicated that without repair a voltage inverter has a 6-month reliability of 0.87; make a rough estimate of the MTTR that must be achieved if the inverter is to operate with an availability of 0.95. (Assume revealed failures and a constant failure rate.)

8.17 A device has a constant failure rate, and the failures are unrevealed. It is found that with a test interval of 6 months the interval availability is 0.98; use the "rare-event" approximation to estimate the failure rate. (Neglect test and repair times.)

8.18 An auxiliary feedwater pump has an availability of 0.960 under the following conditions: The failures are unrevealed; periodic testing is carried out on a monthly (30-day) basis; and testing and repair require that the system be shut down for 8 hr.

(*a*) What will the availability be if the shutdown time can be reduced to 2 hr?

(*b*) What will the availability be if the tests are performed once per week, with the 8-hr shutdown time?

(*c*) Given the 8-hr shutdown time, what is the optimal test interval?

8.19 Unrevealed bearing failures follow a Weibull distribution with $m = 2$ and $\theta = 5000$ operating hours. How frequently must testing and repair take place if bearing availability is to be maintained at least 95%?

8.20 The reliability of a system is represented by the Rayleigh distribution

$$R(t) = e^{-(t/\theta)^2}.$$

Suppose that all failures are unrevealed. The system is tested and repaired to an as-good-as-new condition at intervals of T_0. Neglecting the times required for test and repair, and assuming perfect maintenance:

(*a*) Derive an expression for the asymptotic availability $A^*(\infty)$.
(*b*) Find an approximation for $A^*(\infty)$ when $T_0 << \theta$.
(*c*) Evaluate $A^*(\infty)$ for $T_0/\theta = 0.1, 0.5, 1.0$, and 2.0.

8.21 A pressure relief system consists of two valves in parallel. The system achieves an availability of 0.995 when the valves are tested on a staggered basis, each valve being tested once every 3 months. (*a*) Estimate the failure rate of the valves. (*b*) If the test procedure were relaxed so that each valve is tested once in 6 months, what would the availability be?

8.22 In annual test and replacement procedures 8% of the emergency respirators at a chemical plant are found to be inoperable.

(*a*) What is the availability of the respirators?
(*b*) How frequently must the test and replacement be carried out if an availability of 0.99 is to be reached?
(Assume constant failure rates.)

8.23 Starting with Eqs. 8.124 and 8.129, derive the results for series systems with simultaneous and staggered testing given in Table 8.1.

8.24 Starting with Eqs. 8.125 and 8.130, derive the results for parallel systems with simultaneous and staggered testing given in Table 8.1.

8.25 Consider three units in parallel, each tested at equally staggered intervals of T_0. Assume constant failure rates.

(*a*) What is $A(t)$?
(*b*) Plot $A(t)$.
(*c*) What is $A^*(T_0)$?
(*d*) Find the rare-event approximate for $A^*(T_0)$.

CHAPTER 9

Failure Interactions

9.1 INTRODUCTION

In reliability analysis perhaps the most pervasive technique is that of estimating the reliability of a system in terms of the reliability of its components. In such analysis it is frequently assumed that the component failure and repair properties are mutually independent. In reality, this is often not the case. Therefore, it is necessary to replace the simple products of probabilities with more sophisticated models that take into account the interactions of component failures and repairs.

Many component failure interactions—as well as systems with independent failures—may be modeled effectively as Markov processes, provided that the failure and repair rates can be approximated as time-independent. Indeed, we have already examined a particular example of a Markov process, the derivation of the Poisson process contained in Chapter 4. In this chapter we first formulate the modeling of failures as Markov processes and then apply them to simple systems in which the failures are independent. This allows us both to verify that the same results are obtained as in Chapter 7 and to familiarize ourselves with Markov processes. We then use Markov methods to examine failure interactions of two particular types, shared-load systems and standby systems, and follow with demonstrations of how to incorporate such failure dependencies into the analysis of larger systems. Finally, the analysis is generalized to take into account operational dependencies such as those created by shared repair crews.

9.2 MARKOV ANALYSIS

We begin with the Markov formulation by designating all the possible states of a system. A state is defined to be a particular combination of operating and failed components. Thus, for example, if we have a system consisting of three components, we may easily show that there are eight different combinations of operating and failed components and therefore eight states. These are enumerated in Table 9.1, where O indicates an operational component and X a failed component. In general, a system with N components will have 2^N states so that the number of states increases much faster than the number of components.

For the analysis that follows we must know which of the states correspond to system failure. This, in turn, depends on the configuration in which the components are used. For example, three components might be arranged in any of the three configurations shown in Fig. 9.1. If all the components are in series, as in Fig. 9.1*a*, any combination of one or more component failures will cause system failure. Thus states 2 through 8 in Table 9.1 are failed system states. Conversely, if the three components are in parallel as in Fig. 9.1*b*, all three components must fail for the system to fail. Thus only state 8 is a system failure state. Finally, for the configuration shown in Fig. 9.1*c* both components 1 and 2 or component 3 must fail for the system to fail. Thus states 4 through 8 correspond to system failure.

The object of Markov analysis is to calculate $P_i(t)$, the probability that the system is in state i at time t. Once this is known, the system reliability can be calculated as a function of time from

$$R(t) = \sum_{i \varepsilon O} P_i(t), \tag{9.1}$$

where the sum is taken over all the operating states (i.e., over those states for which the system is not failed). Alternately, the reliability may be calculated from

$$R(t) = 1 - \sum_{i \varepsilon X} P_i(t), \tag{9.2}$$

where the sum is over the states for which the system is failed.

In what follows, we designate state 1 as the state for which all the components are operating, and we assume that at $t = 0$ the system is in state 1. Therefore,

$$P_1(0) = 1, \tag{9.3}$$

and

$$P_i(0) = 0, \qquad i \neq 1. \tag{9.4}$$

Since at any time the system can only be in one state, we have

$$\sum_i P_i(t) = 1, \tag{9.5}$$

where the sum is over all possible states.

To determine the $P_i(t)$, we derive a set of differential equations, one for each state of the system. These are sometimes referred to as state tran-

TABLE 9.1 Markov States of Three-Component Systems

Component	State # 1	2	3	4	5	6	7	8
a	*O*	X	*O*	*O*	X	X	*O*	X
b	*O*	*O*	X	*O*	X	*O*	X	X
c	*O*	*O*	*O*	X	*O*	X	X	X

Note: *O* = operating; *X* = failed.

sition equations because they allow the $P_i(t)$ to be determined in terms of the rates at which transitions are made from one state to another. The transition rates consist of superpositions of component failure rates, repair rates, or both. We illustrate these concepts first with a very simple system, one consisting of only two independent components, *a* and *b*.

Two Independent Components

A two-component system has only four possible states, those enumerated in Table 9.2. The logic of the changes of states is best illustrated by a state transition diagram shown in Fig. 9.2. The failure rates λ_a and λ_b for components *a* and *b* indicate the rates at which the transitions are made between states. Since $\lambda_a\,\Delta t$ is the probability that a component will fail between times t and $t + \Delta t$, given that it is operating at t (and similarly for λ_b), we may write the net change in the probability that the system will be in state 1 as

$$P_1(t + \Delta t) - P_1(t) = -\lambda_a\,\Delta t\,P_1(t) - \lambda_b\,\Delta t\,P_1(t), \tag{9.6}$$

or in differential form

$$\frac{d}{dt}P_1(t) = -\lambda_a P_1(t) - \lambda_b P_1(t). \tag{9.7}$$

To derive equations for state 2, we first observe that for every transition out of state 1 by failure of component *a*, there must be an arrival in state 2. Thus the number of arrivals during Δt is $\lambda_a\,\Delta t\,P_1(t)$. Transitions can also

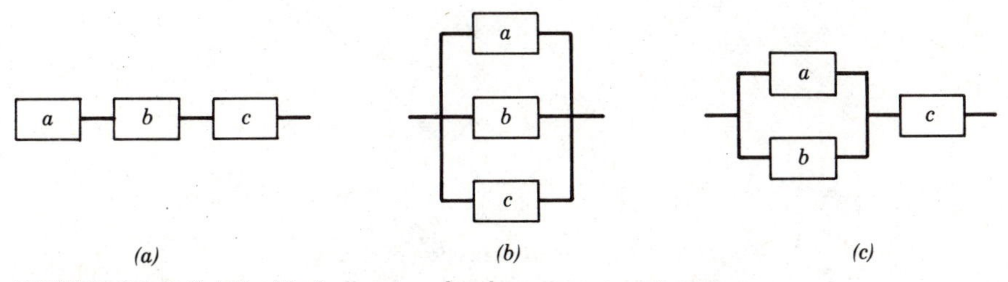

FIGURE 9.1 Reliability block diagrams for three-component systems.

TABLE 9.2 Markov States of Two-Component Systems

Component	State # 1	2	3	4
a	O	X	O	X
b	O	O	X	X

be made out of state 2 during Δt; these will be due to failures of component b, and they will make a contribution of $-\lambda_b\, \Delta t\, P_2(t)$. Thus the net increase in the probability that the system will be in state 2 is given by

$$P_2(t + \Delta t) - P_2(t) = \lambda_a\, \Delta t\, P_1(t) - \lambda_b\, \Delta t\, P_2(t), \tag{9.8}$$

or dividing by Δt and taking the derivative, we have

$$\frac{d}{dt} P_2(t) = \lambda_a P_1(t) - \lambda_b P_2(t). \tag{9.9}$$

Identical arguments can be used to derive the equation for $P_3(t)$. The result is

$$\frac{d}{dt} P_3(t) = \lambda_b P_1(t) - \lambda_a P_3(t). \tag{9.10}$$

We may derive one more differential equation, for state 4. We note from the diagram that the transitions into state 4 may come either as a failure of component b from state 2 or as a failure of component a from

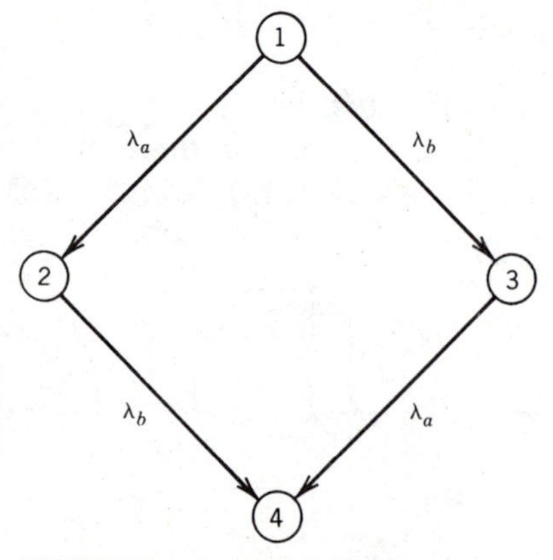

FIGURE 9.2 State transition diagram with independent failures.

state 3; the transitions during Δt are $\lambda_b \, \Delta t \, P_2(t)$ and $\lambda_a \, \Delta t \, P_3(t)$, respectively. Consequently, we have

$$P_4(t + \Delta t) - P_4(t) = \lambda_b \, \Delta t \, P_2(t) + \lambda_a \, \Delta t \, P_3(t) \tag{9.11}$$

or, correspondingly,

$$\frac{d}{dt} P_4(t) = \lambda_b P_2(t) + \lambda_a P_3(t). \tag{9.12}$$

State 4 is called an absorbing state, since there is no way to get out of it. The other states are referred to as nonabsorbing states.

From the foregoing derivation we see that we must solve four coupled ordinary differential equations in time in order to determine the $P_i(t)$. We begin with Eq. 9.7 for $P_1(t)$, since it does not depend on the other $P_i(t)$. By substitution, it is clear that the solution to Eq. 9.7 that meets the initial condition, Eq. 9.3, is

$$P_1(t) = e^{-(\lambda_a + \lambda_b)t}. \tag{9.13}$$

To find $P_2(t)$, we first insert Eq. 9.13 into Eq. 9.9,

$$\frac{d}{dt} P_2(t) = \lambda_a e^{-(\lambda_a + \lambda_b)t} - \lambda_b P_2(t), \tag{9.14}$$

yielding an equation in which only $P_2(t)$ appears. Moving the last term to the left-hand side, and multiplying by an integrating factor $e^{\lambda_b t}$, we obtain

$$\frac{d}{dt} [e^{\lambda_b t} P_2(t)] = \lambda_a e^{-\lambda_a t}. \tag{9.15}$$

Multiplying by dt, and integrating the resulting equation from time equals zero to t, we have

$$[e^{\lambda_b t} P_2(t)]_0^t = \lambda_a \int_0^t e^{-\lambda_a t'} \, dt'. \tag{9.16}$$

Carrying out the integral on the right-hand side, utilizing Eq. 9.4 on the left-hand side, and solving for $P_2(t)$, we obtain

$$P_2(t) = e^{-\lambda_b t} - e^{-(\lambda_a + \lambda_b)t}. \tag{9.17}$$

Completely analogous arguments can be applied to the solution of Eq. 9.10. The result is

$$P_3(t) = e^{-\lambda_a t} - e^{-(\lambda_a + \lambda_b)t}. \tag{9.18}$$

We may now solve Eq. 9.11 for $P_4(t)$. However, it is more expedient to note that it follows from Eq. 9.5 that

$$P_4(t) = 1 - \sum_{i=1}^{3} P_i(t), \tag{9.19}$$

Therefore, inserting Eqs. 9.13, 9.17, and 9.18 into this expression yields the desired solution

$$P_4(t) = 1 - e^{-\lambda_a t} - e^{-\lambda_b t} + e^{-(\lambda_a+\lambda_b)t}. \tag{9.20}$$

With the $P_i(t)$ known, we may now calculate the reliability. This, of course, depends on the configuration of the two components, and there are only two possibilities, series and parallel. In the series configuration any failure causes system failure. Hence

$$R_s(t) = P_1(t) \tag{9.21}$$

or

$$R_s(t) = e^{-(\lambda_a+\lambda_b)t}. \tag{9.22}$$

Since, for the parallel configuration both components a and b must fail to have system failure,

$$R_p(t) = P_1(t) + P_2(t) + P_3(t), \tag{9.23}$$

or, using Eq. 9.19, we have

$$R_p(t) = 1 - P_4(t). \tag{9.24}$$

Therefore,

$$R_p(t) = e^{-\lambda_a t} + e^{-\lambda_b t} - e^{-(\lambda_a+\lambda_b)t}. \tag{9.25}$$

This analysis assumes that the failure rate of each component is independent of the state of the other component. As can be seen from Fig. 9.2, the transitions $1 \rightarrow 2$ and $3 \rightarrow 4$, which involve the failure of component a, have the same failure rate, even though one takes place with component b in operating order and the other with component b failed. The same argument applies in comparing the transitions $1 \rightarrow 3$ and $2 \rightarrow 4$. Since the failure rates—and therefore the failure probabilities—are independent of the system state, they are mutually independent. Therefore, the expressions derived in Chapter 7 should still be valid. That this is the case may be seen from the following. For constant failure rates the component reliabilities derived in Chapter 4 are

$$R_l(t) = e^{-\lambda_l t}, \qquad l = a, b. \tag{9.26}$$

Thus the series expression, Eq. 9.22, reduces to

$$R_s(t) = R_a(t)R_b(t), \tag{9.27}$$

and the parallel expression, Eq. 9.25, is

$$R_p(t) = R_a(t) + R_b(t) - R_a(t)\, R_b(t). \tag{9.28}$$

These are just the expressions derived earlier for independent components, without the use of Markov methods.

Load-Sharing Systems

The primary value of Markov methods appears in situations in which component failure rates can no longer be assumed to be independent of the system state. One of the common cases of dependence is in load-sharing components, whether they be structural members, electric generators, or mechanical pumps or valves. Suppose, for example, that two electric generators share an electric load that either generator has enough capacity to meet. It is nevertheless true that if one generator fails, the additional load on the second generator is likely to increase its failure rate.

To model load-sharing failures, consider once again two components, *a* and *b*, in parallel. We again have a four-state system, but now the transition diagram appears as in Fig. 9.3. Here λ_a^* and λ_b^* denote the increased failure rates brought about by the higher loading after one failure has taken place.

The Markov equations can be derived as for independent failures if the changes in failure rates are included. Comparing Fig. 9.2 with 9.3, we see that the resulting generalizations of Eqs. 9.7, 9.9, 9.10, and 9.12 are

$$\frac{d}{dt} P_1(t) = -(\lambda_a + \lambda_b)P_1(t), \tag{9.29}$$

$$\frac{d}{dt} P_2(t) = \lambda_a P_1(t) - \lambda_b^* P_2(t), \tag{9.30}$$

$$\frac{d}{dt} P_3(t) = \lambda_b P_1(t) - \lambda_a^* P_3(t) \tag{9.31}$$

and

$$\frac{d}{dt} P_4(t) = \lambda_b^* P_2(t) + \lambda_a^* P_3(t). \tag{9.32}$$

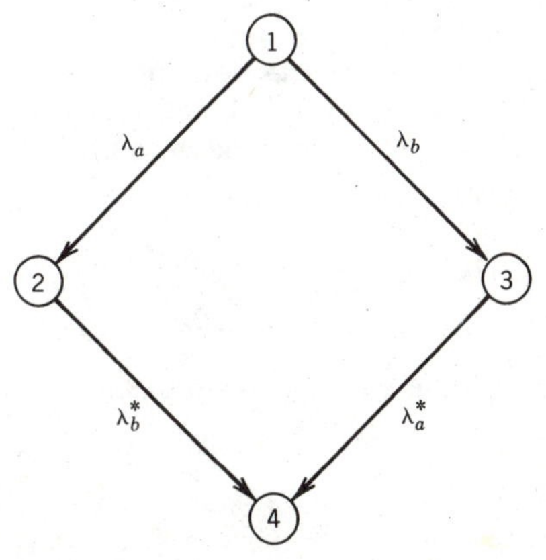

FIGURE 9.3 State transition diagram with load sharing.

The solution procedure is also completely analogous. The results are

$$P_1(t) = e^{-(\lambda_a+\lambda_b)t}, \tag{9.33}$$

$$P_2(t) = e^{-\lambda_b^* t} - e^{-(\lambda_a+\lambda_b^*)t}, \tag{9.34}$$

$$P_3(t) = e^{-\lambda_a^* t} - e^{-(\lambda_a^*+\lambda_b)t} \tag{9.35}$$

and

$$P_4(t) = 1 - e^{-\lambda_a^* t} - e^{-\lambda_b^* t} - e^{-(\lambda_a+\lambda_b)t} + e^{-(\lambda_a+\lambda_b^*)t} + e^{-(\lambda_a^*+\lambda_b)t}. \tag{9.36}$$

Finally, since both components must fail for the system to fail, the reliability is equal to $1 - P_4(t)$, yielding

$$R_p(t) = e^{-\lambda_a^* t} + e^{-\lambda_b^* t} + e^{-(\lambda_a+\lambda_b)t} - e^{-(\lambda_a+\lambda_b^*)t} - e^{-(\lambda_a^*+\lambda_b)t}. \tag{9.37}$$

It is easily seen that if $\lambda_a^* = \lambda_a$ and $\lambda_b^* = \lambda_b$, there is no dependence between failure rates, and Eq. 9.37 reduces to Eq. 9.25. The effects of increased loading on a load-sharing redundant system can be seen graphically by considering the situation in which the two components are identical: $\lambda_a = \lambda_b = \lambda$ and $\lambda_a^* = \lambda_b^* = \lambda^*$. Equation 9.37 then reduces to

$$R(t) = 2e^{-\lambda^* t} + e^{-2\lambda t} - 2e^{-(\lambda+\lambda^*)t}. \tag{9.38}$$

In Fig. 9.4 we have plotted $R(t)$ for the two-component parallel system, while varying the increase in failure rate caused by increased loading (i.e., the ratio λ^*/λ). The two extremes are the system in which the two components are independent, $\lambda^* = \lambda$, and the totally dependent system in which the failure of one component brings on the immediate failure of the other, $\lambda^* = \infty$. Notice that these two extremes correspond to Eqs. 9.25 and 9.22, for independent failures of parallel and series configurations, respectively.

EXAMPLE 9.1

Two diesel generators of known MTTF are hooked in parallel. Because the failure of one of the generators will cause a large additional load on the other, the design engineer estimates that the failure rate will double for the remaining generator. For how many MTTF can the generator system be run without the reliability's dropping below 0.95?

Solution Take $\lambda^* = 2\lambda$. Then Eq. 9.37 is

$$R = 0.95 = 2e^{-2\lambda t} + e^{-2\lambda t} - 2e^{-3\lambda t},$$

where t is the time at which the reliability drops below 0.95. Let $x = e^{-\lambda t}$. Then

$$2x^3 - 3x^2 + 0.95 = 0.$$

The solution must lie in the interval $0 < x < 1$. By plotting the left-hand side of the equation, we may show that the equation is satisfied at only one place, at

$$x = 0.8647.$$

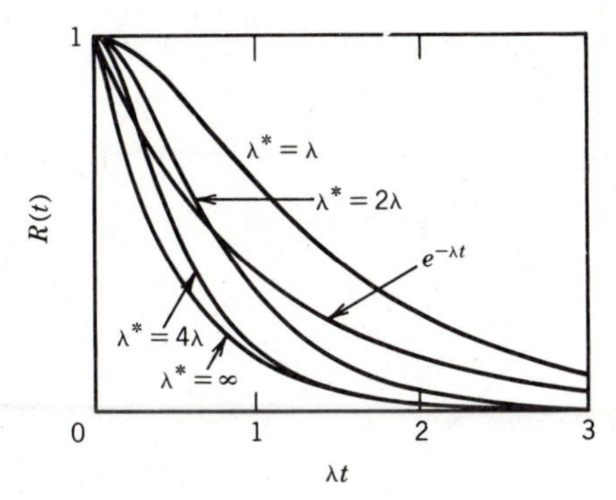

FIGURE 9.4 Reliability of load-sharing systems.

Therefore, $\lambda t = \ln(1/x) = 0.1454$. Since $\lambda = 1/\text{MTTF}$ for the diesel generators, the maximum time of operation is $t = 0.1454/\lambda = 0.1454$ MTTF. ◀ Note that if only a single generator had been used, it could have operated for only $t = \ln(1/R)/\lambda = 0.0513$ MTTF without violating the criterion.

9.3 RELIABILITY WITH STANDBY SYSTEMS

Standby or backup systems are a widely applied type of redundancy in fault-tolerant systems, whether they be in the form of extra logic chips, navigation components, or emergency power generators. They differ, however, from the parallel systems discussed in Chapter 7 in that one of the units is held in reserve and only brought into operation in the event that the first unit fails. For this reason they are often referred to as passive parallel systems, just as the parallel system with both units operating from the start is referred to as an active parallel system. By their nature standby systems involve dependency between components; they are nicely analyzed by Markov methods.

Idealized System

We first consider an idealized standby system consisting of a primary unit a and a backup unit b. If the states are numbered according to Table 9.2, the system operation is described by the transition diagram, Fig. 9.5. When the primary unit fails, there is a transition $1 \rightarrow 2$, and then when the backup unit fails, there is a transition $2 \rightarrow 4$, with state 4 corresponding to system failure. Note that there is no possibility of the system's being in state 3, since we have assumed that the backup unit does not fail while in the standby state. Hence $P_3(t) = 0$. Later we consider the possibility of failure in this standby state as well as the possibility of failures during the switching from primary to backup unit.

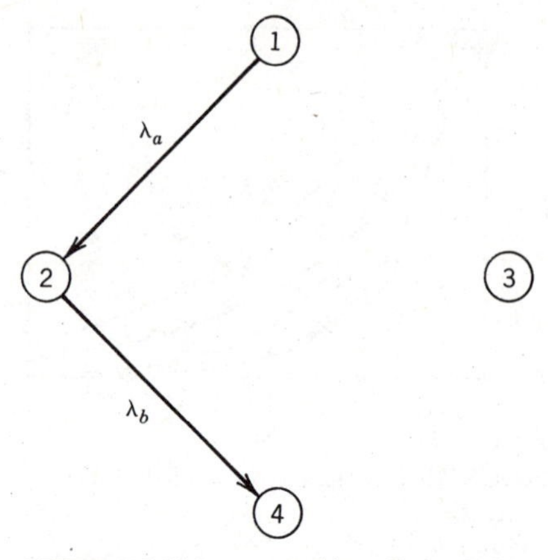

FIGURE 9.5 State transition diagram for a standby configuration.

From the transition diagram we may construct the Markov equations for the three states quite easily. For state 1 there is only a loss term from the transition $1 \rightarrow 2$. Thus

$$\frac{d}{dt} P_1(t) = -\lambda_a P_1(t). \tag{9.39}$$

For state 2 we have one source term, from the $1 \rightarrow 2$ transition, and one loss term from the $2 \rightarrow 4$ transition. Thus

$$\frac{d}{dt} P_2(t) = \lambda_a P_1(t) - \lambda_b P_2(t). \tag{9.40}$$

Since state 4 results only from the transition $2 \rightarrow 4$, we have

$$\frac{d}{dt} P_4(t) = \lambda_b P_2(t). \tag{9.41}$$

The foregoing equations may be solved sequentially in the same manner as those of the preceding sections. We obtain

$$P_1(t) = e^{-\lambda_a t}, \tag{9.42}$$

$$P_2(t) = \frac{\lambda_a}{\lambda_b - \lambda_a}(e^{-\lambda_a t} - e^{-\lambda_b t}), \tag{9.43}$$

$$P_3(t) = 0 \tag{9.44}$$

and

$$P_4(t) = \frac{1}{\lambda_b - \lambda_a}(\lambda_b e^{-\lambda_a t} - \lambda_a e^{-\lambda_b t}), \tag{9.45}$$

where we have again used the initial conditions, Eqs. 9.3 and 9.4. Since state

4 is the only state corresponding to system failure, the reliability is just

$$R(t) = P_1(t) + P_2(t), \tag{9.46}$$

or

$$R(t) = e^{-\lambda_a t} + \frac{\lambda_a}{\lambda_b - \lambda_a}(e^{-\lambda_a t} - e^{-\lambda_b t}). \tag{9.47}$$

This, in turn, may be simplified to

$$R(t) = \frac{1}{\lambda_b - \lambda_a}(\lambda_b e^{-\lambda_a t} - \lambda_a e^{-\lambda_b t}). \tag{9.48}$$

The properties of standby systems are nicely illustrated by comparing their reliability versus time with that of an active parallel system. For brevity we consider the situation $\lambda_a = \lambda_b = \lambda$. In this situation we must be careful in evaluating the reliability, for both Eqs. 9.47 and 9.48 contain $\lambda_b - \lambda_a$ in the denominator. We begin with Eq. 9.47 and rewrite the last term as

$$R(t) = e^{-\lambda_a t} + \frac{\lambda_a}{(\lambda_b - \lambda_a)} e^{-\lambda_a t}\,[1 - e^{-(\lambda_b - \lambda a)t}]. \tag{9.49}$$

Then, going to the limit as λ_b approaches λ_a, we have $(\lambda_b - \lambda_a)t << 1$, and we can expand

$$e^{-(\lambda_b - \lambda_a)t} = 1 - (\lambda_b - \lambda_a)t + \tfrac{1}{2}(\lambda_b - \lambda_a)^2 t^2 - \cdots. \tag{9.50}$$

Combining Eqs. 9.49 and 9.50, we have

$$R(t) = e^{-\lambda_a t} + \lambda_a e^{-\lambda_a t}[t - \tfrac{1}{2}(\lambda_a - \lambda_b)t^2 + \cdots]. \tag{9.51}$$

Thus as λ_b and λ_a become equal, only the first two terms remain, and we have for $\lambda_b = \lambda_a = \lambda$:

$$R(t) = (1 + \lambda t)e^{-\lambda t}. \tag{9.52}$$

In Fig. 9.6 are compared the reliabilities of active and standby parallel

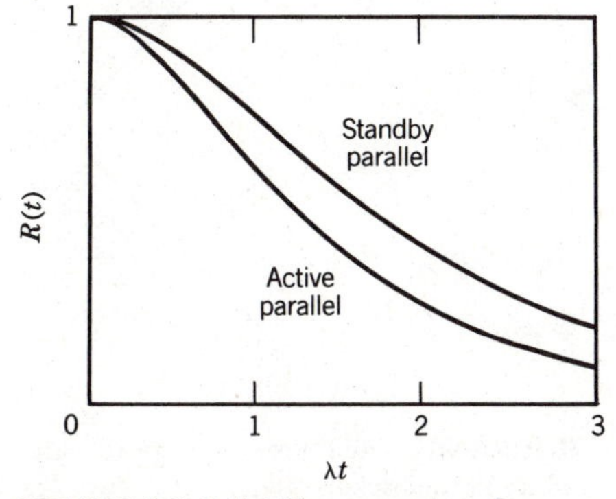

FIGURE 9.6 Reliability comparison for standby and active parallel systems.

systems whose two components have identical failure rates. Note that the standby parallel system is more reliable than the active parallel system because the backup unit cannot fail before the primary unit, even though the reliability of the primary unit is not affected by the presence of the backup unit.

The gain in reliability is further indicated by the increase in the system MTTF for the standby configuration, relative to that for the active configuration. Substituting Eq. 9.52 into Eq. 4.22, we have for the standby parallel system

$$\text{MTTF} = 2/\lambda \tag{9.53}$$

compared to a value of

$$\text{MTTF} = 3/2\lambda \tag{9.54}$$

given by Eq. 7.7 for the active parallel system.

Failures in the Standby State

As anyone who has tried to start a car after it has been idle for several months knows, equipment can fail even while it is not operating. Thus a major consideration with standby systems is that the backup unit does not deteriorate to the point that it is incapable of being operated when required. To model this possibility, we generalize the state transition diagram as shown in Fig. 9.7.

The failure rate λ_b^+ represents failure of the backup unit while it is inactive; state 3 represents the situation in which the primary unit is operating, but there is an undetected failure in the backup unit.

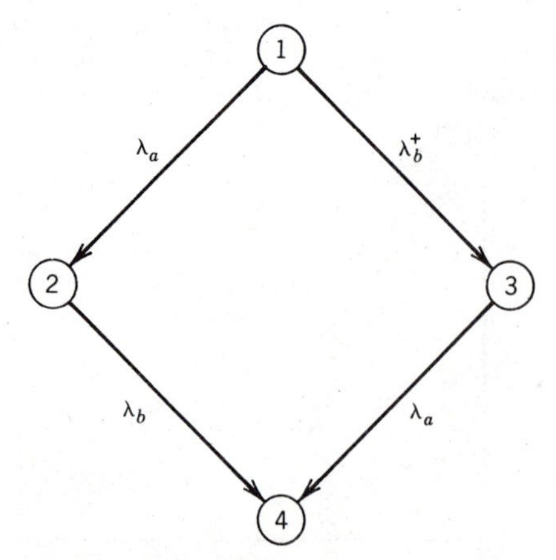

FIGURE 9.7 State transition diagram with failure in the backup mode.

There are now two paths for transition out of state 1. Thus for $P_1(t)$ we have

$$\frac{d}{dt}P_1(t) = -\lambda_a P_1(t) - \lambda_b^+ P_1(t). \tag{9.55}$$

The equation for state 2 is unaffected by the additional failure path; as in Eq. 9.40, we have

$$\frac{d}{dt}P_2(t) = \lambda_a P_1(t) - \lambda_b P_2(t). \tag{9.56}$$

We must now set up an equation to determine $P_3(t)$. This state is entered through the $1 \rightarrow 3$ transition with rate λ_b^+ and exited through the $3 \rightarrow 4$ transition with rate λ_a. Thus

$$\frac{d}{dt}P_3(t) = \lambda_b^+ P_1(t) - \lambda_a P_3(t). \tag{9.57}$$

Finally, state 4 is entered from either states 2 or 3:

$$\frac{d}{dt}P_4(t) = \lambda_b P_2(t) + \lambda_a P_3(t). \tag{9.58}$$

The Markov equations may be solved in the same manner as before. We obtain, with the initial conditions Eqs. 9.3 and 9.4,

$$P_1(t) = e^{-(\lambda_a + \lambda_b^+)t}, \tag{9.59}$$

$$P_2(t) = \frac{\lambda_a}{\lambda_a + \lambda_b^+ - \lambda_b}[e^{-\lambda_b t} - e^{-(\lambda_a + \lambda_b^+)t}] \tag{9.60}$$

and

$$P_3(t) = e^{-\lambda_a t} - e^{-(\lambda_a + \lambda_b^+)t}. \tag{9.61}$$

There is no need to solve for $P_4(t)$, since once again it is the only state for which there is system failure, and therefore,

$$R(t) = P_1(t) + P_2(t) + P_3(t), \tag{9.62}$$

yielding

$$R(t) = e^{-\lambda_a t} + \frac{\lambda_a}{\lambda_a + \lambda_b^+ - \lambda_b}[e^{-\lambda_b t} - e^{-(\lambda_a + \lambda_b^+)t}]. \tag{9.63}$$

Once again it is instructive to examine the case $\lambda_a = \lambda_b = \lambda$ and $\lambda_b^+ = \lambda^+$, in which Eq. 9.63 reduces to

$$R(t) = \left(1 + \frac{\lambda}{\lambda^+}\right)e^{-\lambda t} - \frac{\lambda}{\lambda^+}e^{-(\lambda + \lambda^+)t}. \tag{9.64}$$

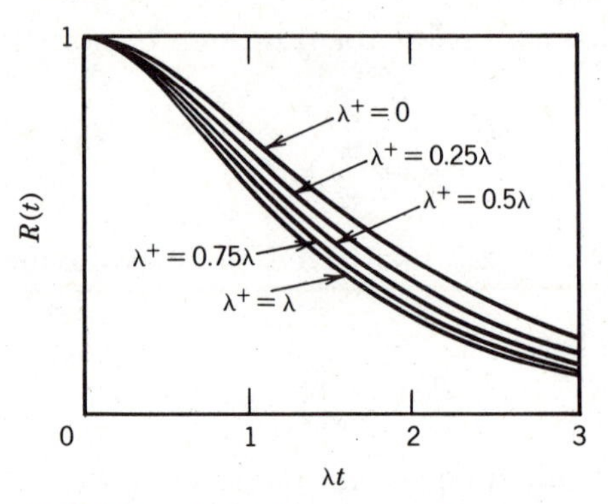

FIGURE 9.8 Reliability of a standby system with different rates of failure in the backup mode.

In Fig. 9.8 the results are shown, having values of λ^+ ranging from zero to λ. The deterioration of the reliability is seen with increasing λ^+. The system MTTF may be found easily by inserting Eq. 9.64 into Eq. 4.22. We have

$$\text{MTTF} = \frac{1}{\lambda} + \frac{1}{\lambda^+} - \frac{\lambda}{\lambda^+}\frac{1}{\lambda + \lambda^+}. \tag{9.65}$$

When $\lambda^+ = \lambda$, the foregoing results reduce to those of an active parallel system. This is sometimes referred to as a "hot-standby system," since both units are then running and only a switch from one to the other is necessary. Fault-tolerant control systems, which can use only the output of one device at a time, but which cannot tolerate the time required to start up the backup unit, operates in this manner. Unlike active parallel systems, however, they must switch from primary unit to backup unit. We consider switching failures next.

EXAMPLE 9.2

A fuel pump with an MTTF of 3000 hr is to operate continuously on a 500-hr mission. (*a*) What is the mission reliability? (*b*) Two such pumps are put in a standby parallel configuration. If there are no failures of the backup pump while in the standby mode, what is the system MTTF and the mission reliability? (*c*) If the standby failure rate is 15% of the operational failure rate, what is the system MTTF and the mission reliability?

Solution (*a*) The component failure rate is $\lambda = 1/3000 = 0.333 \times 10^{-3}$/hr. Therefore, the mission reliability is

$$R(T) = \exp\left(-\frac{1}{3000} \times 500\right) = 0.846. \blacktriangleleft$$

(*b*) In the absence of standby failures, the system MTTF is found from Eq. 9.53 to be

$$\text{MTTF} = \frac{2}{\lambda} = 2 \times 3000 = 6000 \text{ hr.}$$

The system reliability is found from Eq. 9.52 to be

$$R(500) = \left(1 + \frac{1}{3000} \times 500\right) \times \exp\left(-\frac{1}{3000} \times 500\right) = 0.988.$$

(*c*) We find the system MTTF from Eq. 9.65 with $\lambda^+ = 0.15/3000 = 0.5 \times 10^{-4}/$ hr:

$$\text{MTTF} = \frac{1}{0.333 \times 10^{-3}} + \frac{1}{0.5 \times 10^{-4}} - \frac{0.333 \times 10^{-3}}{0.5 \times 10^{-4}} \frac{1}{0.333 \times 10^{-3} + 0.5 \times 10^{-4}}$$

$$\text{MTTF} = 5609 \text{ hr.}$$ ◀

From Eq. 9.64 the system reliability for the mission is $R(500) = 0.986$. ◀

Switching Failures

A second difficulty in using standby systems stems from the switch from the primary unit to the backup. This switch may take action by electric relays, hydraulic valves, electronic control circuits, or other devices. There is always the possibility that the switching device will have a demand failure probability p large enough that switching failures must be considered. For brevity we do not consider backup unit failure while it is in the standby mode.

The state transition diagram with these assumptions is shown in Fig. 9.9. Note that the transition out of state 1 in Fig. 9.5 has been divided into two paths. The primary failure rate is multiplied by $1 - p$ to get the successful transition into state 2, in which the backup system is operating. The second path with rate $p\lambda_a$ indicates a transition directly to the failed-system state that results when there is a demand failure on the switching mechanism.

For the situation depicted in Fig. 9.9, state 1 is still described by Eq. 9.39. Now, however, the $1 \rightarrow 2$ transition is decreased by a factor $1 - p$, and so instead of Eq. 9.40 state 2 is described by

$$\frac{d}{dt} P_2(t) = (1 - p)\lambda_a P_1(t) - \lambda_b P_2(t) \tag{9.66}$$

and state 4 is described by

$$\frac{d}{dt} P_4(t) = \lambda_b P_2(t) + p\lambda_a P_1(t). \tag{9.67}$$

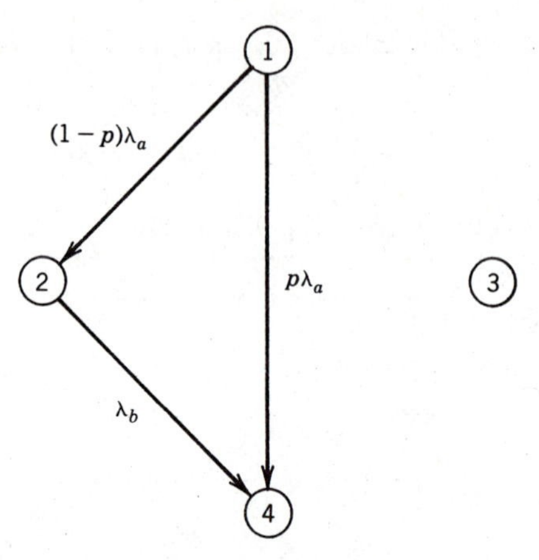

FIGURE 9.9 State transition diagram with standby switching failures.

Since $P_1(t)$ is again given by Eq. 9.42, we need solve only Eq. 9.66 to obtain

$$P_2(t) = (1 - p)\frac{\lambda_a}{\lambda_b - \lambda_a}(e^{-\lambda_a t} - e^{-\lambda_b t}). \tag{9.68}$$

Accordingly, since state 4 is the only failed state and $P_3(t) = 0$, we may write

$$R(t) = P_1(t) + P_2(t), \tag{9.69}$$

or inserting Eqs. 9.42 and 9.68, we obtain for the reliability

$$R(t) = e^{-\lambda_a t} + \frac{(1 - p)\lambda_a}{\lambda_b - \lambda_a}(e^{-\lambda_a t} - e^{-\lambda_b t}). \tag{9.70}$$

Once again it is instructive to consider the case $\lambda_a = \lambda_b = \lambda$, for which we obtain

$$R(t) = [1 + (1 - p)\lambda t]e^{-\lambda t}. \tag{9.71}$$

Clearly, as p increases, the value of the backup system becomes less and less, until finally if p is one (i.e., certain failure of the switching system), the backup system has no effect on the system reliability.

EXAMPLE 9.3

An annunciator system has a mission reliability of 0.9. Because reliability is considered too low, a redundant annunciator of the same design is to be installed. The design engineer must decide between an active parallel and a standby parallel configuration. The engineer knows that failures in standby have a negligible effect, but there is a significant probability of a switching failure. (*a*) How small must the probability of a switching failure be if the standby configuration is to be more reliable than the active configuration? (*b*) Discuss the switching failure requirement of part *a* for very short mission times.

Solution (*a*) Assuming a constant failure rate, we know that for the mission time T,

$$\lambda T = \ln\left[\frac{1}{R(T)}\right] = \ln\left(\frac{1}{0.9}\right) = 0.1054.$$

To find the failure probability, we equate Eq. 9.71 with Eq. 7.8 for the active parallel system:

$$[1 + (1 - p)\lambda T]e^{-\lambda T} = 2e^{-\lambda T} - e^{-2\lambda T}$$

Thus

$$p = 1 - \frac{1}{\lambda T}(1 - e^{-\lambda T})$$

$$= 1 - \frac{1}{0.1054}(1 - e^{-0.1054}) = 0.05. \blacktriangleleft$$

(*b*) For active parallel Eq. 7.46 gives the short mission time approximation:

$$R_a \approx 1 - (\lambda t)^2.$$

For standby parallel we expand 9.71 for small λt:

$$R_{sb} = [1 + (1 - p)\lambda t]e^{-\lambda t} = [1 + (1 - p)\lambda t][1 - \lambda t + \tfrac{1}{2}(\lambda t)^2 \cdots]$$

$$\simeq 1 - p\lambda t - (\tfrac{1}{2} - p)(\lambda t)^2.$$

Then we calculate p for $R_a - R_{sb} = 0$:

$$1 - (\lambda t)^2 - 1 + p\lambda t + (\tfrac{1}{2} - p)(\lambda t)^2 = 0$$

or

$$p = \frac{\frac{1}{2}\lambda t}{1 - \lambda t} \approx \frac{1}{2}\lambda t \blacktriangleleft$$

The shorter the mission, the smaller p must be, or else switching failures will be more probable than the failures of the second annunciator in the active parallel configuration.

The combined effects of failures in the standby mode and switching failures may be included in the foregoing analysis. For two identical units the reliability may be shown to be

$$R(t) = \left[1 + (1 - p)\frac{\lambda}{\lambda^+}\right]e^{-\lambda t} - (1 - p)\frac{\lambda}{\lambda^+}e^{-(\lambda + \lambda^+)t}, \tag{9.72}$$

which reduces to Eq. 9.71 as $\lambda^+ \rightarrow 0$. For a hot standby system in which identical primary and backup systems are both running so that $\lambda^+ = \lambda$, we obtain from Eq. 9.72

$$R(t) = (2 - p)e^{-\lambda t} - (1 - p)e^{-2\lambda t}. \qquad (9.73)$$

Thus the reliability is less than that of an active parallel system because there is a probability of switching failure. As stated earlier, in hot standby systems, such as for control devices, the output of only one unit can be used at a time. If the probability of switching failure is too great, an alternative is to add a third unit and use a 2/3 voting system, as discussed in Chapter 7.

Primary System Repair

Two considerable benefits are to be gained by using redundant system components. The first is that more than one failure must occur in order for the system to fail. A second is that components can be repaired while the system is on line. Much higher reliabilities are possible if the failed component has a high probability of being repaired before a second one fails.

Component repair increases the reliability of either active parallel or standby parallel systems. Moreover, either system may be analyzed using Markov methods. In what follows we derive the reliability for a system consisting of a primary and a backup unit. We assume that the primary unit can be repaired on line. For clarity, we assume that failure of the backup unit in standby mode and switching failures can be neglected.

The state transition diagram shown in Fig. 9.10 differs from Fig. 9.5

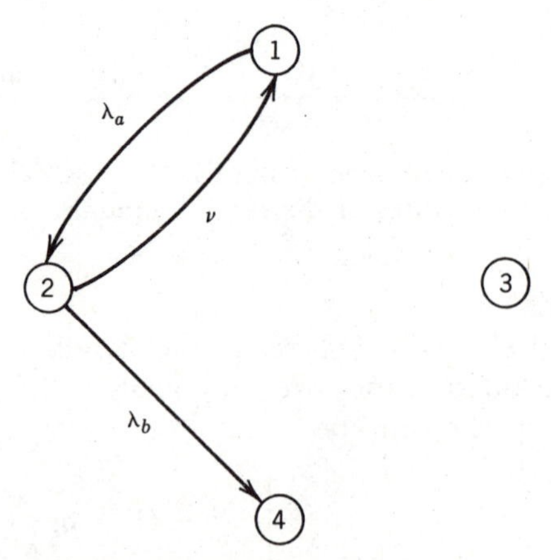

FIGURE 9.10 State transition diagram with primary system repair.

only in that the repair transition has been added. This creates an additional source term of $\nu P_2(t)$ in Eq. 9.39,

$$\frac{d}{dt} P_1(t) = -\lambda_a P_1(t) + \nu P_2(t), \tag{9.74}$$

and the corresponding loss term is substracted from Eq. 9.40,

$$\frac{d}{dt} P_2(t) = \lambda_a P_1(t) - (\lambda_b + \nu)\, P_2(t). \tag{9.75}$$

The reliability, once again, is calculated from Eq. 9.46.

The equations can no longer be solved one at a time, sequentially, as in the previous examples, for now $P_1(t)$ depends on $P_2(t)$. Laplace transforms may be used to solve Eqs. 9.74 and 9.75, but to avoid introducing additional nomenclature we use the following technique instead. Suppose that we look for solutions of the form

$$P_1(t) = Ce^{-\alpha t}; \qquad P_2(t) = C'e^{-\alpha t}, \tag{9.76}$$

where C, C', and α are constants. Substituting these expressions into Eqs. 9.74 and 9.75, we obtain

$$-\alpha C = -\lambda_a C + \nu C'; \qquad -\alpha C' = \lambda_a C - (\lambda_b + \nu)C'. \tag{9.77}$$

The constants C and C' may be eliminated between these expressions to yield the form

$$\alpha^2 - (\lambda_a + \lambda_b + \nu)\alpha + \lambda_a\lambda_b = 0 \tag{9.78}$$

Solving this quadratic equation, we find that there are two solutions for α:

$$\alpha_\pm = \frac{(\nu + \lambda_a + \lambda_b)}{2} \pm \tfrac{1}{2}\,[(\nu + \lambda_a + \lambda_b)^2 - 4\lambda_a\lambda_b]^{1/2}. \tag{9.79}$$

Thus our solutions have the form

$$P_1(t) = C_+e^{-\alpha_+ t} + C_-e^{-\alpha_- t}, \tag{9.80}$$

$$P_2(t) = C'_+e^{-\alpha_+ t} + C'_-e^{-\alpha_- t}. \tag{9.81}$$

We must use the initial conditions along with Eq. 9.79 to evaluate $C_\pm$ and $C'_\pm$. Combining Eqs. 9.80 and 9.81 with the initial conditions $P_1(0) = 1$ and $P_2(0) = 0$, we have

$$C_+ + C_- = 1; \qquad C'_+ + C'_- = 0. \tag{9.82}$$

Furthermore, adding Eqs. 9.77, we may write, for α_+ and α_-,

$$\alpha_\pm C_\pm = (\lambda_b - \alpha_\pm)C'_\pm. \tag{9.83}$$

These four equations can be solved for $C_\pm$ and $C'_\pm$. Then, after some algebra, we may add Eqs. 9.80 and 9.81 to obtain from Eq. 9.46

$$R(t) = \frac{\alpha_+}{\alpha_+ - \alpha_-} e^{-\alpha_- t} - \frac{\alpha_-}{\alpha_+ - \alpha_-} e^{-\alpha_+ t}. \tag{9.84}$$

The improvement in reliability with standby systems is indicated in Fig. 9.11, where the two units are assumed to be identical, $\lambda_a = \lambda_b = \lambda$, and plots are shown for different ratios of ν/λ. In the usual case, where $\nu >> \lambda$, it is easily shown that $\alpha_+ >> \alpha_-$, so that the second term in Eq. 9.84 can be neglected, and that $\alpha_- \approx -\lambda_a\lambda_b/\nu$. Hence we may write, approximately,

$$R(t) \approx \exp\left(-\frac{\lambda_a\lambda_b}{\nu} t\right). \tag{9.85}$$

In the situation in which $\nu >> \lambda_a, \lambda_b$, the deterioration of reliability is likely to be governed not by the possibility that the backup system will fail before the primary system is repaired, but rather by one of two other possibilities: (1) that switching to the backup system will fail, or (b) that the backup system has failed. These failures are dealt with either by improving the switching and standby mode reliabilities or by utilizing an active parallel system with repairable components. Then the switching is obviated, and the configuration is more likely to be designed so that failures in either component are revealed immediately.

9.4 MULTICOMPONENT SYSTEMS

The models described in the two preceding sections concern the dependencies between only two components. In order to make use Markov methods in realistic situations, however, it is often necessary to consider dependencies between more than two components or to build the dependency models

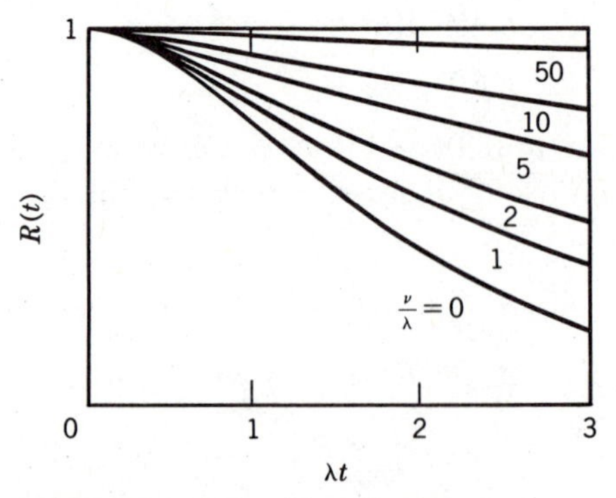

FIGURE 9.11 The effect of primary system repair rate on the reliability of a standby system.

into many-component systems. In this section we first undertake to generalize Markov methods for the consideration of dependencies between more than two components. We then examine how to build dependency models into larger systems in which some of the component failures are independent of the others.

Multicomponent Markov Formulations

The treatment of larger sets of components by Markov methods is streamlined by expressing the coupled set of state transition equations in matrix form. Moreover, the resulting coefficient matrix can be used to check on the formulation's consistency and to gain some insight into the physical processes at play. To illustrate, we first put one of the two-component, four-state systems discussed earlier into matrix form. The generalization to larger systems is then obvious.

Consider the backup configuration shown in Fig. 9.7, in which we allow for failure of the unit in the standby mode. The four equations for the $P_i(t)$ are given by Eqs. 9.55 through 9.58. If we define a vector $\mathsf{P}(t)$, whose components are $P_1(t)$ through $P_4(t)$, we may write the set of simultaneous differential equations as

$$\frac{d}{dt}\begin{bmatrix} P_1(t) \\ P_2(t) \\ P_3(t) \\ P_4(t) \end{bmatrix} = \begin{bmatrix} -\lambda_a - \lambda_b^+ & 0 & 0 & 0 \\ \lambda_a & -\lambda_b & 0 & 0 \\ \lambda_b^+ & 0 & -\lambda_a & 0 \\ 0 & \lambda_b & \lambda_a & 0 \end{bmatrix} \begin{bmatrix} P_1(t) \\ P_2(t) \\ P_3(t) \\ P_4(t) \end{bmatrix}. \tag{9.86}$$

Consider next a system with three components in parallel, as shown in Fig. 9.1*b*. Suppose that this is a load-sharing system in which the component failure rate increases with each component failure:

λ_1 = component failure rate with no component failures,
λ_2 = component failure rate with one component failure,
λ_3 = component failure rate with two component failures.

If we again enumerate the possible system states in Table 9.1, the state transition diagram will appear as in Fig. 9.12. From this diagram we may construct the equations for the $P_i(t)$. In matrix form they are

$$\frac{d}{dt}\begin{bmatrix} P_1(t) \\ P_2(t) \\ P_3(t) \\ P_4(t) \\ P_5(t) \\ P_6(t) \\ P_7(t) \\ P_8(t) \end{bmatrix} = \begin{bmatrix} -3\lambda_1 & 0 & 0 & 0 & 0 & 0 & 0 & 0 \\ \lambda_1 & -2\lambda_2 & 0 & 0 & 0 & 0 & 0 & 0 \\ \lambda_1 & 0 & -2\lambda_2 & 0 & 0 & 0 & 0 & 0 \\ \lambda_1 & 0 & 0 & -2\lambda_2 & 0 & 0 & 0 & 0 \\ 0 & \lambda_2 & \lambda_2 & 0 & -\lambda_3 & 0 & 0 & 0 \\ 0 & \lambda_2 & 0 & \lambda_2 & 0 & -\lambda_3 & 0 & 0 \\ 0 & 0 & \lambda_2 & \lambda_2 & 0 & 0 & -\lambda_3 & 0 \\ 0 & 0 & 0 & 0 & \lambda_3 & \lambda_3 & \lambda_3 & 0 \end{bmatrix} \begin{bmatrix} P_1(t) \\ P_2(t) \\ P_3(t) \\ P_4(t) \\ P_5(t) \\ P_6(t) \\ P_7(t) \\ P_8(t) \end{bmatrix}, \tag{9.87}$$

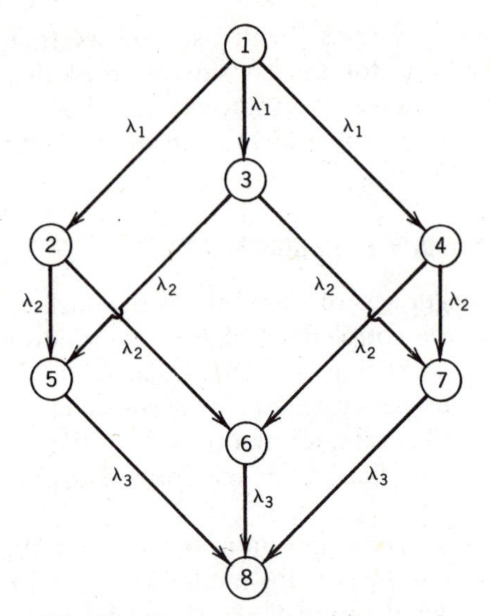

FIGURE 9.12 State transition diagram for a three-component parallel system.

where there are now $2^3 = 8$ states in all. The generalization to more components is straightforward, provided that the logical structure of the dependencies is understood.

Equations 9.86 and 9.87 may be used to illustrate an important property of the coefficient matrix, one which serves as an aid in constructing the set of equations from the state transition diagram. Each transition out of a state must terminate in another state. Thus, for each negative entry in the coefficient matrix, there must be a positive entry in the same column, and the sum of the elements in each column must be zero. Thus the matrix may be constructed systematically by considering the transitions one at a time. If the transition originates from the ith state, the failure rate is subtracted from the ith diagonal element. If the transition is to the jth state, the failure rate is then added to the jth row of the same column.

A second feature of the coefficient matrix involves the distinction between operational and failed states. In reliability calculations we do not allow a system to be repaired once it fails. Hence there can be no way to leave a failed state. In the coefficient matrix this is indicated by the zero in the diagonal element of each failed state. This is not the case, however, when availability rather than reliability is being calculated. Availability calculations are discussed in the following section.

For larger systems of equations it is often more convenient to write Markov equations in the matrix form

$$\frac{d}{dt} \mathsf{P}(t) = \mathsf{M}\mathsf{P}(t), \tag{9.88}$$

where $\mathbf{P}$ is a column vector with components $P_1(t)$, $P_2(t)$, . . ., and $\mathbf{M}$ is referred to as the Markov transition matrix. Instead of repeating the entire set of equations, as in Eqs. 9.86 and 9.87, we need write out only the matrix. Thus, for example, the matrix for Eq. 9.86 is

$$\mathbf{M} = \begin{bmatrix} -\lambda_a - \lambda_b^+ & 0 & 0 & 0 \\ \lambda_a & -\lambda_b & 0 & 0 \\ \lambda_b^+ & 0 & -\lambda_a & 0 \\ 0 & \lambda_b & \lambda_a & 0 \end{bmatrix}. \tag{9.89}$$

The dimension of the matrix increases as 2^N, where N is the number of components. For larger systems, particularly those whose components are repaired, the simple solution algorithms discussed earlier become intractable. Instead, more general Laplace transform techniques may be required. If there are added complications, such as time-dependent failure rates, the equations may require solution by numerical integration or by Monte Carlo simulation.

EXAMPLE 9.4

A 2/3 system is constructed as follows. After the failure of either component a or c, whichever comes first, component b is switched on. The system fails after any two of the components fail. The components are identical with failure rate λ. (*a*) Draw a state transition diagram for the system. (*b*) Write the corresponding Markov transition matrix. (*c*) Find the system reliability $R(t)$. (*d*) Determine the reliability when time is set equal to the MTTF one component.

Solution For this three-component system, there are eight states. We define these according to Table 9.1.
(*a*) The state transition diagram is shown in Fig. 9.13. Note that states 3 and 8 are not reachable.
(*b*) The Markov transition matrix is

$$\mathbf{M} = \begin{bmatrix} -2\lambda & 0 & 0 & 0 & 0 & 0 & 0 & 0 \\ \lambda & -2\lambda & 0 & 0 & 0 & 0 & 0 & 0 \\ 0 & 0 & 0 & 0 & 0 & 0 & 0 & 0 \\ \lambda & 0 & 0 & -2\lambda & 0 & 0 & 0 & 0 \\ 0 & \lambda & 0 & 0 & 0 & 0 & 0 & 0 \\ 0 & \lambda & 0 & \lambda & 0 & 0 & 0 & 0 \\ 0 & 0 & 0 & \lambda & 0 & 0 & 0 & 0 \\ 0 & 0 & 0 & 0 & 0 & 0 & 0 & 0 \end{bmatrix} \quad ◀$$

(*c*) The reliability is given by $R(t) = P_1(t) + P_2(t) + P_4(t)$; thus only three of the eight equations need be solved. First, $dP_1/dt = -2\lambda P_1$, with $P_1(0) = 1$ yields $P_1(t) = e^{-2\lambda t}$. The equations for $P_2 + P_4$ are the same:

$$\frac{dP_n}{dt} = \lambda P_1 - 2\lambda P_n, \quad P_n(0) = 0; \qquad n = 2, 4.$$

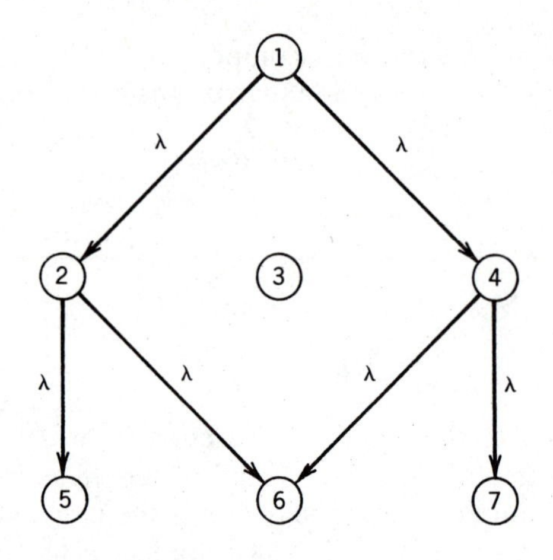

FIGURE 9.13 State transition diagram for Example 9.4.

Therefore,

$$\frac{dP_n}{dt} = \lambda e^{-2\lambda t} - 2\lambda P_n.$$

We use the integrating factor $e^{2\lambda t}$ to obtain

$$\frac{d}{dt}(P_n e^{-2\lambda t}) = \lambda.$$

Then integrating between 0 and t, we obtain

$$P_n(t)e^{2\lambda t} - P_n(0) = \lambda t.$$

Thus

$$P_n(t) = \lambda t e^{-2\lambda t}, \qquad n = 2, 4.$$

Substituting into $R(t) = P_1 + P_2 + P_4$ yields

$$R(t) = (1 + 2\lambda t)e^{-2\lambda t}. \blacktriangleleft$$

(*d*) $t = \text{MTTF} \equiv 1/\lambda$. Then

$$R(\text{MTTF}) = (1 + 2 \times 1)e^{-2\times 1} = 0.406. \blacktriangleleft$$

Combinations of Subsystems

In principle, we can treat systems of many components using Markov methods. However, with 2^N equations the solutions soon become unmanageable. A more efficient approach is to define one or more subsystems containing the components with dependencies between them. These subsystems can then be treated as single blocks in a reliability block diagram, and the system reliability can be calculated using the techniques of Chapter 7, since the failures in the subsystem defined in this way are independent of one another.

To understand this procedure, consider the system configurations shown in Fig. 9.14. In Fig. 9.14*a* is shown the convention for drawing a two-component standby system of the type discussed in the preceding section as a reliability block diagram. In Fig. 9.14*b* the standby parallel subsystem, consisting of components *a* and *b*, is in series with two other components. The reliability of the standby subsystem (with no switching errors) is given by Eq. 9.63. Therefore, we define the reliability of the standby subsystem as

$$R_{sb}(t) = e^{-\lambda_a t} + \frac{\lambda_a}{\lambda_a + \lambda_b^+ - \lambda_b}\left[e^{-\lambda_b t} - e^{-(\lambda_a+\lambda_b^+)t}\right]. \tag{9.90}$$

Then, if the failures in components *c* and *d* are independent of those in the standby subsystem, the system reliability can be calculated using the product rule

$$R(t) = R_{sb}(t)R_c(t)R_d(t). \tag{9.91}$$

Generalization of this technique to more complex configurations is straightforward.

The configuration in Fig. 9.14*c* illustrates a somewhat different situation. Here the primary and standby subsystems themselves each consist of two components, *a* and *c*, and *b* and *d*, respectively. Here we may simplify the Markov analysis by first combining the four components into two subsystems, each having a composite failure rate. Thus we define

$$\lambda_{ac} = \lambda_a + \lambda_c, \tag{9.92}$$

$$\lambda_{bd} = \lambda_b + \lambda_d, \tag{9.93}$$

and

$$\lambda_{bd}^+ = \lambda_b^+ + \lambda_d^+. \tag{9.94}$$

We may again apply Eq. 9.90 to calculate the system reliability if we replace λ_a, λ_b, and λ_b^+ with λ_{ac}, λ_{bd}, and λ_{bd}^+, respectively.

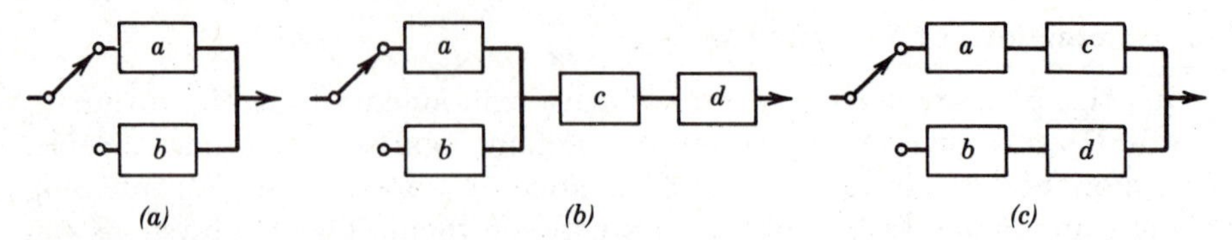

FIGURE 9.14 Standby configurations.

9.5 AVAILABILITY

In availability, as well as in reliabiity, there are situations in which the component failures cannot be considered independent of one another. These include shared-load and backup systems in which all the components are repairable. They may also include a variety of other situations in which the dependency is introduced by the limited number of repair personnel or by replacement parts that may be called on to put components into working order. Thus, for example, the repair of two redundant components cannot be considered independent if only one crew is on station to carry out the repairs.

The dependencies between component failure and repair rates may be approached once more with Markov methods, provided that the failures are revealed, and that the failure and repair rates are time-independent. Although we have already treated the repair of components in reliability calculations, there is a fundamental difference in the analysis that follows. In reliability calculations components can be repaired only as long as the system has not failed; the analysis terminates with the first system failure. In availability calculations we continue to repair components after a system failure in order to bring the system back on line, that is, to make it available once again.

The differences between Markov reliability and availability calculations for systems with repairable components can be illustrated best in terms of the matrix notion developed in the preceding section. For this reason we first illustrate an availability calculation with a system for which the reliability was calculated in the preceding section, standby redundance. We then illustrate the limitation placed on the availability of an active parallel configuration by the availability of only one repair crew.

Standby Redundancy

Suppose that we consider the reliability of a two-component system, consisting of a primary and a backup unit. We assume that switching failures and failure in the standby mode can be neglected. In the preceding section the analysis of such a system is carried out assuming that the primary unit can be repaired with a rate ν. Since there are only three states with nonzero

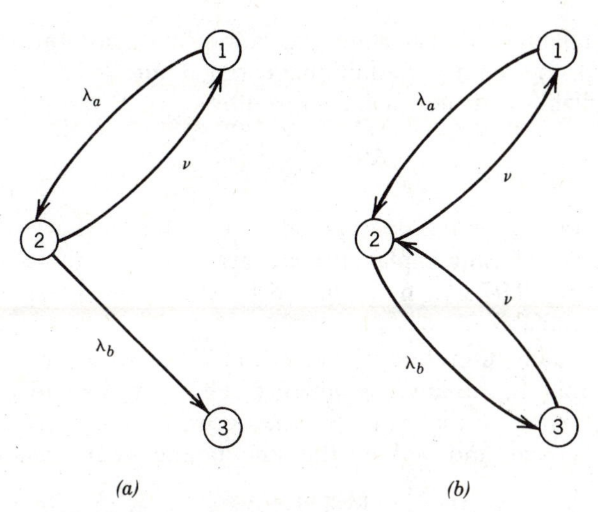

FIGURE 9.15 State transition diagrams for a standby system: (*a*) for reliability, (*b*) for availability.

probabilities as indicated in Table 9.3, the state transition diagram may be drawn as in Fig. 9.15*a*, where state 3 is the failed state. The transition matrix for Eq. 9.88 is then given by

$$\mathbf{M} = \begin{bmatrix} -\lambda_a & \nu & 0 \\ \lambda_a & -\lambda_b - \nu & 0 \\ 0 & \lambda_b & 0 \end{bmatrix}. \tag{9.95}$$

The estimate of the availability of this system involves one additional state transition. In order for the system to go back into operation after both units have failed, we must be able to repair the backup unit. This requires an added repair transition from state 3 to state 2, as indicated in Fig. 9.15*b*. This repair transition is represented by two additional terms in the Markov transition matrix. We have

$$\mathbf{M} = \begin{bmatrix} -\lambda_a & \nu & 0 \\ \lambda_a & -\lambda_b - \nu & \nu \\ 0 & \lambda_b & -\nu \end{bmatrix}. \tag{9.96}$$

Here we assume that when both units have failed, the backup unit will be repaired first; we also assume that the repair rates are equal. More general cases may also be considered.

An important difference can be seen in the structures of Eqs. 9.95 and 9.96. In Eq. 9.96 all the diagonal elements are nonzero. This is a fundamental difference from reliability calculations. In availability calculations the system must always be able to recover from any failed state. Thus there can be no zero diagonal elements, for these would represent an absorbing or inescapable failed state; transitions can always be made out of operating states through the failure of additional components.

The availability of the system is given by

$$A(t) = \sum_{i \in O} P_i(t), \tag{9.97}$$

where the sum is over the operational states. The Markov equations, Eq. 9.88, may be solved using Laplace transforms or other methods to determine the $\mathsf{P}(t)$, and Eq. 9.97 may be evaluated for the detailed time dependence of the point availability.

We are usually interested in the asymptotic or steady-state availability, $A(\infty)$, rather than in the time dependence. This quantity may be calculated more simply. We note that as $t \rightarrow \infty$, the derivative on the right-hand side of Eq. 9.88 vanishes and we have the time-independent relationship

$$\mathsf{MP}(\infty) = 0. \tag{9.98}$$

In our problem this represents the three simultaneous equations

$$-\lambda_a P_1(\infty) + \nu P_2(\infty) = 0, \tag{9.99}$$

$$\lambda_a P_1(\infty) - (\lambda_b + \nu) P_2(\infty) + \nu P_3(\infty) = 0, \tag{9.100}$$

and

$$\lambda_b P_2(\infty) - \nu P_3(\infty) = 0. \tag{9.101}$$

This set of three equations is not sufficient to solve for the $P_i(\infty)$. For all Markov transition matrices are singular; that is, the equations are linearly dependent, yielding only $N - 1$ (in our case two) independent relationships. This is easily seen, for adding Eqs. 9.99 and 9.101 yields Eq. 9.100. The needed piece of additional information is the condition that all of the probabilities must sum to one:

$$\sum_i P_i(\infty) = 1. \tag{9.102}$$

In the situation in which we take $\lambda_a = \lambda_b = \lambda$, our problem is easily solved. Combining Eqs. 9.99, 9.101, and 9.102, we obtain

$$P_1(\infty) = \left[1 + \frac{\lambda}{\nu} + \left(\frac{\lambda}{\nu}\right)^2\right]^{-1}, \tag{9.103}$$

$$P_2(\infty) = \left[1 + \frac{\lambda}{\nu} + \left(\frac{\lambda}{\nu}\right)^2\right]^{-1} \frac{\lambda}{\nu}, \tag{9.104}$$

and

$$P_3(\infty) = \left[1 + \frac{\lambda}{\nu} + \left(\frac{\lambda}{\nu}\right)^2\right]^{-1} \left(\frac{\lambda}{\nu}\right)^2. \tag{9.105}$$

The steady-state availability may be found by setting $t = \infty$ in Eq. 9.97:

$$A(\infty) = 1 - \left[1 + \frac{\lambda}{\nu} + \left(\frac{\lambda}{\nu}\right)^2\right]^{-1}\left(\frac{\lambda}{\nu}\right)^2. \tag{9.106}$$

If we further assume that $\lambda/\nu << 1$, we may write

$$A(\infty) \approx 1 - \left(\frac{\lambda}{\nu}\right)^2. \tag{9.107}$$

EXAMPLE 9.5

Suppose that the system availability for standby systems must be 0.9. What is the maximum acceptable value of the failure to repair rate ratio λ/ν?

Solution Let $x = \lambda/\nu$ in Eq. 9.106. Then

$$A(\infty) = 1 - (1 + x + x^2)^{-1}(x^2).$$

Converting to a quadratic equation, we have $x^2 - \gamma x - \gamma = 0$, where

$$\gamma = \frac{1 - A}{A} = \frac{1 - 0.9}{0.9} = \frac{1}{9}$$

and

$$\frac{\lambda}{\nu} \equiv x = \frac{+\gamma + \gamma\sqrt{1 + 4/\gamma}}{2} = 0.393. \blacktriangleleft$$

If instead the rare-event approximation is used,

$$\frac{\lambda}{\nu} \approx \sqrt{1 - A(\infty)} = \sqrt{1 - 0.9} = 0.316. \blacktriangleleft$$

Other configurations are also possible. If two repair crews are available, repairs may be carried out on the primary and backup units simultaneously; the result is the four-state system of Table 9.2. As indicated in Fig. 9.16*a*, it is possible to get the primary unit running before the backup unit is repaired. In this situation states 1, 2, and 3 are operating states and must be included in the sum in Eq. 9.97. The Markov matrix now becomes

$$\mathbf{M} = \begin{bmatrix} -\lambda_a & \nu & \nu & 0 \\ \lambda_a & -\nu - \lambda_b & 0 & \nu \\ 0 & 0 & -\nu - \lambda_a & \nu \\ 0 & \lambda_b & \lambda_a & -2\nu \end{bmatrix}. \tag{9.108}$$

Other possibilities may also be added. For example, if switching failures and failures of the backup unit while in standby are not negligible, the state transition diagram is modified as shown in Fig. 9.16*b*, where p represents the probability of failure in switching from the primary to the backup, and

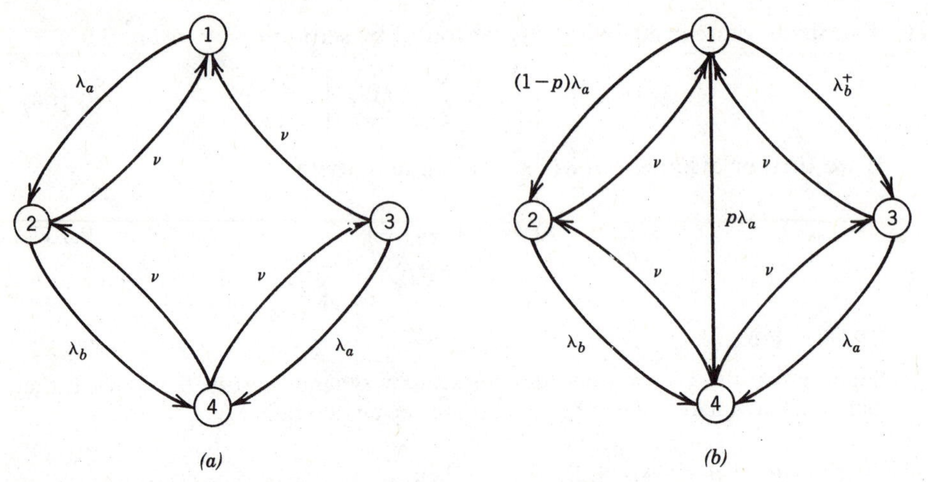

FIGURE 9.16 State transition diagrams for repairable standby systems.

λ_b^+ the standby failure rate of the backup unit. The Markov transition matrix corresponding to Fig. 9.16*b* is

$$\mathsf{M} = \begin{bmatrix} -\lambda_a - \lambda_b^+ & \nu & \nu & 0 \\ (1-p)\lambda_a & -\lambda_b - \nu & 0 & \nu \\ \lambda_b^+ & 0 & -\lambda_a - \nu & \nu \\ p\lambda_a & \lambda_b & \lambda_a & -2\nu \end{bmatrix}. \tag{9.109}$$

To recapitulate, steady-state availability problems are solved by the same procedure. Any $N - 1$ of the N equations represented by Eq. 9.98 are combined with the condition, Eq. 9.102, that the probabilities must add to one, to solve for the components of $\mathsf{P}(\infty)$. These are then substituted into Eq. 9.97 with the sum taken over all operating states to obtain the availability.

Shared Repair Crews

We conclude with the analysis of an active parallel system consisting of two identical units. We assume that the failure rates are identical and that they are independent of the state of the other unit. We also assume that the repair rates for the two units are the same. In this situation the failures and repairs of the two units are independent, provided that each unit has its own repair crew. The availability is then given by Eq. 8.112. The dependency is introduced not by a hardware failure, as in the case of standby redundance, but by an operational decision to provide a single repair crew that can handle only one unit at a time.

The state transition diagram for the system using two repair crews is shown in Fig. 9.17*a*. Since the availability can be calculated from the component availabilities, as in Eq. 8.112, we shall not pursue the Markov solution

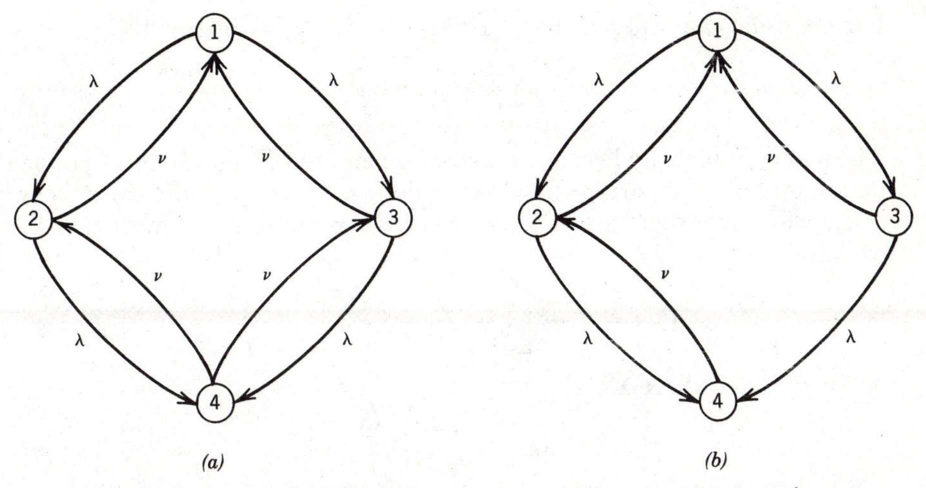

FIGURE 9.17 State transition diagrams for an active parallel system: (*a*) two repair crews, (*b*) one repair crew.

further. Our attention is directed to the system using one repair crew, indicated by the state transition diagram given in Fig. 9.17*b*.

The transition matrix corresponding to Fig. 9.17*b* is

$$\mathbf{M} = \begin{bmatrix} -2\lambda & \nu & \nu & 0 \\ \lambda & -\lambda - \nu & 0 & \nu \\ \lambda & 0 & -\lambda - \nu & 0 \\ 0 & \lambda & \lambda & -\nu \end{bmatrix}. \tag{9.110}$$

We solve the equations obtained from this matrix along with Eq. 9.102 to yield, after some algebra,

$$P_1(\infty) = \left[1 + 2\frac{\lambda}{\nu} + 2\left(\frac{\lambda}{\nu}\right)^2\right]^{-1}, \tag{9.111}$$

$$P_2(\infty) + P_3(\infty) = \left[1 + 2\frac{\lambda}{\nu} + 2\left(\frac{\lambda}{\nu}\right)^2\right]^{-1} \frac{2\lambda}{\nu}, \tag{9.112}$$

and

$$P_4(\infty) = \left[1 + 2\frac{\lambda}{\nu} + 2\left(\frac{\lambda}{\nu}\right)^2\right]^{-1} \frac{2\lambda^2}{\nu^2}. \tag{9.113}$$

Substitution of the results into Eq. 9.97 then yields for the steady-state availability

$$A(\infty) = 1 - \left[1 + 2\frac{\lambda}{\nu} + 2\left(\frac{\lambda}{\nu}\right)^2\right]^{-1} \frac{2\lambda^2}{\nu^2}. \tag{9.114}$$

For the usual case where $\lambda/\nu << 1$, this may be approximated by

$$A(\infty) \simeq 1 - 2\left(\frac{\lambda}{\nu}\right)^2. \tag{9.115}$$

The loss in availability because a second repair crew is not on hand can be determined by comparing these expressions to those obtained for system availability when there are two repair crews. From Eq. 8.112, with $N = 2$, we have

$$A(\infty) = 1 - \left[1 + 2\frac{\lambda}{\nu} + \left(\frac{\lambda}{\nu}\right)^2\right]^{-1}\left(\frac{\lambda}{\nu}\right)^2, \tag{9.116}$$

or for the case where $\lambda/\nu << 1$,

$$A(\infty) \simeq 1 - \left(\frac{\lambda}{\nu}\right)^2. \tag{9.117}$$

Thus the unavailability is roughly doubled if only one repair crew is present.

EXAMPLE 9.6

A system has an availability of 0.90. Two such systems, each with its own repair crew, are placed in parallel. What is the availability (*a*) for a standby parallel configuration with perfect switching and no failure of the unit in standby; (*b*) for an active parallel configuration? (*c*) What is the availability if only one repair crew is assigned to the active parallel configuration?

Solution The system availability is given by $A(\infty) = \nu/(\nu + \lambda)$. Therefore $\nu/\lambda = A(\infty)/[1 - A(\infty)] = 0.9/(1 - 0.9) = 9$; $\lambda/\nu = 0.1111$.
(*a*) From Eq. 9.106,

$$A(\infty) = 1 - \frac{(0.1111)^2}{1 + 0.1111 + (0.1111)^2} = 0.989. \blacktriangleleft$$

(*b*) From Eq. 9.116,

$$A(\infty) = 1 - \frac{(0.1111)^2}{1 + 2 \times 0.1111 + (0.1111)^2} = 0.990. \blacktriangleleft$$

(*c*) From Eq. 9.114,

$$A(\infty) = 1 - \frac{2 \times (0.1111)^2}{1 + 2 \times 0.1111 + 2 \times (0.1111)^2} = 0.980. \blacktriangleleft$$

Bibliography

Barlow, R. E., and F. Proschan, *Mathematical Theory of Reliability,* Wiley, New York, 1965.

Green, A. E., and A. J. Bourne, *Reliability Technology,* Wiley, New York, 1972.

Henley, E. J., and H. Kumamoto, *Reliability Engineering and Risk Assessment,* Prentice-Hall, Englewood Cliffs, NJ, 1981.

McCormick, N. J., *Reliability and Risk Analysis,* Academic Press, New York, 1981.

Sandler, G. H., *System Reliability Engineering,* Prentice-Hall, Englewood Cliffs, NJ, 1963.

EXERCISES

9.1 Enumerate the 16 possible states of a four-component system by writing a table similar to Table 9.1. For the following configurations which are the failed states?

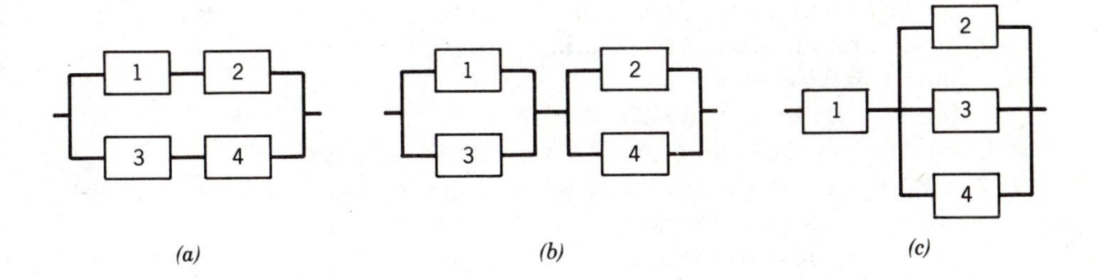

9.2 Repeat Exercise 9.1 for the standby configurations shown in Fig. 9.14.

9.3 Two stamping machines operate in parallel positions on an assembly line, each with the same MTTF at the rated speed. If one fails, the other takes up the load by doubling its operating speed. When this happens, however, the failure rate also doubles. Assuming no repair, how many MTTF for a machine at the rated speed will elapse before the system reliability drops below (*a*) 0.99, (*b*) 0.95, (*c*) 0.90?

9.4 Consider a system consisting of two identical units in an active parallel configuration. The units cannot be repaired. Moreover, because they share loads, the failure rate λ^* of the remaining unit is substantially larger than the unit failure rates when both are operating.

(*a*) Find an approximation for the system reliability for a short period of time (i.e., $\lambda t << 1$ and $\lambda^* t << 1$).

(*b*) How large must the ratio of λ^*/λ become before the MTTF of the system is no greater than that for a single unit with failure rate λ?

9.5 For the idealized standby system for which the reliability is given by Eq. 9.52,

(*a*) Calculate the MTTF in terms of λ.

(*b*) Plot the time-dependent failure rate $\lambda(t)$ and compare your results to the active parallel system depicted in Fig. 7.3*b*.

9.6 Verify Eqs. 9.42 through 9.45.

9.7 Derive Eq. 9.52 assuming that $\lambda_b = \lambda_a$ from the beginning.

9.8 Calculate the variance for the time to failure for two identical units, each with a failure rate λ, placed in standby parallel configuration, and compare your results to the variance of the same two units placed in active parallel configuration. (Ignore switching failures and failures in the standby mode.)

9.9 Show that Eq. 9.64 reduces to Eq. 9.52 as $\lambda^+ \to 0$.

9.10 Verify Eq. 9.68.

9.11 Under a specified load the failure rate of a turbogenerator is decreased by 30% if the load is shared by two such generators. A designer must decide whether to put two such generators in active or standby parallel configuration. Assuming that there are no switching failures or failures in the standby mode,

(*a*) Which system will yield the larger MTTF?

(*b*) What is the ratio of MTTF for the two systems?

9.12 Consider a standby system in which there is a switching failure probability p and a failure rate in the standby mode of λ_b^+.

(*a*) Draw the transition diagram.

(*b*) Write the Markov equations.

(*c*) Solve for the system reliability.

(*d*) Reduce the reliability to the situation in which the units are identical, $\lambda_a = \lambda_b = \lambda$, $\lambda_b^+ = \lambda$.

9.13 Consider the following configuration consisting of four identical units with failure rate λ and with negligible switching and standby failure rates. There is no repair.

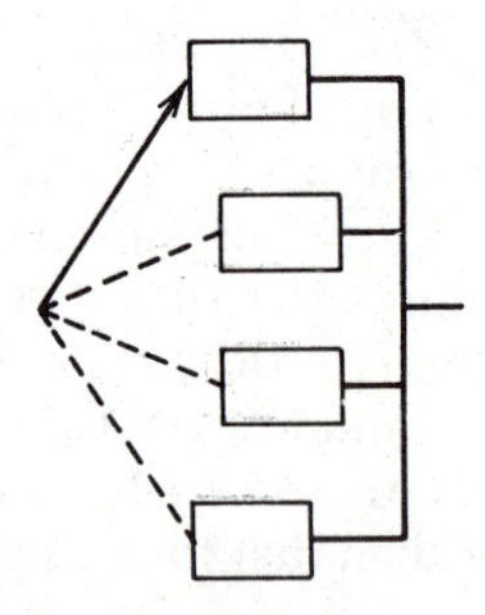

(*a*) Show that the reliability can be expressed in terms of the Poisson distribution discussed in Chapter 4.

(*b*) Evaluate the reliability in the rare-event approximation for small λt.

(*c*) Compare the result from part *b* to the rare-event approximation for four identical units in active parallel configuration, as developed in Chapter 7, and evaluate the reliabilities for $\lambda t = 0.1$.

9.14 For the following systems, assume unit failure rates λ, no repair, and no switching or standby failures.

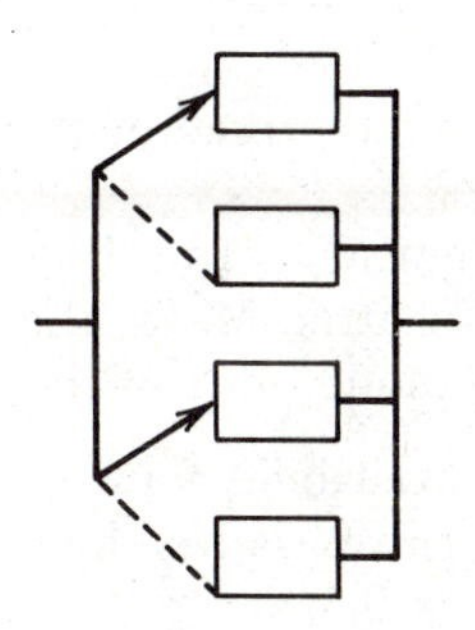

(*a*) Calculate the reliability.

(*b*) Approximate the result in part *a* by the rare-event approximation for small λt, and compare your result to that for four units in active parallel configuration.

9.15 Derive Eq. 9.72.

9.16 A design team is attempting to optimize the reliability of a navigation device. The choices for the rate gyroscopes are (*a*) a hot standby system consisting of two gyroscopes, and (*b*) a 2/3 voting system consisting of three gyroscopes. The mission time is 20 hr, and the gyroscope failure rate is 3×10^{-5}/hr. What is the greatest probability of switching failure in the hot standby system for which mission reliability is greater than that of the $\frac{2}{3}$ system? Assume that failures in logic on the 2/3 system can be neglected. (*Hint*: Assume rare-event approximations for the gyroscope failures.)

9.17 Derive Eqs. 9.82 and 9.83.

9.18 Derive Eqs. 9.103 through 9.105.

9.19 Consider the 2/3 standby configuration shown on the following page. It consists of three identical units; two units are required for operation. If either unit *a* or *c* fails, unit *b* is switched on. Ignore switching failures and repair, but assume failure rate λ and λ^* in the operating and standby modes.

(*a*) Enumerate the possible system states and draw a transition diagram.

(*b*) Write the Markov equations for the system.

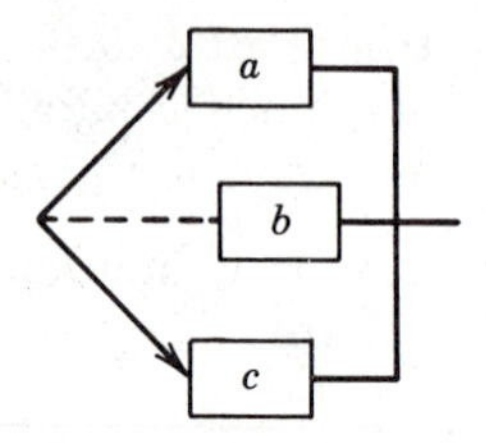

9.20 Assume that the units in the preceding problem have identical repair rates ν.
(*a*) Enumerate the system states and draw a transition diagram.
(*b*) Write the transition matrix, **M,** for the Markov equations.
(*c*) Determine the asymptotic value of the system availability.

9.21 (*a*) Find the asymptotic availability for a standby system with two repair crews; the Markov matrix is given by Eq. 9.108. Assume that $\lambda_a = \lambda_b = 0.01/\text{hr}$ and $\nu = 0.5/\text{hr}$.
(*b*) Evaluate the asymptotic availability for a standby system for the same data, except that there is only one repair crew. The Markov matrix is given by Eq. 9.96.

9.22 A system has an asymptotic availability of 0.93. A second redundant system is added, but only the original repair crew is retained. Assuming that all failures are revealed, estimate the asymptotic availability.

9.23 Assume that the units in Exercise 9.13 all have failure and repair rates λ and ν. A single crew repairs the most recently failed unit first.
(*a*) Determine the asymptotic availability in terms of ν and λ.
(*b*) Approximate your result for the case $\lambda/\nu << 1$.
(*c*) Compare your result to that for some of the same units in active parallel configuration when $\lambda/\nu = 0.02$.

9.24 Two ventilation units are in active parallel configuration. Each has an MTTF of 120 hr. Each is attended by a repair crew, and the MTTR is known to be 8 hr.
(*a*) Calculate the availability, assuming that either unit can provide adequate ventilation.
(*b*) The units are replaced by new models with an MTTF of 200 hr. Can the staff be reduced to one repair crew without a net loss of availability? (Assume that the MTTR remains the same.)

9.25 Consider the system in Exercise 9.19 with the assumption that $\lambda^* = 0$. If each unit has a repair rate ν,
(*a*) Find the expression for the asymptotic availability, that there are two repair crews.
(*b*) Reduce the result of part *a* to a rare-event approximation in which $\lambda/\nu << 1$.

CHAPTER 10

System Safety Analysis

10.1 INTRODUCTION

In this chapter the discussion of system safety analysis presents a different emphasis from the more general reliability considerations contained in Chapters 1 through 9. Although all system failures are included in the determination of reliability, our attention now is turned specifically to the failures that may create safety hazards. The analysis of such hazards is often difficult, for with reasonable precautions taken in design and manufacture, failures that cause safety problems usually have small probabilities of occurring, therefore complicating the collection of data needed for analysis and for making system improvements.

The problems inherent in system safety analysis may be seen by examining its application to mass-produced items such as household appliances. Although reliability testing is invariably done on such items, breakdowns for the most part do not endanger people but are simply failures to operate. In contrast, failures of appliances that have grave consequences—such as electrocution—happen only very rarely when the equipment has shortcomings or is used in unanticipated environments or for inappropriate purposes. Conversely, for systems whose every failure endangers safety, as in an artificial respirator or other life support systems, the design criteria are likely to require such a high degree of reliability that failures occur only in circumstances not anticipated by the designer.

System safety analysis as a discipline has derived much of its importance from being applied to industrial activities that may engender accidents of grave consequences. Such activities are carried out by systems for which safety analysis is very difficult. If we examine, in detail, historic accidents such as the 1985 crash of the Boeing 747 in Japan, the 1979 meltdown of the reactor core at Three Mile Island, or the disastrous chemical leak at

Bhopal, India in 1984, some of the difficulties begin to become apparent. The first problem is that the system is likely to have very small probabilities of catastrophic failure, because it has redundant configurations of critical components. It then follows that the events that are to be avoided have either never occurred, or if they have, only rarely. There are few if any statistics on the probabilities of failure of the systems as a whole, and reliability testing at the system level would be futile. A second problem with the testing of such systems is that whatever accidents have occurred have rarely been the result of component failures of a type that would be easy to predict through reliability testing. Rather, the web of events is usually a complex of equipment failure, faulty maintenance, instrumentation and control problems, and human errors.

A thorough understanding of the functioning of the particular system being considered and an intimate knowledge of how failures are likely to occur are prerequisites for applying any of the formalized safety analysis procedures. Such knowledge must include the many reliability considerations covered in the preceding chapters. But, in addition, considerable effort must be devoted to the study of the physical phenomena that may lead to failures of specific devices. By and large, the detailed knowledge of the relevent failure mechanisms requires a thorough grounding in one or more specific engineering disciplines that goes well beyond what can be covered in a broad-based book on reliability engineering.

In addition to a thorough understanding of equipment failures, we must have a basis for assessing the role played by operators, maintenance crews, and other personnel in potentially hazardous situations. Because of its importance to system safety analysis, we devote the following section to a discussion of human error.

In order to understand how accidents may happen, to estimate their probabilities, and—more important—to reduce the likelihood of their happening, a number of analytical methods have been developed. We introduce three of the more important of these in Section 10.3. Such methods provide formalisms for systematizing the analyst's knowledge, drawing attention to gaps in it, and pointing out areas where improvements in design, manufacture, or operating procedures are likely to be the most fruitful. The methods may be either qualitative or quantitative. From qualitative analysis the logical structure of the failure modes and their relation to one another may become clearer, giving clues to how the possibility of certain accidents may be eliminated or greatly reduced. In quantitative analysis we use failure rate data, repair time estimates, human error estimates, and other available data to predict the probabilities of accidents.

Whether qualitative or quantitative, the methods fall into two broad categories, inductive and deductive.* Inductive methods are those that pos-

* H. R. Roberts, W. E. Vesley, D. F. Haust, and F. F. Goldberg, *Fault Tree Handbook*, U.S. Nuclear Regulatory Commission, NUREG-0492, 1981.

tulate a particular failure or error and then attempt to assess its consequence on the system as a whole. These methods are often referred to as "what if?" analysis, and historically they have been carried out, at least on an informal basis, as an integral part of both equipment design and the setting up of operational and maintenance procedures. In contrast, deductive methods have historically been more closely associated with after-the-fact investigations of accidents, for they begin with a particular undesirable event, the accident, and then look for the root causes or combinations of causes that may have led to it.

After examining briefly the rudiments of some of the more commonly used inductive and deductive methods in Section 10.3, we concentrate our efforts for the remainder of the chapter on a more detailed exposition of fault-tree analysis. Fault trees have become a central tool of safety analysis in a variety of different applications. Fault trees, along with event trees, are also central to the broader discipline of probabilistic risk assessment (PRA), in which quantitative estimates are made of overall risks. Probabilistic risk assessment necessarily involves estimating the consequences of major accidents, as well as the probabilities that they will occur. For some circumstances the determination of consequences is relatively straightforward; for example, if a heart-lung machine fails, the patient will die. In releases of toxic chemicals, nuclear reactor accidents, and a variety of other situations, consequence modeling is far more complex and depends on the details of the particular accident. For this reason PRA can normally be considered only within the framework of the physics, chemistry, or biology of the particular application at hand. For this reason consequence models, and therefore PRA in a broader sense, lie outside the scope of this book.

10.2 HUMAN ERROR

All engineering is a human endeavor, and in the broadest sense most failures are due to human causes, whether they be ignorance, negligence, or limitations of vigilance, strength, and manual dexterity. Designers may fail to understand fully system characteristics or to anticipate properly the nature and magnitudes of the loading to which a system may be subjected or the environmental conditions under which it must operate. Indeed, much of engineering education is devoted to understanding these and related phenomena. Similarly, errors committed during manufacture or construction are attributable either to the personnel involved or to the engineers responsible for the setup of the manufacturing process. Quality assurance programs have a central role in detecting and eliminating such errors in manufacture and construction.

We shall consider here only human errors that are committed after design and manufacture, those that are committed in the operation and maintenance of a system. This is a convenient separation, since design and manufacturing errors, whether they are considered human or not, appear in the as-built system as shortcomings in the reliability of the hardware.

Even with our attention confined to human errors appearing in the operation and maintenance of a system, we find that the uncertainties involved are generally much greater than in the analysis of hardware reliability. As discussed in Chapter 3, there are three categories of uncertainty. First, the natural variability of human performance is considerable. Not only do the capabilities of people differ, but the day-to-day and hour-to-hour performance of any one individual also varies. Second, there is a great deal of uncertainty about how to model probabilistically the variability of human performance, since the interactions with the environment, with stress, and with fellow workers are extremely complex and to a large extent psychological. Third, even when tractable models for limited aspects of human performance can be formulated, the numerical probabilities or model parameters that must be estimated in order to apply them are usually only very approximate, and the range of situations to which they apply is relatively narrow.

It is, nevertheless, necessary to include the effects of human error in the safety analysis of any complex system. For as the consequences of accidents become more serious and more emphasis is put on reliable hardware and highly redundant configurations, an increasing proportion of the risk is likely to come from human error, or more accurately from complex interactions of human shortcomings and equipment problems. Even though accurate predictions of failure probabilities are problemmatical, a great deal may be gained from studying the characteristics of human reliability and contrasting them with those of hardware. From such study comes an insight into how systems may be designed and operated in order to minimize and mitigate accidents in which the operating and maintenance staff may play an important role.

It has been pointed out* that increasingly there is a centralization of systems, whether they be larger-capacity power and chemical plants, aircraft carrying greater number of passengers, or structures with larger capacities. Since human error in the operation of many such centralized systems may lead to accidents of major consequence to life and property, there has been an increased emphasis on plant automation. There are certainly limitations on such automation, particularly when the uncertainty of how an operator may react to a situation is overriden by the need for human adaptability in dealing with conditions that have not or could not be incorporated into the automated control system. Moreover, automated operation does not tend to eliminate humans from consideration, but rather to remove them to tasks of two quite dissimilar varieties; routine tasks of maintaining, testing, and

* J. Rasmussen, "Human Factors in High Risk Technology," in *High Risk Technology,* A. E. Green (ed.) Wiley, New York, 1982.

calibrating equipment; and protective tasks of watching for plant malfunctions and preventing their accident propagation. These two classes of tasks tend to enter system safety considerations in different ways. When humans err in routine testing, maintenance, and repair work, they may introduce latently risky conditions into the plant. Any errors that they make in taking protective actions under emergency conditions may increase the severity of an accident.

The problems inherent in maximizing human reliability for the two classes of tasks may be viewed graphically in Fig. 10.1. Generally, there is an optimum level of psychological stress for human performance. When the level is too low, humans are bored and make careless errors; too high a level may cause them to make a number of inappropriate, near-panic responses to a situation. To illustrate, consider the example of flying a commercial airliner. The pilot's monitoring of controls during level, uneventful flight in a highly automated aircraft would fall on the low level of the curve. The principal danger here is carelessness or lack of attention. Normal take-offs and landings are likely to be closer to the optimum stress level for attentive behavior. At the other extreme pilot reaction to major inflight emergencies, such as onboard fires or power failures, is likely to be degraded by the high stress level present. Because of the quite different factors that come into play, we shall now consider human reliability and its degradation under the two limiting situations of very routine tasks and tasks performed in emergency situations.

Routine Operations

For purposes of analysis it is useful to classify human errors as random, systematic, or sporadic. These classes may be illustrated by considering the simple example, shown in Fig. 10.2, of the ability to hit a target.* Random errors are dispersed about the desired value without bias; that is, they have the true mean value (in x and y), but the variance may be too large. These errors may be corrected if they are attributable to an inappropriate tool or man–machine interface. For example, if it is not possible to read instruments finely enough or to adjust setting precisely enough, such improvements are in order. Similarly, training in the particular task may reduce the dispersion of random errors. Figure 10.2*b* illustrates systematic errors whose dispersion is sufficiently small, but with a bias departing from the mean value. Such bias may be caused by tools or instruments that are out of calibration, or it may come from incorrect performance of a procedure. In either case corrective measures may be taken. More subtle psychological factors—such as the desire of an inspector not to miss any faulty parts, and thus declaring a good many faulty even though they are not—may also cause bias errors.

* H. R. Guttmann, unpublished lecture notes, Northwestern University, 1982.

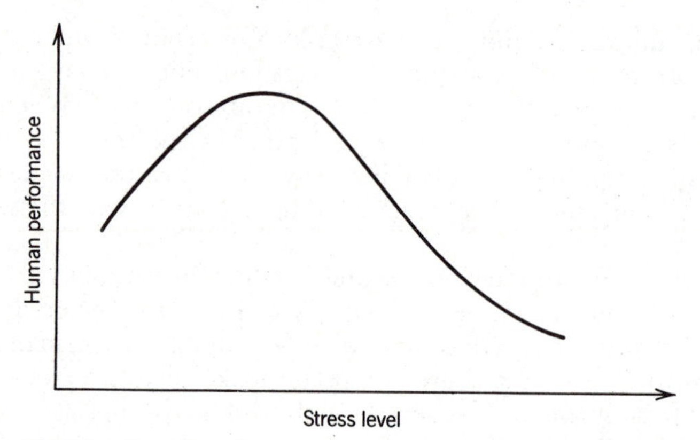

FIGURE 10.1 The effect of stress level on human performance.

Perhaps sporadic errors, pictured in Fig. 10.2*c*, are the most difficult to deal with, for they rarely show observable patterns. They are committed when the person acts in an extreme or careless way: forgetting to do something altogether, performing an action that was not called for, or reversing the order in which things are done. For example, a meter reader might, in taking a series of meter readings, read a wrong meter. Again, careful design of the man–machine interface can minimize the number of sporadic errors. Color, shape, and other means can be used to differentiate instruments and control and to minimize confusion. Sporadic errors, in particular, are amplified by the carelessness inherent in low-stress situations, as well as by the confusion of high-stress situations.

Let us first examine sporadic errors made in routine situations. Certainly, under any circumstances, errors are minimized by a well-designed work environment. Such design would take into account all the standard considerations or human factors engineering: comfortable seating, adequate

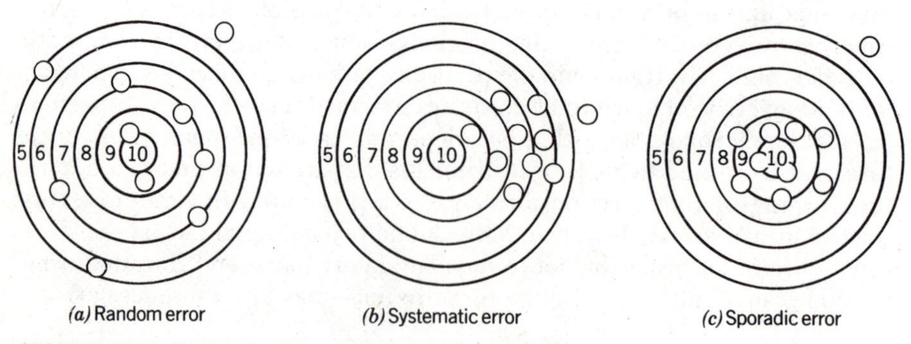

FIGURE 10.2 Classes of human error.

light, temperature and humidity control, and well-designed control and instrument panels to minimize the possibilities for confusion. The attention span that can be expected for routine tasks is still limited. As indicated in Fig. 10.3, attention spans for detailed monitoring tend to deteriorate rapidly after about half an hour, indicating the need for frequent rotation of such duties for optimal performance. The same deterioration may be expected for very repetitive tasks, unless there is careful checking or other intervention to insure that such deterioration does not take place.

Probably one of the most important ways in which system reliability is degraded is through the dependencies introduced between redundant components during the course of routine maintenance, testing, and repair. We may cite the turning off of both of the redundant auxiliary feedwater systems at the Three Mile Island reactor. The point is that if technicians perform a task incorrectly on one piece of equipment, they are likely to do it incorrectly on all like pieces of equipment. This problem may be countered, at least in part, by a variety of techniques. Diversity of equipment is one, for just as the hardware will not be subjected to the same failure modes, the maintenance procedures will also be different. Staggering the times or the personnel doing maintenance on redundant equipment also tends to reduce dependencies, although some smaller degree of dependency may remain through the use of common tools or incorrect training procedures.

Independent checking of procedures also decreases both the probability of failure and the degree of dependency. Even here, however, psychological factors limit effectiveness. When the inspector and the person performing the maintenance have worked with each other for an extended period of time, the inspector may tend to become less careful as he or she grows more confident of the colleague's abilities. Similarly, if two independent checks

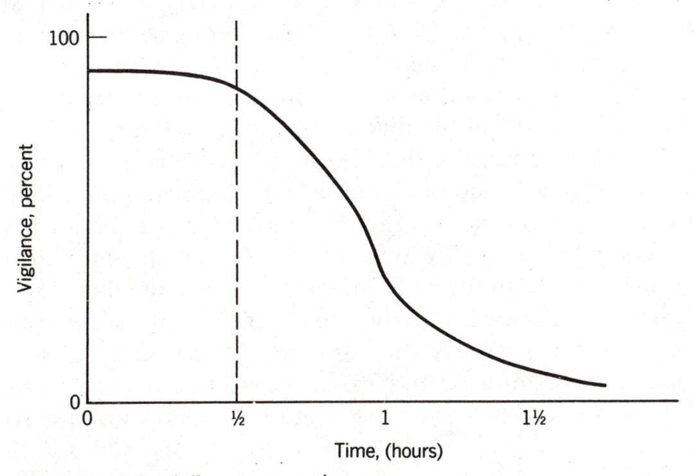

FIGURE 10.3 Vigilance versus time.

are to be performed, they are unlikely to be truly independent, for often the very knowledge that a procedure is being checked twice will tend to decrease the care with which it is done.

Reliability is also degraded when operating and maintenance personnel inappropriately modify or make shortcuts in operating and maintenance procedures. Often operating and maintenance personnel gain an understanding of the system that was not available at the time of design and modify procedures to make them more efficient and safer. The danger is that, without a thorough design review, new loadings and environment degradation may be introduced, and component dependencies may increase inadvertently. For example, in the 1979 crash of the DC-10 in Chicago, it is thought that a modified procedure for removing the engines for inspection and preventive maintenance led to excessive fatigue stresses on the engine support pylon, causing the engine to break off during takeoff.

Although the methodology is not straightforward, data are available on the errors committed in the course of routine tasks. Extensive efforts have been made to develop task analysis and simulation methods.* Failure probabilities are first estimated for rudimentary functions. Then, by combining these factors, we can estimate probabilities that more extensive procedures will engender errors.

Emergency Operations

At the high-stress end of the spectrum shown in Fig. 10.1 are the protective tasks that must be performed by operations personnel under emergency conditions to prevent potentially dangerous situations from getting completely out of hand. Here a well-designed, man–machine interface, clear-cut procedures, and thorough training are critical, for in such situations actions that are not familiar from routine use must be taken quickly, with the knowledge that mistakes may be disastrous. Moreover, since such situations are likely to be caused by subtle combinations of malfunctions, they may be confusing and call for diagnostic and problem-solving ability, not just the skill and rule-based actions exercised for routine tasks.

Under emergency conditions conflicting information may well confuse operators who then act in ways that further propagate the accident. With proper training and the ability to function under psychological stress, however, they may be able to solve the problem and save the day. For example, the confusion of the operators at the Three Mile Island reactor caused them to turn off the emergency core-cooling system, thus worsening the accident. In contrast, the pilot of a Boeing 767 managed to make use of his earlier experience as an amateur glider pilot and safely land his aircraft after a

* A. D. Swain, and H. R. Guttmann, *Handbook of Human Reliability Analysis with Emphasis on Nuclear Power Plant Applications,* U.S. Nuclear Regulatory Commission, NUREG/CR-1287, 1980.

series of equipment failures and maintenance errors had caused the plane to run out of fuel while in flight over Canada.

There are a number of common responses to emergency situations that must be taken into consideration when designing systems and establishing operating procedures. Perhaps the most important is the incredulity response. In the rare event of a major accident, it is common for an operator not to believe that an accident is taking place. The operator is more likely to think that there is a problem with the instruments or alarms, causing them to produce spurious signals. At installations that have been subjected to substantial numbers of false alarms, a real one may very well be disbelieved. Systems should be carefully designed to keep spurious alarms to a minimum, and straightforward checks to distinguish accidents from faulty instrument performance should be provided. In some situations it is desirable to mandate that safety actions be taken, even though the operator may feel that faulty instruments are the cause of the problem.

A second common reaction to emergencies is reverting to stereotype. The operator reverts to the stereotypical response of the population of which he or she is a part, even though more recent training has been to the contrary. For example, in the United States turning a light or other switch "up" means that it is "on." (In Europe, however, "down" is "on.") Thus, although Americans may be trained to put a particular switch down to turn it on, under the time pressure of an emergency they are likely to revert to the population stereotype and try to put the switch up. The obvious solution to this problem is to take great care in human factors engineering not to violate population stereotypes in the design of instrumentation and control systems. This problem may be aggravated if operators from one culture are transferred to another, or if care is not taken in the use of imported equipment.

Finally, once a mistake is made, such as placing a switch in the wrong position, in a panic an operator is likely to repeat the mistake rather than think through the problem. This reaction, as well as other inappropriate emergency responses, must be considered when deciding the extent to which emergency actions should or can be automated. On the one hand, when there is extreme time pressure, automated protection systems may eliminate the errors discussed. At the same time, such systems do not have the flexibility and problem-solving ability of human operators, and these advantages may be of overwhelming importance, assuming that there is time for the situation to be properly assessed.

In summary, to ensure a high degree of human reliability in emergency situations, control rooms, whether they be aircraft cockpits or chemical plant control installations, must be carefully designed according to good human factors practice. It is also important that the procedures for all anticipated situations are readily understandable, and finally, that operators are drilled at frequent intervals on emergency procedures, preferably with simulators that model the real conditions.

Even though we may characterize human behavior under emergency conditions and suggest actions that will improve human reliability, it is difficult indeed to obtain quantitative data on failure probabilites. As we have indicated, such situations happen only infrequently and often they are not well documented. Moreover, it is difficult to obtain a realistic response from simulator experiments when the subjects know that they are in an experiment and not a life-threatening situation.

10.3 METHODS OF ANALYSIS

Probably the most important task in eliminating or reducing the probability of accidents is to identify the mechanisms by which they may take place. The ability to make such identifications in turn requires that the analyst have a comprehensive understanding of the system under consideration, both in how it operates and in the limitations of its components. Even the most knowledgeable analysts are in danger of missing critical failure modes, however, unless the analysis is carried out in a very systematic manner. For this reason a substantial number of formal approaches have been developed for safety analysis. In this section we introduce three of the most widely used: failure modes and effects analysis, event trees, and fault trees. In later sections the use of fault trees is developed in more detail.

Failure Modes and Effects Analysis

Failure modes and effects analysis, usually referred to by the acronym FMEA, is one of the most widely employed techniques for enumerating the possible modes by which components may fail and for tracing through the characteristics and consequences of each mode of failure on the system as a whole. The method is primarily qualitative in nature, although some estimates of failure probabilities are often included.

Although there are many variants of FMEA, its general characteristics can be illustrated with the analysis of a rocket shown in Fig. 10.4. In the left-hand column the major components or subsystems are listed; then, in the next column the physical modes by which each of the components may fail are given. This is followed, in the third column, by the possible causes of each of the failure modes. The fourth column lists the effects of the failure. The method becomes more quantitative if an estimate of the probability of each failure mode is made. Criticality or an alternative ranking of the failure's importance is usually included to separate failure modes that are catastrophic from those that merely cause inconvenience or moderate economic loss. The final column in most FMEA charts is a listing of possible remedies.

In a more extensive FMEA the information shown in Figure 10.4 may

FAILURE MODES AND EFFECTS ANALYSIS

1. SUBSYSTEM ________ 2. DWG. NR. ________ 3. PREPARED BY ________ 4. DATE ________

ITEM	FAILURE MODES	CAUSE OF FAILURE	POSSIBLE EFFECTS	PROBABILITY OF OCCURRENCE	CRITICALITY	POSSIBLE ACTION TO REDUCE FAILURE RATE OR EFFECTS
Motor case	Rupture	a. Poor workmanship b. Defective materials c. Damage during transportation d. Damage during handling e. Overpressurization	Destruction of missile	0.0006	Critical	Close control of manufacturing processes to ensure that workmanship meets prescribed standards. Rigid quality control of basic materials to eliminate defectives. Inspection and pressure testing of completed cases. Provision of suitable packaging to protect motor during transportation.
Propellant grain	a. Cracking b. Voids c. Bond separation	a. Abnormal stresses from cure b. Excessively low temperatures c. Aging effects	Excessive burning rate; overpressurization; motor case rupture during otherwise normal operation	0.0001	Critical	Carefully controlled production. Storage and operation only within prescribed temperature limits. Suitable formulation to resist effects of aging.
Liner	a. Separation from motor case b. Separation from motor grain or insulation	a. Inadequate cleaning of motor case after fabrication b. Use of unsuitable bonding material c. Failure to control bonding process properly	Excessive burning rate Overpressurization Case rupture during operation	0.0001	Critical	Strict observance of proper cleaning procedures. Strict inspection after cleaning of motor case to ensure that all contaminants have been removed.

FIGURE 10.4 Failure modes and effects analysis. (From Willie Hammer, *Handbook of System and Product Safety*, © 1972, p. 153, with permission from Prentice-Hall, Englewood Cliffs, NJ.)

be expanded. For example, failures are not categorized as simply critical or not critical but by four levels denoting seriousness.

1. Negligible—loss of function that has no effect on the system.
2. Marginal—a fault that will degrade the system to some extent but will not cause the system to be unavailable, for example, the loss of one of two redundant pumps, either of which can perform a required function.
3. Critical—a fault that will completely degrade system performance, for example, the loss of a component that renders a safety system unavailable.
4. Catastrophic—a fault that will have severe consequences and perhaps cause injuries or fatalities, for example, catastrophic pressure vessel failure.

Additional columns also may be included in FMEA. A list of symptoms or methods of detection of each failure mode may be very important for safe operations. A list of compensating provisions for each failure mode may be provided to emphasize the relative seriousness of the modes. In order to concentrate improvement efforts on eliminating those having the widest effects, it is common also to rank the various causes of a particular mode according to the percentage of the mode's failures that they incur.

The emphasis in FMEA is usually on the basic physical phenomena that can cause a device or component to fail. Therefore, it often serves as a suitable starting point for enumerating and understanding the failure mechanisms before proceeding to one of the other techniques for safety analysis. To understand better the progression of accidents when they pass through several stages and to analyze the effects of component redundancies on system safety, engineers often supplement FMEA with the more graphic event-tree and fault-tree methods for quantifying system behavior during accidents.

Event Trees

In many accident scenarios the initiating event—say, the failure of a component—may have a wide spectrum of results, ranging from inconsequential to catastrophic. The consequences may be determined by how the accident progression is affected by subsequent failure or operation of other components or subsystems, particularly safety or protection devices, and by human errors made in responding to the initiating event. In such situations an inductive method may be very useful. We begin by asking "what if" the initiating event occurs and then follow each of the possible sequences of events that result from assuming failure or success of the components and humans affected as the accident propagates. After such sequences are defined, we may attempt to attach probabilities to them if such a quantitative estimate is needed.

The event tree is a quantitative technique for such inductive analysis. It begins with a specific initiating event, a particular cause of an accident, and then follows the possible progressions of the accident according to the success or failure of other components or pieces of equipment. Event trees are a particular adaptation of the more general decision-tree formalism that is widely employed for business and economic analysis. They are quite useful in analyzing the effects of the functioning or failure of safety systems in response to an accident, particularly when events follow with a particular time progression. The following is a very simple application of event-tree analysis.

Suppose that we want to examine the effects of the power failure in a hospital in order to determine the probability of a blackout, along with other likely consequences. For simplicity we assume that the situations may be analyzed in terms of just three components: (1) the off-site local utility power system that supplies electricity to the hospital; (2) a diesel generator that supplies emergency power, and (3) a voltage-monitoring system that monitors the off-site power supply and, in the event of a failure, transmits a signal that starts the diesel generator.

We are concerned with a sequence of three events. The initiating event is the loss of off-site power. The second event is detection of the loss and subsequent functioning of the voltage-monitoring system; and the third event is the start-up and operation of the diesel generator. This sequence is shown in the event tree in Fig. 10.5. Note that at each event there is a branch corresponding to whether a system operates or fails. By convention, the upward branches signify successful operation, and the lower branches failure.

Note that for a sequence of N events there will be 2^N branches of the tree. The number may be reduced, however, by eliminating impossible branches. For example, the generator cannot start unless the voltage monitor functions. Thus the path is impossible (has a zero probability) and can be pruned from the tree, as in Fig. 10.6.

We may follow an event tree from left to right to find the probabilities and consequences of differing sequences of events. The probabilities of the

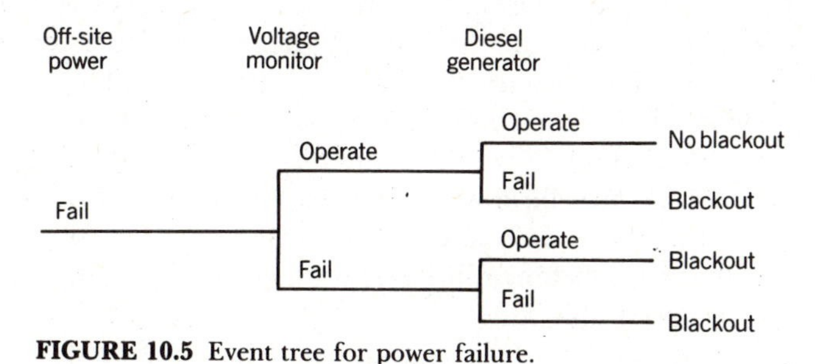

FIGURE 10.5 Event tree for power failure.

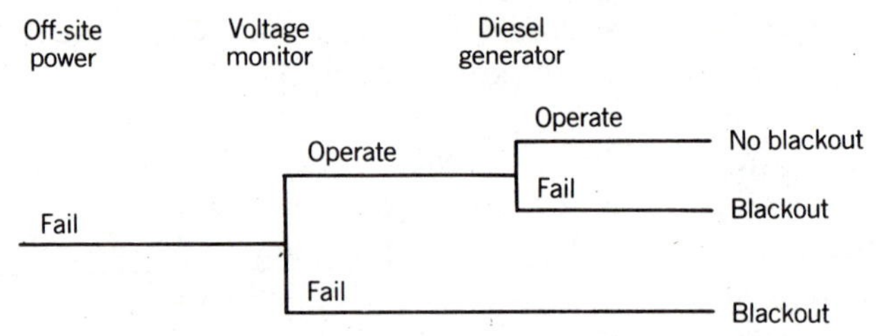

FIGURE 10.6 Reduced event tree for power failure.

various outcomes are determined by attaching a probability to each event on the tree. In our tree the probabilities are P_i for the initial event, P_v for the failure of the voltage monitoring system, and P_g for the failure of the diesel generator. With the assumption that the failures are independent, the probability of a blackout is therefore $P_iP_v + P_i(1 - P_v)P_g$.

Fault Trees

Fault-tree analysis is a deductive methodology for determining the potential causes of accidents, or for system failures more generally, and for estimating the failure probabilities. In its narrowest sense fault-tree analysis may be looked on as an alternative to the use of reliability block diagrams in determining system reliability in terms of the corresponding components. However, fault-tree analysis differs both in the approach to the problem and in the scope of the analysis.

Fault-tree analysis is centered about determining the causes of an undesired event, referred to as the top event, since fault trees are drawn with it at the top of the tree. We then work downward, dissecting the system in increasing detail to determine the root causes or combinations of causes of the top event. Top events are usually failures of major consequence, engendering serious safety hazards or the potential for significant economic loss.

The analysis yields both qualitative and quantitative information about the system at hand. The construction of the fault tree in itself provides the anlayst with a better understanding of the potential sources of failure and thereby a means to rethink the design and operation of a system in order to eliminate many potential hazards. Once completed, the fault tree can be analyzed to determine what combinations of component failures, operational errors, or other faults may cause the top event. Finally, the fault tree may be used to calculate the demand failure probability, unreliability, or unavailability of the system in question. This task of quantitative evaluation is often of primary importance in determining whether a final design is considered to be acceptably safe.

The rudiments of fault-tree analysis may be illustrated with a very simple example. We use the same problem of a hospital power failure treated inductively by event-tree analysis earlier to demonstrate the deductive logic of fault-tree analysis. We begin with blackout as the top event and look for the causes, or combination of causes, that may lead to it. To do this, we construct a fault tree as shown in Fig. 10.7. In examining its causes, we see that both the off-site power system *and* the emergency power supply must fail. This is represented by a $\cap$ gate in the fault tree, as shown. Moving down to the second level, we see that the emergency power supply fails if the voltage monitor *or* the diesel generator fails. This is represented by a $\cup$ gate in the fault tree as shown.

We see that the fault tree consists of a structure of OR and AND gates, with boxes to describe intermediate events. Using the same probabilities as in the event tree, we can determine the probability of a blackout in terms of P_i, P_v, and P_g, the failure probabilities for off-site power, voltage monitor, and diesel generator.

The most straightforward fault trees to draw are those, such as in the preceding example, in which all the significant primary failures are component failures. If a reliability block diagram can be drawn, a fault tree can also be drawn. This can be seen in an additional example.

Consider the system shown in Fig. 7.10. We may look at the system as consisting of an upper subsystem ($a1$, $a2$, and $b1$) and a lower subsystem ($a3$, $a4$, and $b2$), in addition to component c. For a system to fail, either component c must fail or the upper and lower subsystems must fail. Proceeding downward, for the upper subsystem to fail either component $b1$ must fail or both $a1$ and $a2$ must fail. Treating the lower subsystem analogously, we obtain the tree shown in Fig. 10.8.

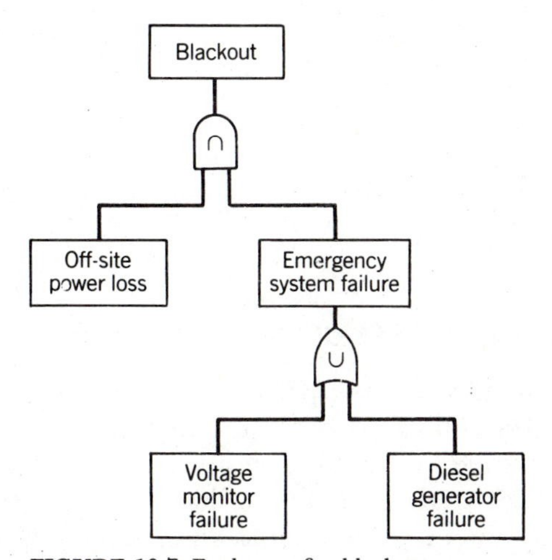

FIGURE 10.7 Fault tree for blackout.

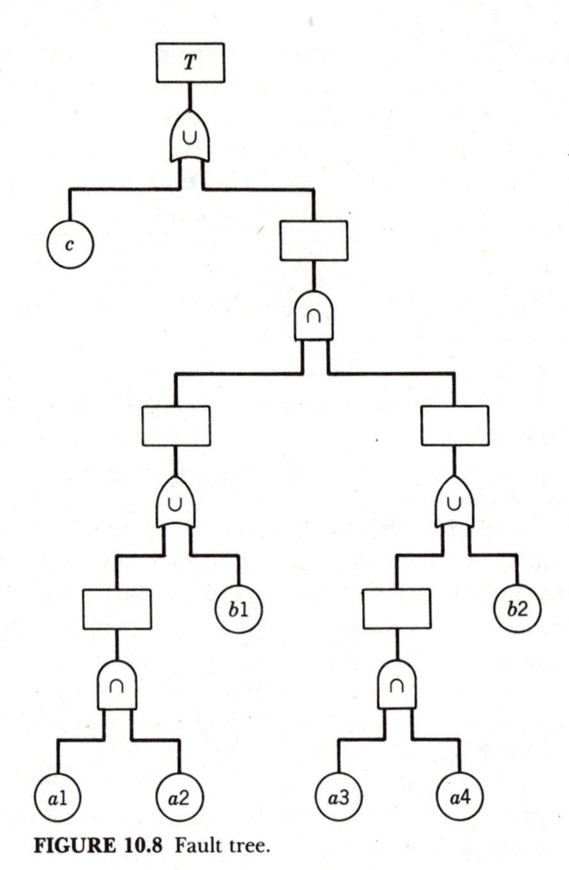

FIGURE 10.8 Fault tree.

EXAMPLE 10.1

Construct a reliability block diagram corresponding to the fault tree in Fig. 10.7.

Solution The reliability block diagram having the same logic and failure probability as the fault tree of Fig. 10.7 is depicted in Fig. 10.9.

10.4 FAULT-TREE CONSTRUCTION

Of the methods discussed in the preceding section, fault-tree analysis has been the most thoroughly developed and is finding increased use for system safety analysis in a wide variety of applications. It is particularly well suited to situations in which tracing a failure to its root causes requires dissecting the system into subsystems, components, and parts to get at the level where

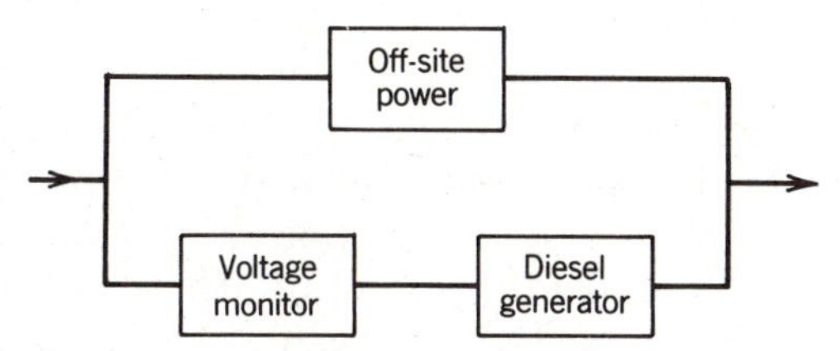

FIGURE 10.9 Reliability block diagram for electrical power.

failure data are available. For example, in the aforetreated hospital blackout we may not have the test data that is required to determine P_v for the voltage monitor or P_g for the diesel generator. We must then delve more deeply and examine the components of these devices; we may need to construct the probability that the voltage monitor will fail from the failure rates of its components.

It may be argued that such dissection can also be done by subdividing the blocks appearing in reliability block diagrams. Although this is true, there are some important differences. Reliabiity block diagrams are success-oriented; that is, all failures are lumped together to obtain the probability that a system will fail. In most reliability studies we are interested only in knowing the reliability (i.e., the probability that the system does not fail). Conversely, in fault-tree analysis we are often interested only in a particular undesirable event (i.e., a failure that leads to a safety hazard) and in calculating the probability that it will happen. Hence failures that do not cause the safety hazard defined by the top event are excluded from consideration.

The difference between reliability analysis and safety analysis may be illustrated by the example of a hot-water heater. In reliability analysis—carried out with a reliability block diagram—failure of any kind will cause failure of the system to supply hot water. Most of these failures have no safety implications: The heater unit fails to turn on, the tank develops a leak, and so on. In safety analysis—using a fault tree—we would be interested in a particular safety hazard such as the explosion of the tank. The other failures listed would not be included in the fault-tree construction.

Because of the increasing importance of fault-tree analysis, the remainder of this chapter is devoted to it. In this section we discuss the construction of fault trees by first giving the standardized nomenclature. Then following a brief discussion of fault classifications, we supply several illustrative examples. In Sections 10.5 and 10.6 fault trees are evaluated. In qualitative evaluation the fault tree is reduced to a logical expression, giving the top event in terms of combinations of primary-failure events. In quantitative evaluation the probability of the top event is expressed in terms of the probabilities of the primary-failure events.

Nomenclature

As we have seen, the fault tree is made up of events, expressed as boxes, and gates. Two types of gates appear, the OR and the AND gate. The OR gate as indicated in Fig. 10.10*a* is used to show that the output event occurs only if one or more of the input events occur. There may be any number of input events of an OR gate. The AND gate as indicated in Fig. 10.10*b* is used to show that the output fault occurs only if all the input faults occur. There may be any number of input faults to an AND gate.

Generally, OR and AND gates are distinguished by their shape. In free-hand drawings, however, it may be desirable to put the ∪ and ∩ symbols on the gates. Or the so-called engineering notation, in which OR is represented by a "+" and AND by "·", may be used. Obviously, if these notations are included, the care with which the shape of the gate is drawn becomes of secondary importance.

In addition to the AND and OR gates, the INHIBIT gate shown in Fig. 10.11*a* is also widely used. It is a special case of the AND gate. The output is caused by a single input, but some qualifying condition must be satisfied before the input can produce the output. The condition that must exist is indicated conventionally by an ellipse, which is located to the right of the gate. In other words the output happens only if the input occurs under the conditions specified within the ellipse. The ellipse may also be used to indicate conditions on OR or AND gates. This is shown in Figs. 10.11*b* and *c*.

The rectangular boxes in the foregoing figures indicate top or intermediate events; they appear as outputs of gates. Shape also distinguishes different types of primary or input events appearing at the bottom of the fault tree. The primary events of a fault tree are events that, for one of a number of reasons, are not developed further. They are events for which probabilities must be provided if the fault tree is to be evaluated quantitatively (i.e., if the probability of the top event is to be calculated).

In general, four different types of primary events are distinguished. These make up part of the list of symbols in Table 10.1. The circle describes

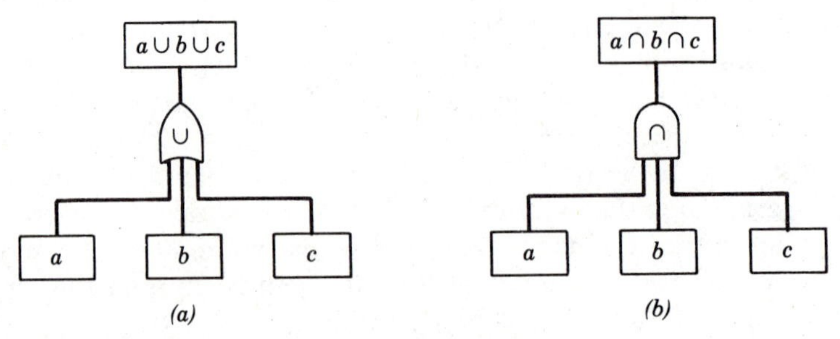

FIGURE 10.10 Fault-tree gates: (*a*) OR, (*b*) AND.

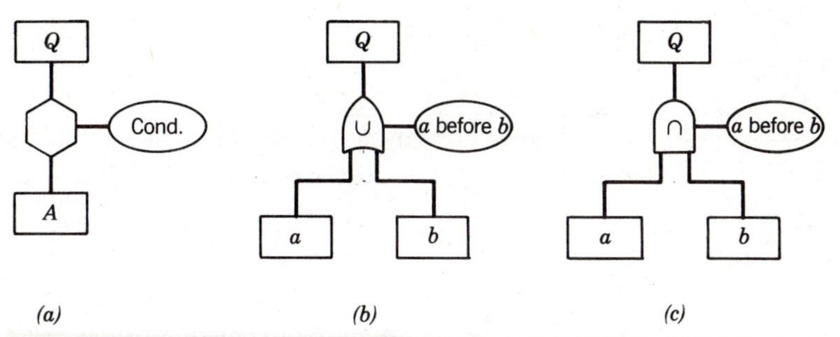

FIGURE 10.11 Fault-tree conditional gates.

a basic event. This is a basic initiating fault event that requires no further development. The circle indicates that the appropriate resolution of the fault tree has been reached.

The undeveloped event is indicated by a diamond. It refers to a specific fault event, although it is not further developed, either because the event is of insufficient consequence or because information relevant to the event is unavailable. In contrast, the external event, signified by a house-shaped figure, indicates an event that is normally expected to occur. Thus house symbol displays are not of themselves faults.

The last symbols in Table 10.1 are the triangles indicating transfers into and out of the fault tree. These are used when more than one page is required to draw a fault tree. A transfer-in triangle indicates that the input to a gate is developed on another page. A transfer-out triangle at the top of a tree indicates that it is the input to a gate appearing on another page.

In fault-tree construction a distinction is made between a fault and a failure. The word *failure* is reserved for basic events such as a burned-out bearing in a pump or a short circuit in an amplifier. The word *fault* is more all-encompassing. Thus, if a valve closes when it should not, this may be considered a valve fault. However, if the valve fault is due to a spurious signal from the shorted amplifier, it is not a valve failure. Thus all failures are faults, but not all faults are failures.

Fault Classification

The dissection of a system to determine what combinations of primary failures may lead to the top event is central to the construction of a fault tree. This dissection is likely to proceed most smoothly when the system can be divided into subsystems, components, or parts in order to associate the faults with discrete pieces of the system. Even then, a great deal of attention must be given to the component interactions, particularly common-mode failures. Beyond decomposing the system into components, however, we

TABLE 10.1 Fault-Tree Symbols Commonly Used

Symbol	Name	Description
	Rectangle	Fault event; it is usually the result of the logical combination of other events.
	Circle	Independent primary fault event.
	Diamond	Fault event not fully developed, for its causes are not known; it is only an assumed primary fault event.
	House	Normally occurring basic event; it is not a fault event.
	OR Gate	The union operation of events; i.e., the output event occurs if one or more of the inputs occur.
	AND Gate	The intersection operation of events; i.e., the output event occurs if and only if all the inputs occur.
A, X	INHIBIT Gate	Output exists when *X* exists and condition *A* is present; this gate functions somewhat like an AND gate and is used for a secondary fault event *X*.
A	Triangle-in	Triangle symbols provide a tool to avoid repeating sections of a fault tree or to transfer the tree construction from one sheet to the next. The triangle-in appears at the bottom of a tree and represents the branch of the tree (in this case *A*) shown someplace else. The trianlge-out appears at the top of a tree and denotes that the tree *A* is a subtree to one shown someplace else.
A	Triangle-out	

Source: Adapted from H. R. Roberts, W. E. Vesley, D. F. Haast, and F. F. Goldberg, *Fault Tree Handbook*, U.S. Nuclear Regulatory Commission, NUREG-0492, 1981.

must also examine which components are more likely to fail and study with care the various modes by which component failure may occur.

In the material already covered, we have examined several ways of classifying failures that are very useful for fault-tree construction. Distinguishing between hardware faults and human error is essential, as is the classification of hardware failures into early, random, and aging, each with its own characteristics and causes. In what follows we discuss briefly two additional classifications. The division of failures into primary, secondary, and command faults is particularly useful in determining the logical structure of a fault tree. The classification of components as passive or active is important in determining which ones are likely to make larger contributions to system failure.

Primary, Secondary, and Command Faults

Failures may be usefully classified as primary, secondary, and command faults.* A primary fault by definition occurs in an environment and under a loading for which the component is qualified. Thus a pressure vessel's bursting at less than the design pressure is classified as a primary fault. Primary faults are most often caused by defective design, manufacture, or construction and are therefore most closely correlated to wearin failures. Primary faults may also be caused by excessive or unanticipated wear, or they may occur when the system is not properly maintained and parts are not replaced on time.

Secondary faults occur in an environment or under loading for which the component is not qualified. For example, if a pressure vessel fails through excessive pressure for which it was not designed, it has a secondary fault. As indicated by the name, the basic failure is not of the vessel but in the excessive loading or adverse environment. Such failures often occur randomly and are characterized by constant failure rates.

Although a component fails when it has primary and secondary faults, it operates correctly when it has a command fault, but at the wrong time or place. Thus, our pressure vessel might lose pressure through the unwanted opening of a relief valve, even though there is no excessive pressure. If the valve opens through an erroneous signal, it has a command fault. For command failures we must look beyond the component failure to find the source of the erroneous command.

Passive and Active Faults

Components may be designated as either passive or active. Passive components include such things as pipes, cables, bearings, welds, and bolts. They function in a more or less static manner, often acting as transmitters of energy, such as a buss bar or cable, or of fluids such as piping. Transmitters of mechanical loads, such as structural members, beams, columns, and so on, and connectors, such as welds, bolts, or other fasteners, are also passive components. A passive component may usually be thought of as a mechanism for transmitting the output of one active component to the input of

* *Fault Tree Handbook,* op. cit.

another. In the broadest sense, the quantity transmitted may be an electric signal, a fluid, mechanical loading, or any number of other quantities.

Active components contribute to the system function in a dynamic manner, altering in some way the system's behavior. For example, pumps and valves modify fluid flow; relays, switches, amplifiers, rectifiers, and computer chips modify electric signals; motors, clutches, and other machinery modify the transmission of mechanical loading.

Our primary reason for distinguishing between active and passive components is that failure rates are normally much higher for active components than for passive components, often by two or three orders of magnitude. The terms *active* and *passive* refer to the primary function of the component. Indeed, an active component may have many passive parts that are prone to failure. For example, a pump and its function are active, but the pump housing is considered passive, even though a housing rupture is one mode by which the pump may fail. In fact, one of the reasons that active components have higher failure rates than passive ones is that they tend to be made up of many nonredundant parts both active and passive.

Examples

We present here four examples of rather simple systems, and ones that are, moreover, readily understandable without specialized knowledge. This is consistent with the philosophy that one should not attempt to construct a fault tree until the design and function of the system is thoroughly understood. The first example is a demand failure, the failure of a motor to start; and the second is the failure of a continuously operating system. The third involves both start-up and operation; in the fourth the top event is a catastrophic failure, and its causes involve faulty procedures and operator actions as well as equipment failures.

EXAMPLE 10.2*

Draw a fault tree for the motor circuit shown in Fig. 10.12. The top event for the fault-tree analysis is simply failure of the motor to operate.

Solution The fault tree is shown in Figure 10.13. Note that failures are distinguished as primary and secondary. For primary failures we would expect data to be available to determine the failure probabilities. If not, further dissection of the component into its parts might be necessary. The secondary faults are either command faults, such as no current to the motor, or excessive loading, such as an overload in the circuits. For these we must delve deeper to locate the causes of the faults.

* Adapted from J. B. Fussel in *Generic Techniques in System Reliability Assessment,* E. J. Henley and J. W. Lynn (eds.), Nordhoff, Leyden, Holland, 1976.

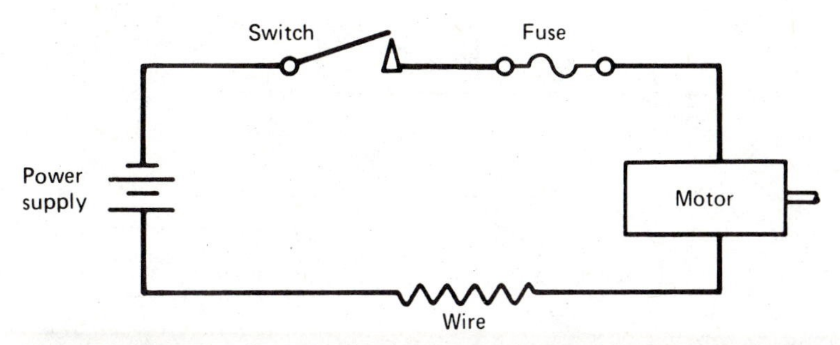

FIGURE 10.12 Electric motor circuit. (From J. B. Fussel, in *Generic Techniques in System Reliability Assessment,* pp. 133–162, E. J. Henley and J. W. Lynn (eds.), Martinus Nijhoff/Dr. Junk Publishers (was Sijthoff Noordhoff), Leyden, 1976, reprinted by permission.)

EXAMPLE 10.3*

Draw a fault tree for the coolant supply system pictured in Fig. 10.14. Here the top event is loss of minimum flow to a heat exchanger.

Solution The fault tree is shown in Fig. 10.15. Not all of the faults at the bottom of the tree are primary failures. Thus it may be desirable to develop some of the faults, such as loss of the pump inlet supply, further. Conversely, the faults may be considered too insignificant to be traced further, or data may be available even though they are not primary failures.

EXAMPLE 10.4†

Consider the sump pump system shown in Fig. 10.16. Redundance is provided by a battery-driven backup system which is activated when the utility power supply fails. Draw a fault tree for the flooding of a basement protected by this system.

Solution The fault tree is shown in Fig. 10.17. The tree accounts for the fact that flooding can occur if the rate of inflow from the storm exceeds the pump capacity. Moreover, flooding can occur from storms within the system's capacity if there are malfunctions of both pumps and the inflow is large enough to fill the sump. Primary pump failures may be caused either by the failure of the pump itself or by loss of ac power. Similarly, the second pump may malfunction or it may be lost through failure of the battery. The battery fails only if all three events at the bottom of the tree take place.

* Adapted from J. A. Burgess, "Spotting Trouble Before It Happens," *Machine Design,* **42,** No. 23, 150 (1970).
† Adapted from A. H-S. Ang and W. H. Tang, *Probability Concepts in Engineering Planning and Design,* Vol. 2, Wiley, New York, 1984.

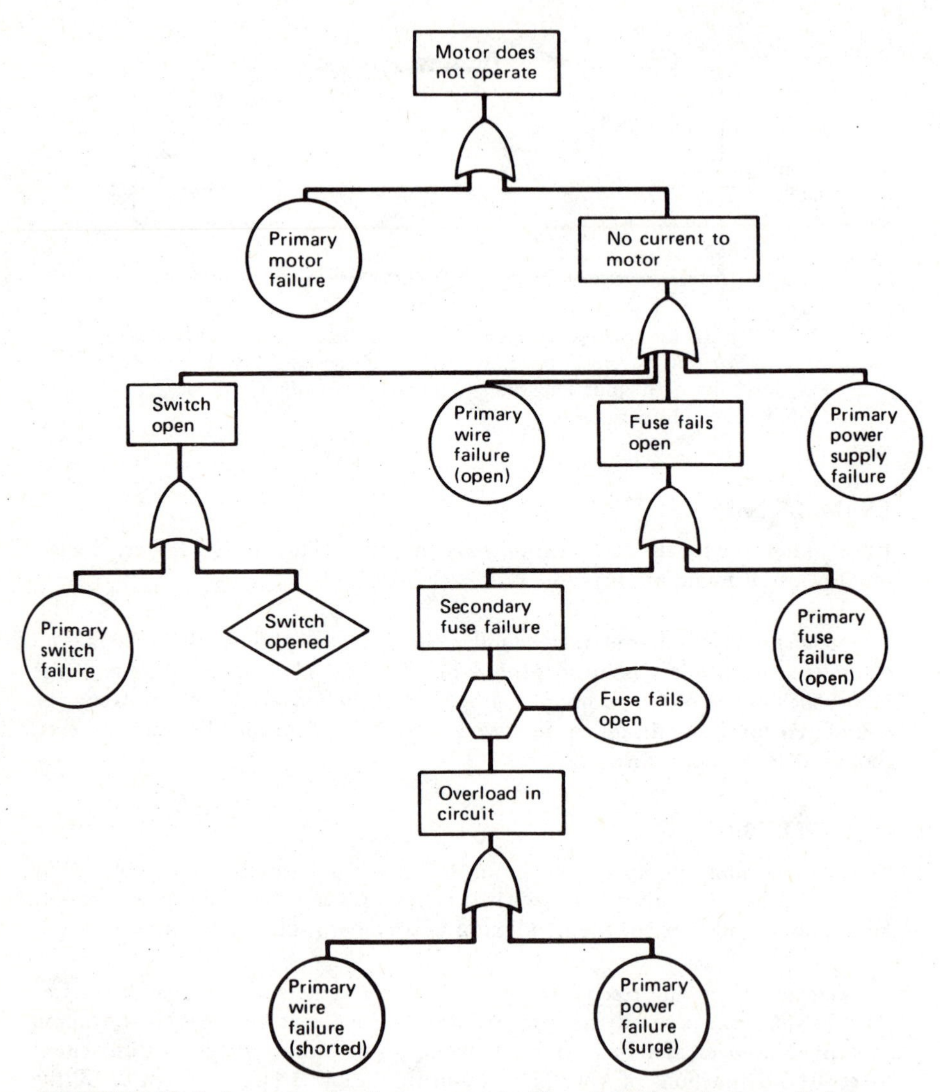

FIGURE 10.13 Fault tree for electric motor circuit. (From J. B. Fussel in *Generic Techniques in System Reliability Assessment,* pp. 133–162, E. J. Henley and J. W. Lynn (eds.), Martinus Nijhoff/Dr. Junk Publishers (was Sijthoff Noordhoff), Leyden, 1976, reprinted by permission.)

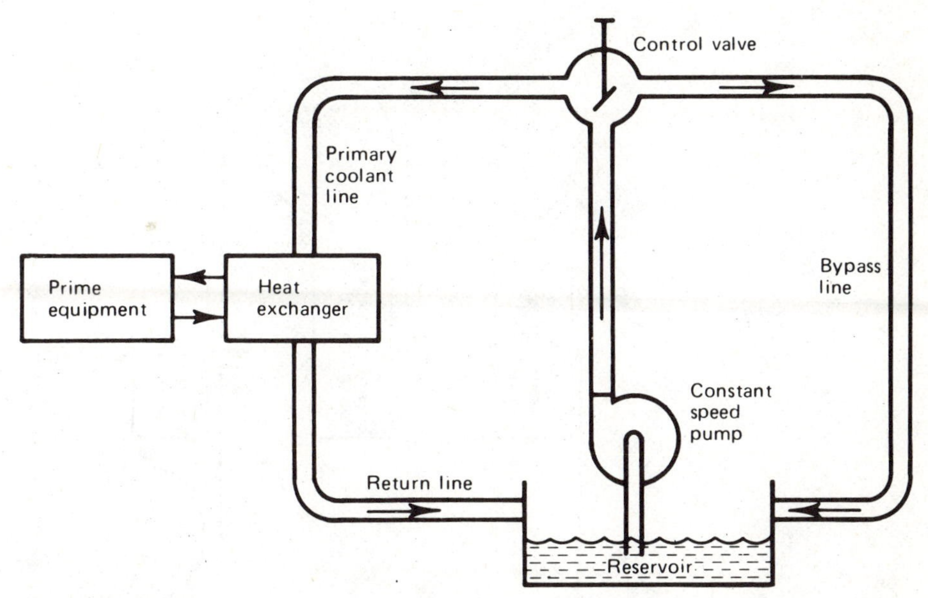

FIGURE 10.14 Coolant supply system. (Reprinted from *Machine Design,* © 1984, by Penton/IPC, Cleveland, Ohio.)

EXAMPLE 10.5*

The final example that we consider is the pumping system shown in Fig. 10.18. The top event here is rupture of the pressure tank. This situation has the added complication that operator errors as well as equipment failures may lead to the top event. Before a fault tree can be drawn, the procedure by which the system is operated must be specified. The tank is filled in 10 min and empties in 50 min. Thus there is a 1-hr cycle time. After the switch is closed, the timer is set to open the contact in 10 min. If there is a failure in the mechanism, the alarm horn sounds. The operator then opens the switch to prevent the tank from overfilling and therefore rupturing.

Solution A fault tree for the tank rupture is shown in Fig. 10.19. Notice how the analyst has used primary (i.e. basic), secondary, and command faults at several points in developing the tree. The operator's actions, a primary failure, are interpreted as the operator's failing to push the button when the alarm sounds. A secondary fault would occur, for example, if the operator is absent or unconscious when the alarm sounded, and the command fault for the operator would take place if the alarm does not sound.

* Adapted from E. J. Henley and H. Kumamoto, ***Reliability Engineering and Risk Assessment,*** Prentice Hall, Englewood Cliffs, NJ, 1981.

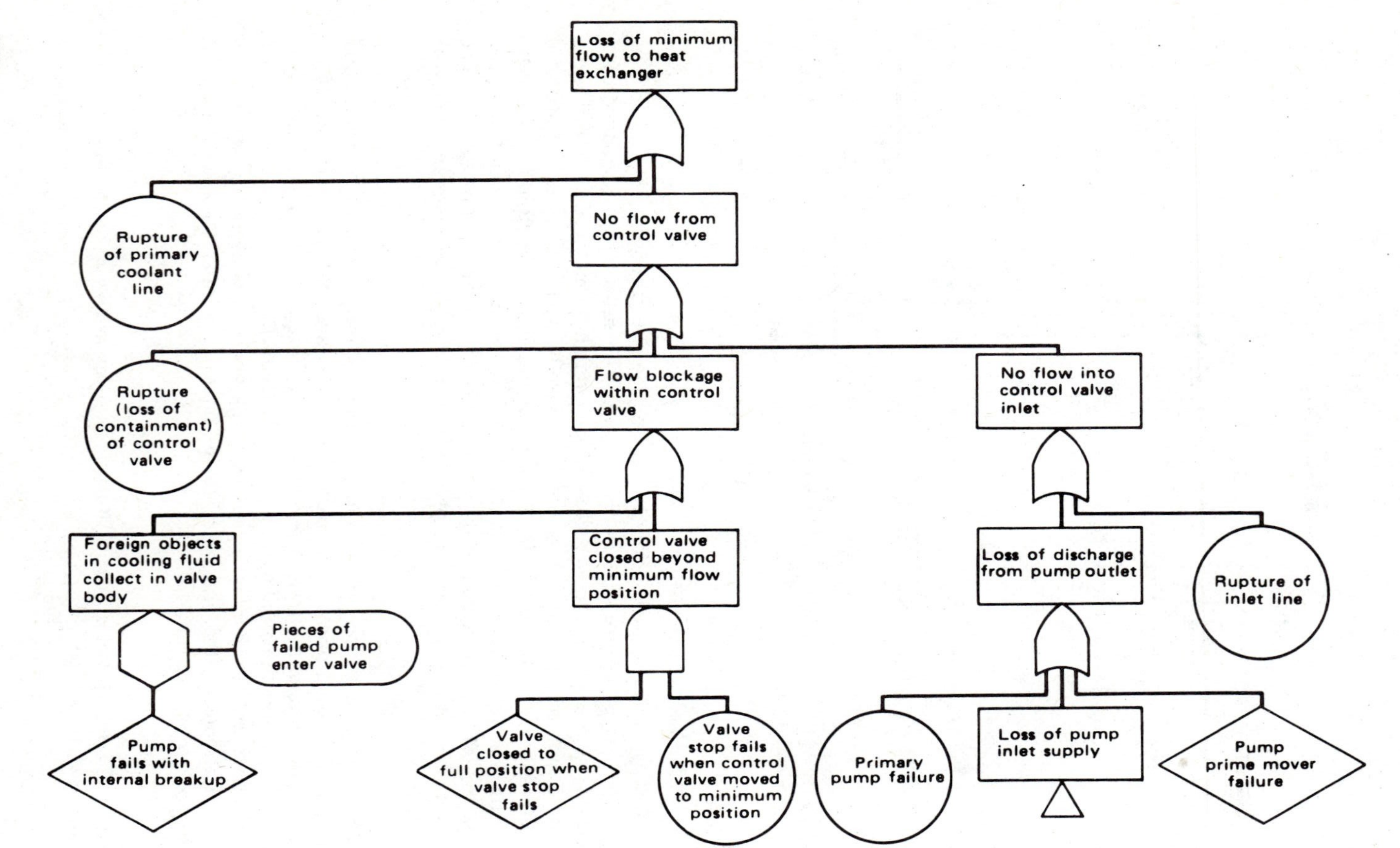

FIGURE 10.15 Fault tree for coolant supply system. (Reprinted from *Machine Design,* © 1984, by Penton/IPC, Cleveland, Ohio.)

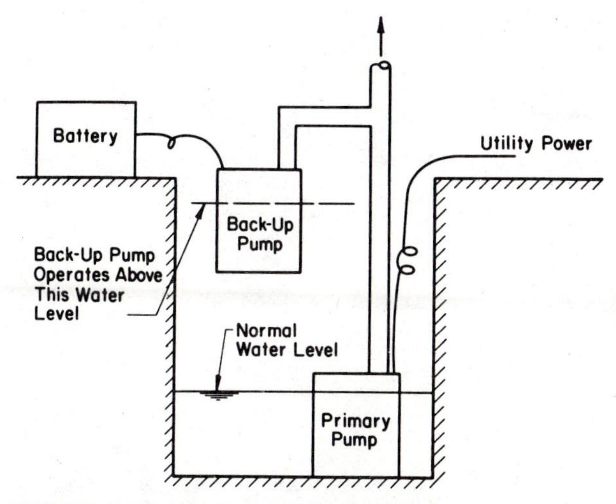

FIGURE 10.16 Sump pump system. (From A. H-S. Ang and W. H. Tang, *Probability Concepts in Engineering Planning and Design,* Vol. 2, p. 496. Copyright © 1984, by John Wiley and Sons, New York. Reprinted by permission.)

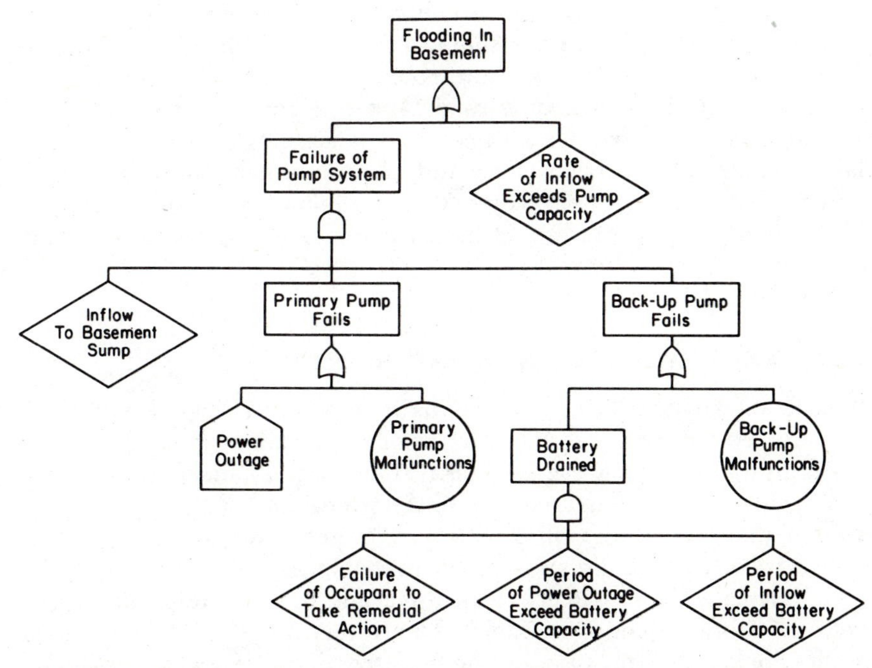

FIGURE 10.17 Fault tree for basement flooding. (From A. H-S. Ang and W. H. Tang, *Probability Concepts in Engineering Planning and Design,* Vol. 2, p. 496. Copyright © 1984, by John Wiley and Sons, New York. Reprinted by permission.)

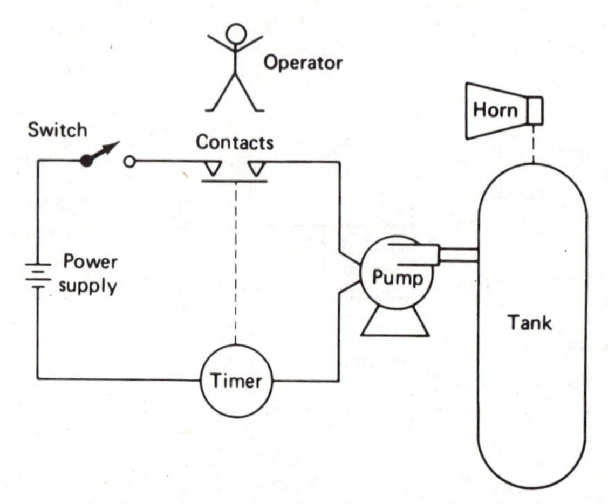

FIGURE 10.18 Schematic diagram for a pumping system. (From Ernest J. Henley and Hiromitsu Kumamoto, *Reliability Engineering and Risk Assessment,* p. 73, © 1981, with permission from Prentice-Hall, Englewood Cliffs, NJ.)

The foregoing examples give some idea of the problems inherent in drawing fault trees. The reader should consult more advanced literature for fault-tree constructions for more complex configurations, keeping in mind that the construction of a valid fault tree for any real system (as opposed to textbook examples) is necessarily a learning experience for the analyst. As the tree is drawn, more and more knowledge must be gained about the details of the system's components, its failure modes, the operating and maintenance procedures and the environment in which the system is to be located.

10.5 DIRECT EVALUATION OF FAULT TREES

The evaluation of a fault tree proceeds in two steps. First, a logical expression is constructed for the top event in terms of combinations (i.e., unions and intersections) of the basic events. This is referred to as qualitative analysis. Second, this expression is used to give the probability of the top event in terms of the probabilities of the primary events. This is referred to as quantitative analysis. Thus, knowing the probabilities of the primary events, we can calculate the probability of the top event. In these steps the rules of Boolean algebra contained in Table 10.2 are very useful. They allow us to simplify the logical expression for the fault tree and thus also to streamline the formula giving the probability of the top event in terms of the primary-failure probabilities.

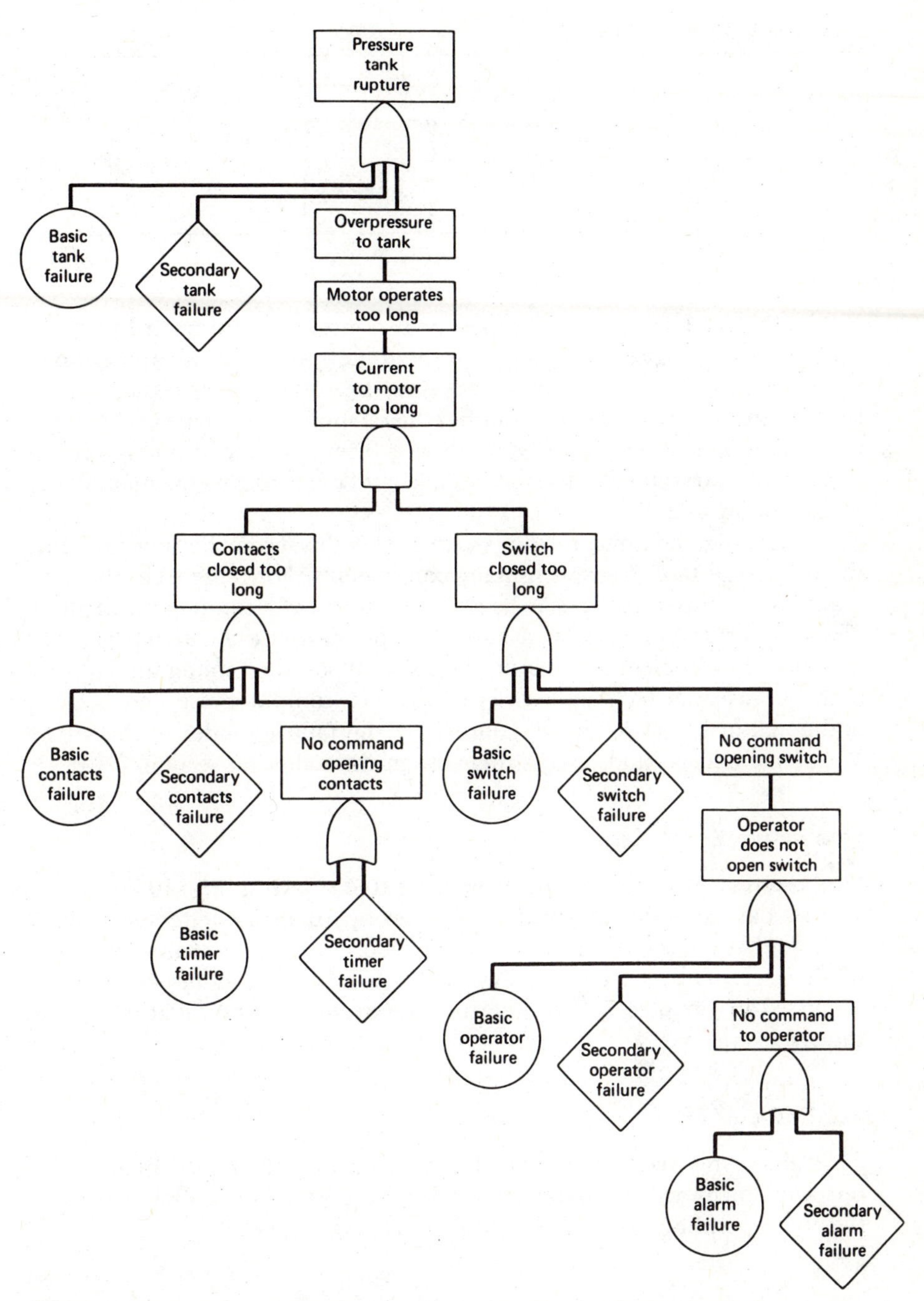

FIGURE 10.19 Fault tree for pumping system. (From Ernest J. Henley and Hiromitsu Kumamoto, *Reliability Engineering and Risk Assessment,* p. 73, © 1981, with permission from Prentice-Hall, Englewood Cliffs, NJ.)

TABLE 10.2 Boolean Logic

A	B	$A \cap B$	$A \cup B$
0	0	0	0
1	0	0	1
0	1	0	1
1	1	1	1

In this section we first illustrate the two most straightforward methods for obtaining a logical expression for the top event, top-down and bottom-up evaluation. We then demonstrate how the resulting expression can be reduced in a way that greatly simplifies the relation between the probabilities of top and basic events. Finally, we discuss briefly the most common forms that the primary-failure probabilities take and demonstrate the quantitative evaluation of a fault tree.

The so-named direct methods discussed in this section become unwieldy for very large fault trees with many components. For large trees the evaluation procedure must usually be cast in the form of a computer algorithm. These algorithms make extensive use of an alternative evaluation procedure in which the problem is recast in the form of so-called minimum cut sets, both because the technique is well suited to computer use and because additional insights are gained concerning the failure modes of the sytem. We define cut sets and discuss their use in the following section.

Qualitative Evaluation

Suppose that we are to evaluate the fault tree shown in Fig. 10.20. In this tree we have signified the primary failures by uppercase letters A through C. Note that the same primary failure may occur in more than one branch of the tree. This is typical of systems with m/N redundancy of the type discussed in Chapter 7. The intermediate events are indicated by E_i, and the top event by T.

Top Down

To evaluate the tree from the top down, we begin at the top event and work our way downward through the levels of the tree, replacing the gates with the corresponding OR or AND symbol. Thus we have

$$T = E_1 \cap E_2 \tag{10.1}$$

at the highest level of the tree, and

$$E_1 = A \cup E_3; \quad E_2 = C \cup E_4 \tag{10.2}$$

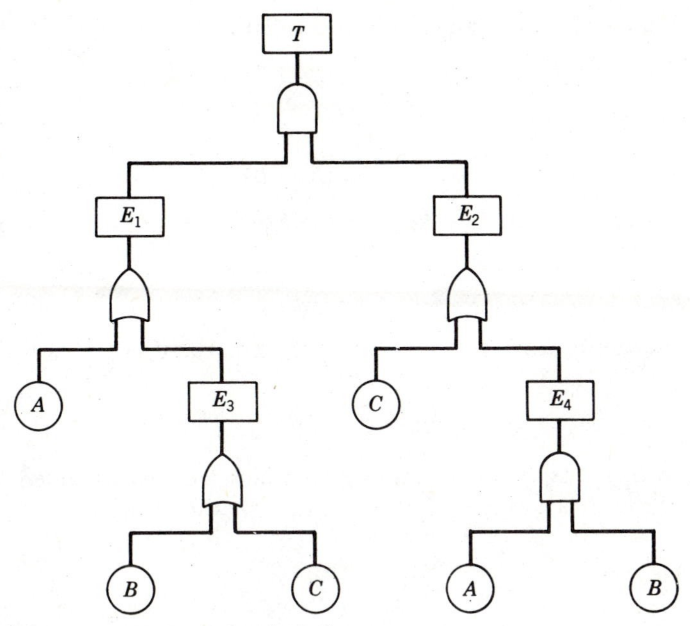

FIGURE 10.20 Example of a fault tree.

at the intermediate level. Substituting Eq. 10.2 into Eq. 10.1, we then obtain

$$T = (A \cup E_3) \cap (C \cup E_4). \tag{10.3}$$

Proceeding downward to the lowest level, we have

$$E_3 = B \cup C; \quad E_4 = A \cap B. \tag{10.4}$$

Substituting these expressions into Eq. 10.3, we obtain as our final result

$$T = [A \cup (B \cup C)] \cap [C \cup (A \cap B)]. \tag{10.5}$$

Bottom Up

Conversely, to evaluate this same tree from the bottom up, we first write the expressions for the gates at the bottom of the fault tree as

$$E_3 = B \cup C; \quad E_4 = A \cap B. \tag{10.6}$$

Then, proceeding upward to the intermediate level, we have

$$E_1 = A \cup E_3; \quad E_2 = C \cup E_4. \tag{10.7}$$

Hence we may substitute Eq. 10.6 into Eq. 10.7 to obtain

$$E_1 = A \cup (B \cup C) \tag{10.8}$$

and

$$E_2 = C \cup (A \cap B). \tag{10.9}$$

We now move to the highest level of the fault tree and express the AND gate appearing there as

$$T = E_1 \cap E_2. \tag{10.10}$$

Then, substituting Eqs. 10.8 and 10.9 into Eq. 10.10, we obtain the final form:

$$T = [A \cup (B \cup C)] \cap [C \cup (A \cap B)]. \tag{10.11}$$

The two results, Eqs. 10.5 and 10.11, which we have obtained with the two evaluation procedures, are not surprisingly the same.

Logical Reduction

For most fault trees, particularly those with one or more primary failures occurring in more than one branch of the tree, the rules of Boolean algebra contained in Table 2.1 may be used to simplify the logical expression for T, the top event. In our example, Eq. 10.11 can be simplified by first applying the associative and then the cummunicative law to write $A \cup (B \cup C) = (A \cup B) \cup C = C \cup (A \cup B)$. Then we have

$$T = [C \cup (A \cup B)] \cap [C \cup (A \cap B)]. \tag{10.12}$$

We then apply the distributive law with $X \equiv C$, $Y \equiv A \cup B$, and $Z \equiv A \cap B$ to obtain

$$T = C \cup [(A \cup B) \cap (A \cap B)]. \tag{10.13}$$

From the associative law we can eliminate the parenthesis on the right. Then, since $A \cap B = B \cap A$, we have

$$T = C \cup [(A \cup B) \cap B \cap A]. \tag{10.14}$$

Now, from the absorption law $(A \cup B) \cap B = B$. Hence

$$T = C \cup (B \cap A). \tag{10.15}$$

This expression tells us that for the fault tree under consideration the failure of the top system is caused by the failure of C or by the failure of both A and B. We then refer to $M_1 = C$ and $M_2 = A \cap B$ as the two failure modes leading to the top event. The reduced fault tree can be drawn to represent the system as shown in Fig. 10.21.

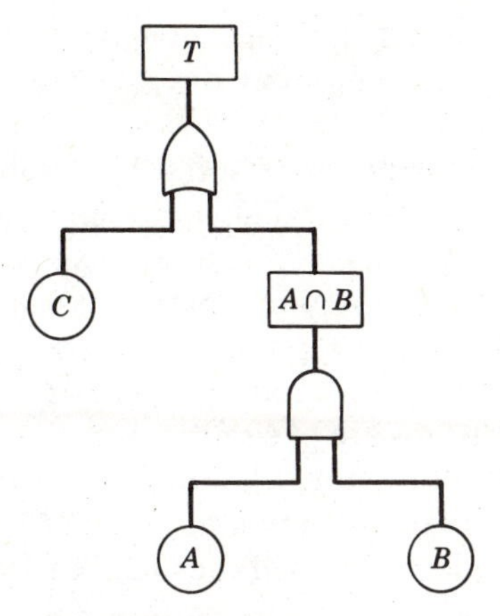

FIGURE 10.21 Fault-tree equivalent to Fig. 10.20.

Quantitative Evaluation

Having obtained, in its simplest form, the logical expression for the top event in terms of the primary failures, we are prepared to evaluate the probability that the top event will occur. The evaluation may be divided into two tasks. First, we must use the logical expression and the rules developed in Chapter 2 for combining probabilities to express the probability of the top event in terms of the probabilities of the primary failures. Second, we must evaluate the primary-failure probabilties in terms of the data available for component unreliabilities, component unavailabilities, and demand-failure probabilities.

Probability Relationships

To illustrate the quantitative evaluation, we again use the fault tree that reduces to Eq. 10.15. Since the top event is the union of C with $B \cap A$, we use Eq. 2.10 to obtain

$$P\{T\} = P\{C\} + P\{B \cap A\} - P\{A \cap B \cap C\}, \tag{10.16}$$

thus expressing the top events in terms of the intersections of the basic events. If the basic events are known to be independent, the intersections may be replaced by the products of basic-event probabilities. Thus, in our example,

$$P\{T\} = P\{C\} + P\{A\}P\{B\} - P\{A\}P\{B\}P\{C\}. \tag{10.17}$$

If there are known dependencies between events, however, we must determine expression for $P\{A \cap B\}$, $P\{A \cap B \cap C\}$, or both through more sophisticated treatments such as the Markov models discussed in Chapter 9. Alternatively, we may be able to apply the β-factor treatment of Chapter 7 for common-mode failures.

Even where independent failures can be assumed, a problem arises when larger trees with many different component failures are considered. Instead of three terms as in Eq. 10.17, there may be hundreds of terms of vastly different magnitudes. A systematic way is needed for making reasonable approximations without evaluating all the terms. Since the failure probabilities are rarely known to more than two or three places of accuracy, often only a few of the terms are of significance. For example, suppose that in Eq. 10.17 the probabilities of A, B, and C are $\sim 10^{-2}$, 10^{-4}, and $\sim 10^{-6}$, respectively. Then the first two terms in Eq. 10.17 are each of the order 10^{-6}; in comparison the last term is of the order of 10^{-12} and may therefore be neglected.

One approach that is used in rough calculations for larger trees is to approximate the basic equation for $P\{X \cup Y\}$ by assuming that both events are improbable. Then, instead of using Eq. 2.10, we may approximate

$$P\{X \cup Y\} \approx P\{X\} + P\{Y\}, \tag{10.18}$$

which leads to a conservative (i.e., pessimistic) approximation for the system failure. For our simple example, we have, instead of Eq. 10.17, the approximation

$$P\{T\} \approx P\{C\} + P\{A\}P\{B\}. \tag{10.19}$$

The combination of this form of the rare-event approximation and the assumption of independence,

$$P\{X \cap Y\} = P\{X\}P\{Y\}, \tag{10.20}$$

often allows a very rough estimate of the top-event probability. We simply perform a bottom-up evaluation, multiplying probabilities at AND gates and adding them at OR gates. Care must be exercised in using this technique, for it is applicable only to trees in which basic events are not repeated—since repeated events are not independent—or to trees that have been logically reduced to a form in which primary failures appear only once. Thus we may not evaluate the tree as it appears in Fig. 10.20 in this way, but we may evaluate the reduced form in Fig. 10.21. More systematic techniques for truncating the prohibitively long probability expressions that arise from large fault trees are an integral part of the minimum cut-set formulation considered in the next section.

Primary-Failure Data

In our discussions we have described fault trees in terms of failure probabilities without specifying the particular types of failure represented either by the top event or by the primary-failure data. In fact, there are three types of top events and, correspondingly, three types of basic events frequently used in conjunction with fault trees. They are (1) the failure on demand, (2) the unreliability for some fixed period of time t, and (3) the unavailability at some time.

When failures on demand are the basic events, a value of p is needed. For the unreliability or unavailability it is often possible to use the following approximations to simplify the form of the data, since the probabilities of failure are expected to be quite small. If we assume a constant failure rate, the unreliability is

$$\tilde{R} \simeq \lambda t. \tag{10.21}$$

Similarly, the most common unavailability is the asymptotic value, for a system with constant failure and repair rates λ and ν. From Eq. 8.73 we have

$$\tilde{A}(\infty) = 1 - \frac{\nu}{\nu + \lambda}. \tag{10.22}$$

But, since in the usual case $\nu >> \lambda$, we may approximate this by

$$\tilde{A}(\infty) \approx \lambda/\nu. \tag{10.23}$$

Often, demand failures, unreliabilities, and unavailabilities will be mixed in a single fault tree. Consider, for example, a very simple fault tree for the failure of a light to go on when the switch is flipped. We assume that the top event, T, is the failure on demand for the light to go on, which is due to

X = bulb burned out,

Y = switch fails to make contact,

Z = power failure to house.

Therefore $T = X \cup Y \cup Z$. In this case, X might be considered an unreliability of the bulb, with the time being that since it was originally installed; Y would be a demand failure, assuming that the cause was a random failure of the switch to make contact; and Z would be the unavailability of power to the circuit. Of course, the tree can be drawn in more depth. Is the random demand failure the only significant reason (a demand failure) for the switch not to make contact, or is there a significant probability that the switch is corroded open (an unreliability)?

10.6 FAULT-TREE EVALUATION BY CUT SETS

The direct evaluation procedures just discussed allow us to assess fault trees with relatively few branches and basic events. When larger trees are considered, both evaluation and interpretation of the results become more difficult and digital computer codes are invariably employed. Such codes are usually formulated in terms of the minimum cut-set methodology discussed in this section. There are at least two reasons for this. First, the techniques lend themselves well to the computer algorithms, and second, from them a good deal of intermediate information can be obtained concerning the combination of component failures that are pertinent to improvements in system design and operations.

The discussion that follows is conveniently divided into qualitative and quantitative analysis. In qualitative analysis information about the logical structure of the tree is used to locate weak points and evaluate and improve system design. In quantitative analysis the same objectives are taken further by studying the probabilities of component failures in relation to system design.

Qualitative Analysis

In these subsections we first introduce the idea of minimum cut sets and relate it to the qualitative evaluation of fault trees. We then discuss briefly how the minimum cut sets are determined for large fault trees. Finally, we discuss their use in locating system weak points, particularly possibilities for common-mode failures.

Minimum Cut-Set Formulation

A minimum cut set is defined as the smallest combination of primary failures which, if they all occur, will cause the top event to occur. It, therefore, is a combination (i.e., intersection) of primary failures sufficient to cause the top event. It is the smallest combination in that all the failures must take place for the top event to occur. If even one of the failures in the minimum cut set does not happen, the top event will not take place.

The terms minimum cut set and failure mode are sometimes used interchangeably. However, there is a subtle difference that we shall observe hereafter. In reliability calculations a failure mode is a combination of component or other failures that cause a system to fail, regardless of the consequences of the failure. A minimum cut set is usually more restrictive, for it is the minimum combination of failures that causes the top event as defined for a particular fault tree. If the top event is defined broadly as system failure, the two are indeed interchangeable. Usually, however, the top event encompasses only the particular subset of system failures that bring about a particular safety hazard.

The origin for using the term cut set may be illustrated graphically using the reduced fault tree in Fig. 10.21. The reliability block diagram corresponding to the tree is shown in Fig. 10.22. The idea of a cut set comes originally from the use of such diagrams for electric apparatus, where the signal enters at the left and leaves at the right. Thus a minimum cut set is

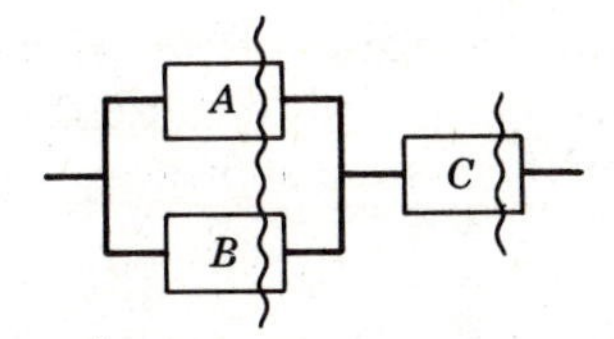

FIGURE 10.22 Minimum cut sets on a reliability block diagram.

the minimum number of components that must be cut to prevent the signal flow. There are two minimum cut sets, M_1, consisting of components A and B, and M_2, consisting of component C.

For a slightly more complicated example, consider the redundant system of Fig. 7.10, for which the equivalent fault tree appears in Fig. 10.8. In this system there are five cut sets, as indicated in the reliability block diagram of Fig. 10.23.

For larger systems, particularly those in which the primary failures appear more than once in the fault tree, the simple geometrical interpretation becomes problemmatical. However, the primary characteristics of the concept remain valid. It permits the logical structure of the fault tree to be represented in a systematic way that is amenable to interpretation in terms of the behavior of the minimum cut sets.

Suppose that the minimum cut sets of a system can be found. The top event, system failure, may then be expressed as the union of these sets. Thus, if there are N minimum cut sets,

$$T = M_1 \cup M_2 \cup \cdots \cup M_N. \tag{10.24}$$

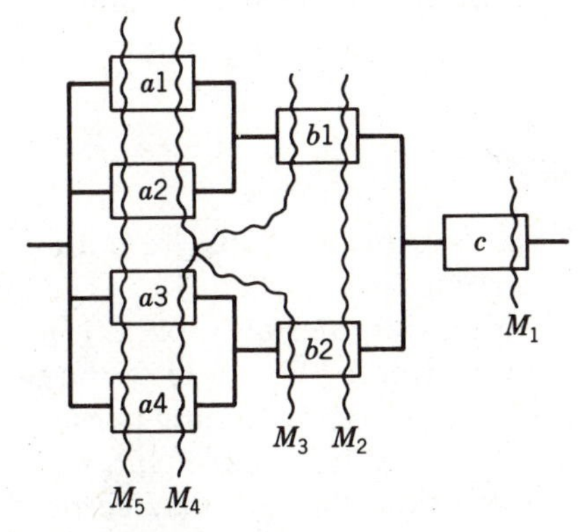

FIGURE 10.23 Minimum cut sets on a reliability block diagram of a seven-component system.

Each minimum cut set then consists of the intersection of the minimum number of primary failures required to cause the top event. For example, the minimum cut sets for the system shown in Figs. 10.8 and 10.23 are

$$M_1 = c \qquad M_3 = a1 \cap a2 \cap b2$$
$$M_2 = b1 \cap b2 \qquad M_4 = a3 \cap a4 \cap b1 \tag{10.25}$$
$$M_5 = a1 \cap a2 \cap a3 \cap a4.$$

Before proceeding, it should be pointed out that there are other cut sets that will cause the top event, but they are not minimum cut sets. These need not be considered, however, because they do not enter the logic of the fault tree. By the rules of Boolean algebra contained in Table 2.1, they are absorbed into the minimum cut sets. This can be illustrated using the configuration of Fig. 10.23 again. Suppose that we examine the cut set $M_0 = b1 \cap c$, which will certainly cause system failure, but it is not a minimum cut set. If we include it in the expression for the top event, we have

$$T = M_0 \cup M_1 \cup M_2 \cup \cdots \cup M_N. \tag{10.26}$$

Now suppose that we consider $M_0 \cup M_1$. From the absorption law of Table 2.1, however, we see that

$$M_0 \cup M_1 = (b1 \cap c) \cup c = c. \tag{10.27}$$

Thus the nonminimum cut set is eliminated from the expression for the top event. Because of this property, minimum cut sets are often referred to simply as cut sets, with the minimum implied.

Since we are able to write the top event in terms of minimum cut sets as in Eq. 10.24, we may express the fault tree in the standardized form shown in Fig. 10.24. In this X_{mn} is the nth element of the mth minimum cut set. Note from our example that the same primary failures may often be expected to occur in more than one of the minimum cut sets. Thus the minimum cut sets are not generally independent of one another.

Cut-Set Determination

In order to utilize the cut-set formulations, we must express the top event as the union of minimum cut sets, as in Eq. 10.24. For small fault trees this can be done by hand, using the rules of Table 2.1, just as we reduced the top-event expression for T given by Eq. 10.11 to the two-cut-set expression given by Eq. 10.15. For larger trees, containing perhaps 20 or more primary failures, this procedure becomes intractable, and we must resort to digital computer evaluation. Even then the task may be prodigious, for a larger tree with a great deal of redundancy may have a million or more minimum cut sets.

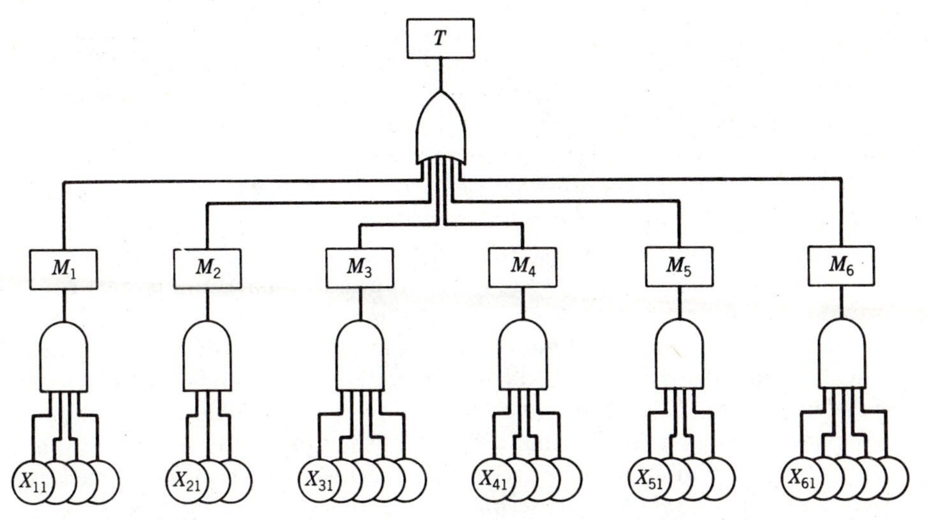

FIGURE 10.24 Generalized minimum cut-set representation of a fault tree.

The computer codes for determining the cut sets* do not typically apply the rules of Boolean algebra to reduce the expression for the top set to the form of Eq. 10.24. Rather, a search is performed for the minimum cut sets; in this, a failure is represented by 1 and a success by 0. Then each expression for the top event is evaluated using the outcome shown in Table 10.2 for the union and intersection of the events. A number of different procedures may be used to find the cut sets. In exhaustive searches, all single failures are first examined, and then all combinations of two primary failures, and so on. In general, there are 2^N, where N is the number of primary failures that must be examined. Other methods involve the use of random number generators in Monte Carlo simulation to locate the minimum cut sets.

When millions of minimum cut sets are possible, the search procedures are often truncated, for cut sets requiring many primary failures to take place are so improbable that they will not significantly affect the overall probability of the top event. Moreover, simulation methods must be terminated after a finite number of trials.

Cut-Set Interpretations

Knowing the minimum cut sets for a particular fault tree can provide valuable insight concerning potential weak points of complex systems, even when it is not possible to calculate the probability that either a particular

* See, for example, N. J. McCormick, *Reliability and Risk Analysis,* Academic Press, New York, 1981.

cut set or the top event will occur. Three qualitative considerations, in particular, may be very useful: the ranking of the minimal cut sets by the number of primary failures required, the importance of particular component failures to the occurrence of the minimum cut sets, and the susceptibility of particular cut sets to common-mode failures.

Minimum cut sets are normally categorized as singlets, doublets, triplets, and so on, according to the number of primary failures in the cut set. Emphasis is then put on eliminating cut sets corresponding to small numbers of failures, for ordinarily these may be expected to make the largest contributions to system failure. In fact, the common design criterion, that no single component failure should cause system failure is equivalent to saying that all singlets must be removed from the fault tree for which the top event is system failure. Indeed, if component failure probabilities are small and independent, then provided that they are of the same order of magnitude, doublets will occur much less frequently than singlets, triplets much less frequently than doublets, and so on.

A second application of cut-set information is in assessing qualitatively the importance of a particular component. Suppose that we wish to evaluate the effect on the system of improving the reliability of a particular component, or conversely, to ask whether, if a particular component fails, the system-wide effect will be considerable. If the component appears in one or more of the low-order cut sets, say singlets or doublets, its reliability is likely to have a pronounced effect. On the other hand, if it appears only in minimum cut sets requiring several independent failures, its importance to system failure is likely to be small.

These arguments can rank minimum cut-set and component importance, assuming that the primary failures are independent. If they are not, that is, if they are susceptible to common-mode failure, the ranking of cut-set importance may be changed. If five of the failures in a minimum cut set with six failures, for example, can occur as the result of a common cause, the probability of the cut set's occurring is more comparable to that of a doublet.

Extensive analysis is often carried out to determine the susceptibility of minimum cut sets to common-cause failures. In an industrial plant one cause might be fire. If the plant is divided into several fire-resistant compartments, the analysis might proceed as follows. All the primary failures of equipment located in one of the compartments that could be caused by fire are listed. Then these components would be eliminated from the minimum cut sets (i.e., they would be assumed to fail). The resulting cut sets would then indicate how many failures—if any—in addition to those caused by the fire, would be required for the top event to happen. Such analysis is critical for determining the layout of the plant that will best protect it from a variety of sources of damage: fire, flooding, collision, earthquake, and so on.

Quantitative Analysis

With the minimum cut sets determined, we may use probability data for the primary failures and proceed with quantitative analysis. This normally includes both an estimate of the probability of the top event's occurring and quantitative measures of the importance of components and cut sets to the top event. Finally, studies of uncertainty about the top event's happening because the probability data for the primary failures are uncertain are often needed to assess the precision of the results.

Top-Event Probability

To determine the probability of the top event, we must calculate

$$P\{T\} = P\{M_1 \cup M_2 \cup \cdots \cup M_N\}. \tag{10.28}$$

As indicated in Section 2.2, the union can always be eliminated from a probability expression by writing it as a sum of terms, each one of which is the probability of an intersection of events. Here the intersections are the minimum cut sets. Probability theory provides the expansion of Eq. 10.28 in the following form

$$\begin{aligned} P\{T\} = &\sum_{i=1}^{N} P\{M_i\} - \sum_{i=2}^{N}\sum_{j=1}^{i-1} P\{M_i \cap M_j\} \\ &+ \sum_{i=3}^{N}\sum_{j=2}^{i-1}\sum_{k=1}^{j-1} P\{M_i \cap M_j \cap M_k\} - \cdots \\ &+ (-1)^{N-1} P\{M_1 \cap M_2 \cap \cdots \cap M_N\}. \end{aligned} \tag{10.29}$$

This is sometimes referred to as the inclusion–exclusion principle.

The first task in evaluating this expression is to evaluate the probabilities of the individual minimum cut sets. Suppose that we let X_{im} represent the mth basic event in minimum cut set i. Then

$$P\{M_i\} = P\{X_{i1} \cap X_{i2} \cap X_{i3} \cap \cdots \cap X_{iM}\}. \tag{10.30}$$

If it may be proved that the primary failures in a given cut set are independent, we may write

$$P\{M_i\} = P\{X_{i1}\}P\{X_{i2}\} \cdots P\{X_{iM}\}. \tag{10.31}$$

If they are not, a Markov model or some other procedure must be used to relate $P\{M_i\}$ to the properties of the primary failures.

The second task is to evaluate the intersections of the cut-set probabilities. If the cut sets are independent of one another, we have simply

$$P\{M_i \cap M_j\} = P\{M_i\}P\{M_j\}, \tag{10.32}$$

$$P\{M_i \cap M_j \cap M_k\} = P\{M_i\}P\{M_j\}P\{M_k\}, \tag{10.33}$$

and so on. More often than not, however, these conditions are not valid, for in a system with redundant components, a given component is likely to appear in more than one minimum cut set: If the same primary failure appears in two minimum cut sets, they cannot be independent of one another. Thus an important point is to be made. Even if the primary events are independent of one another, the minimum cut sets are unlikely to be. For example, in the fault trees of Figs. 10.8 and 10.23 the minimum cut sets $M_1 = c$ and $M_2 = b1 \cap b2$ will be independent of one another if the primary failures of components $b1$ and $b2$ are independent of c. In this system, however, M_2 and M_3 will be dependent even if all the primary failures are independent because they contain the failure of component $b2$.

Although minimum cut sets may be dependent, calculation of their intersections is greatly simplified if the primary failures are all independent of one another, for then the dependencies are due only to the primary failures that appear in more than one minimum cut set. To evaluate the intersection of minimum cut sets, simply take the product of probabilities that appear in one or more of the minimal cut sets:

$$P\{M_i \cap M_j\} = P\{X_{1ij}\}P\{X_{2ij}\} \cdots P\{X_{Nij}\}, \tag{10.34}$$

where $X_{1ij}, X_{2ij}, \cdots, X_{Nij}$ is the list of the failures that appear in M_i, M_j, or both.

That the foregoing procedure is correct is illustrated by a simple example. Suppose that we have two minimal cut sets $M_1 = A \cap B$, $M_2 = B \cap C$, where the primary failures are independent. We then have

$$M_1 \cap M_2 = (A \cap B) \cap (B \cap C) = A \cap B \cap B \cap C, \tag{10.35}$$

but $B \cap B = B$. Thus

$$P\{M_1 \cap M_2\} = P\{A \cap B \cap C\} = P\{A\}P\{B\}P\{C\}. \tag{10.36}$$

In the general notation of Eq. 10.34 we would have

$$X_{112} = A, \quad X_{212} = B, \quad X_{312} = C. \tag{10.37}$$

With the assumption of independent primary failures, the series in Eq. 10.29 may in principle be evaluated exactly. When there are thousands or even millions of minimum cut sets to be considered, however, the task may be both prohibitive and unwarranted, for many of the terms in the series are likely to be completely negligible compared to the leading one or two terms.

The true answer may be bracketed by taking successive terms, and it is rarely necessary to evaluate more than the first two or three terms. If $P\{T\}$ is the exact value, it may be shown that*

$$P_1\{T\} \equiv \sum_{i=1}^{N} P\{M_i\} > P\{T\}, \tag{10.38}$$

$$P_2\{T\} \equiv P_1\{T\} - \sum_{i=2}^{N} \sum_{j=1}^{i-1} P\{M_i \cap M_j\} < P\{T\}, \tag{10.39}$$

$$P_3\{T\} \equiv P_2\{T\} + \sum_{i=3}^{N} \sum_{j=2}^{i-1} \sum_{k=1}^{j-1} P\{M_i \cap M_j \cap M_k\} > P\{T\}. \tag{10.40}$$

and so on, with $P_4\{T\} < P\{T\}$.

Often the first-order approximation $P_1\{T\}$ gives a result that is both reasonable and pessimistic. The second-order approximation might be evaluated to check the accuracy of the first. And rarely would more than the third-order approximation be used.

Even taking only a few terms in Eq. 10.38 may be difficult, and wasteful, if a million or more minimum cut sets are present. Thus, as mentioned in the preceding subsection, we often truncate the number of minimum cut sets to include only those that contain fewer than some specified number of primary failures. If all the failure probabilities are small, say < 0.1, the cut-set probabilities should go down by more than an order of magnitude as we go from singlets to doublets, doublets to triplets, and so on.

Importance

As in qualitative analysis, it is not only the probability of the top event that normally concerns the analyst. The relative importance of single components and of particular minimum cut sets must be known if designs are to be optimized and operating procedures revised.

Two measures of importance† are particularly simple but useful in system analysis. In order to know which cut sets are the most likely to cause the top event, the cut-set importance is defined as

$$I_{M_i} = \frac{P\{M_i\}}{P\{T\}} \tag{10.41}$$

for the minimum cut set i. Generally, we would also like to determine the relative importance of different primary failures in contributing to the top

* W. E. Vesely, "Time Dependent Methodology for Fault Tree Evaluation," *Nucl. Eng. Design*, **13**, 337–357 (1970).

† See, for example, E. J. Henley and H. Kumamoto, op. cit., Chapter 10.

event. To accomplish this, the simplest measure is to add the probabilities of all the minimum cut sets to which the primary failure contributes. Thus the importance of component X_i is

$$I_{X_i} = \frac{1}{P\{T\}} \sum_{X_i \in M_l} P\{M_l\}. \tag{10.42}$$

Other more sophisticated measures of importance have also found applications.

Uncertainty

What we have obtained thus far are point or best estimates of the top event's probability. However, there are likely to be substantial uncertainties in the basic parameters—the component failure rates, demand failures, and other data—that are input to the probability estimates. Given these considerable uncertainties, it would be very questionable to accept point estimates without an accompanying interval estimate by which to judge the precision of the results. To this end the component failure rates and other data may themselves be represented as random variables with a mean or best-estimate value and a variance to represent the uncertainty. The lognormal distribution has been very popular for representing failure data in this manner. For small fault trees a number of analytical techniques may be applied to determine the sensitivity of the results to the data uncertainty. For larger trees the Monte Carlo method has found extensive use.*

Bibliography

Roberts, H. R., Vesley, W. E., Haast, D. F., and Goldberg, F. F., *Fault Tree Handbook,* U.S. Nuclear Regulatory Commission, NUREG-0492, 1981.

Green, A. E. (ed.) *High Risk Safety Technology,* Wiley, New York, 1982.

Green, A. E., *Safety Systems Analysis,* Wiley, New York, 1983.

Henley, E. J., and H. Kumamoto, *Reliability Engineering and Risk Assessment,* Prentice-Hall, Englewood Cliffs, NJ, 1981.

McCormick, E. J., *Human Factors in Engineering Design,* McGraw-Hill, New York, 1976.

McCormick, N. J., *Reliability and Risk Analysis,* Academic Press, New York, 1981.

PRA Procedures Guide, Vol. 1, U.S. Nuclear Regulatory Commission, NUREG/CR-2300, 1983.

Swain, A. D., and H. R. Guttmann, *Handbook of Human Reliability Analysis with Emphasis on Nuclear Power Plant Applications,* U.S. Nuclear Regulatory Commission, NUREG/CR-1287, 1980.

* See, for example, E. J. Henley and H. Kumanoto, op. cit, Chapter 12.

Waller, R. A., and V. C. Covello (eds.), *Low-Probability High Consequence Risk Analysis*, Plenum Press, New York, 1984.

EXERCISES

10.1 Make a list of six population stereotypical responses.

10.2 Find the fault tree for system failure for the following configurations.

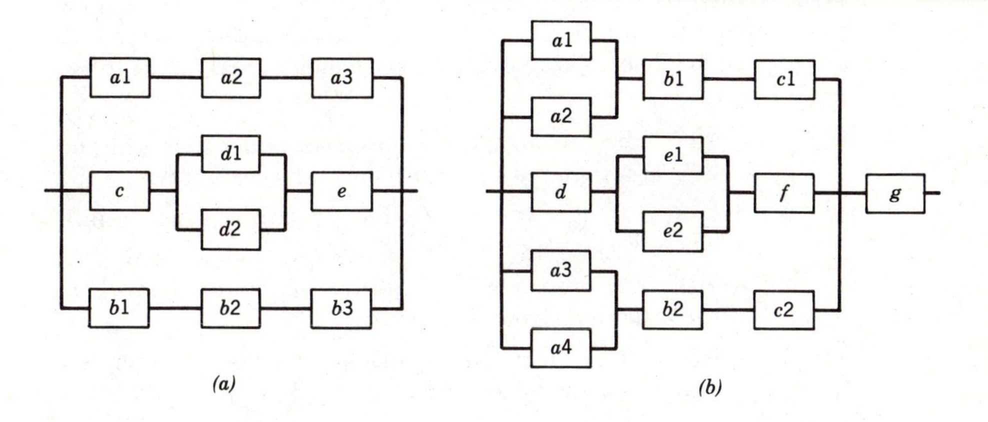

10.3 Draw a fault tree corresponding to the reliability block diagram in Exercise 7.27.

10.4 Construct a fault tree for which the top event is your failure to arrive on time for the final exam of a reliability engineering course. Include only the primary failures that you think have probabilities large enough to significantly affect the result.

10.5 Classify each of the failures in Fig. 10.15 as (*a*) passive, (*b*) active, or (*c*) either.

10.6 Suppose that a system consists of two subsystems in parallel. Each has a mission reliability of 0.9.
(*a*) Draw a fault tree for mission failure and calculate the probability of the top event.
(*b*) Assume that there are common-mode failures described by the β-factor method (Chapter 7) with $\beta = 0.1$. Redraw the fault tree to take this into account and recalculate the top event.

10.7 The following system is designed to deliver emergency cooling to a nuclear reactor.

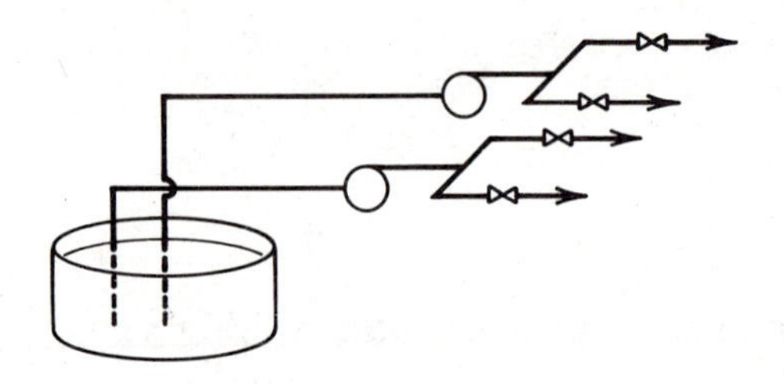

In the event of an accident the protection system delivers an actuation signal to the two identical pumps and the four identical valves. The pumps then start up; the valves open; and liquid coolant is delivered to the reactor. The following failure probabilities are found to be significant:

$p_{ps} = 10^{-5}$ the probability that the protection system will not deliver a signal to the pump and valve actuators.

$p_p = 2 \times 10^{-2}$ the probability that a pump will fail to start when the actuation signal is received.

$p_v = 10^{-1}$ the probability that a valve will fail to open when the actuation signal is received.

$p_r = 0.5 \times 10^{-5}$ the probability that the reservoir will be empty at the time of the accident.

(*a*) Draw a fault tree for the failure of the system to deliver any coolant to the primary system in the event of an accident.
(*b*) Evaluate the probability that such a failure will take place in the event of an accident.

10.8 Find the minimum cut sets of the tree on the following page.

10.9 Develop a logical expression for the fault trees in Fig. 10.13 in terms of the nine root causes. Find the minimum cut sets.

10.10 The logical expression for a fault tree is given by

$$T = A \cap (B \cup C) \cap [D \cup (E \cap F \cap G)].$$

(*a*) Construct the corresponding fault tree.
(*b*) Find the minimum cut sets.
(*c*) Construct an equivalent reliability block diagram.

10.11 Construct the fault trees for system failure for the low- and high-level redundant systems shown in Fig. 7.8. Then find the minimum cut sets.

10.12 Suppose that a fault tree has three minimum cut sets. The basic failures are independent and do not appear in more than one cut

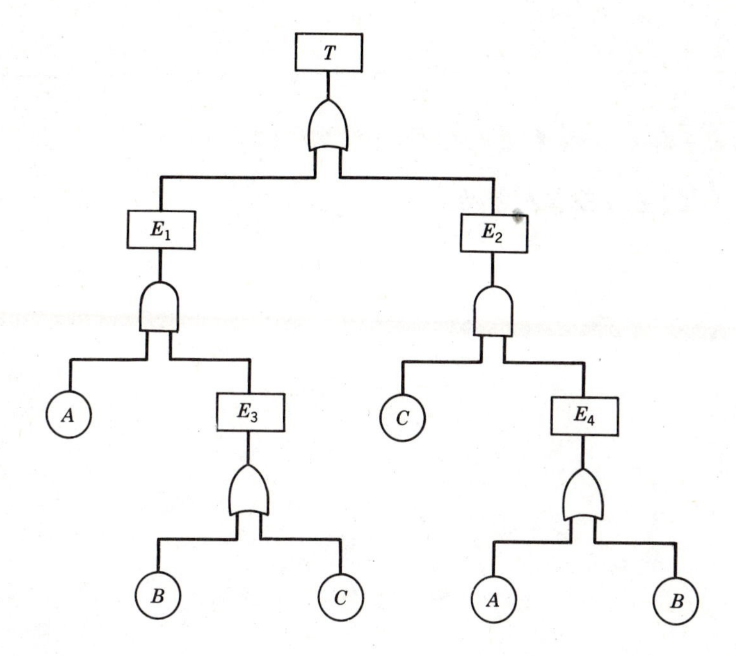

set. Assume that $P\{M_1\} = 0.03$, $P\{M_2\} = 0.12$ and $P\{M_3\} = 0.005$. Estimate $P\{T\}$ by the three successive estimates given in Eqs. 10.38, 10.39, and 10.40.

10.13 Suppose that for the fault tree given in Fig. 10.21 $P\{A\} = 0.15$, $P\{B\} = 0.20$, and $P\{C\} = 0.05$.
(*a*) Calculate the cut-set importances.
(*b*) Calculate the component importances.
(Assume independent failures.)

10.14 From the reliability block diagram shown in Figure 10.23, draw a fault tree for system failure in minimum cut-set form. Assume that the failure probabilities for component types *a*, *b*, and *c* are, respectively, 0.1, 0.02, and 0.005. Assuming independent failures, calculate (*a*) $P\{T\}$, the probability of the top event; (*b*) the importance of components $a1$, $b1$ and c; (*c*) the importance of each of the five minimum cut sets.

APPENDIX A

Useful Mathematical Relationships

A.1 INTEGRALS

Definite Integrals

$$\int_0^\infty e^{-ax}\,dx = \frac{1}{a}, \qquad a > 0.$$

$$\int_0^\infty x^n e^{-ax}\,dx = \frac{n!}{a^{n+1}}, \qquad n = \text{integer} > 0,\ a > 0.$$

$$\int_0^\infty e^{-a^2x^2}\,dx = \frac{\sqrt{\pi}}{2a}, \qquad a > 0.$$

$$\int_0^\infty x e^{-x^2}\,dx = \tfrac{1}{2}.$$

$$\int_0^\infty x^2 e^{-x^2}\,dx = \frac{\sqrt{\pi}}{4}.$$

$$\int_0^\infty x^{2n} e^{-ax^2}\,dx = \frac{1\cdot 3\cdot 5\cdots(2n-1)}{2^{n+1}a^n}\sqrt{\pi/a},$$

$$n = \text{integer} > 0,\ a > 0.$$

Integration by Parts

$$\int_a^b f(x)\frac{d}{dx}g(x)\,dx = f(b)g(b) - f(a)g(a) - \int_a^b g(x)\frac{d}{dx}f(x)\,dx.$$

Derivative of an Integral

$$\frac{d}{dc}\int_p^q f(x, c)\,dx = \int_p^q \frac{d}{dc}f(x, c)\,dx + f(q, c)\frac{dq}{dc} - f(p, c)\frac{dp}{dc}.$$

A.2 EXPANSIONS

Integer Series

$$1 + 2 + 3 + \cdots + n = \frac{n}{2}(n + 1).$$

$$1^2 + 2^2 + 3^2 + \cdots + n^2 = \frac{n}{6}(2n^2 + 3n + 1).$$

$$1^3 + 2^3 + 3^3 + \cdots + n^3 = \frac{n^2}{4}(n + 1)^2.$$

$$1 + 3 + 5 + \cdots + (2n - 1) = n^2.$$

Binomial Expansion

$$(p + q)^N = \sum_{n=0}^{N} C_n^N\, p^n q^{N-n}.$$

$$C_n^N \equiv \frac{N!}{(N - n)!n!}.$$

Geometric Progression

$$\frac{1 - p^n}{1 - p} = 1 + p + p^2 + p^3 + \cdots + p^{n-1}.$$

Infinite Series

$$e^x = 1 + \frac{x}{1!} + \frac{x^2}{2!} + \frac{x^3}{3!} + \cdots, \qquad x^2 < \infty.$$

$$\log(1 + x) = x - \frac{x^2}{2} + \frac{x^3}{3} - \frac{x^4}{4} + \cdots, \qquad x^2 < 1.$$

A.3 SOLUTION OF A FIRST-ORDER LINEAR DIFFERENTIAL EQUATION

$$\frac{d}{dx}y(x) + \alpha(x)y(x) = S(x).$$

Note that

$$\frac{d}{dx}y(x)\exp\left[\int_{x_0}^{x}\alpha(x')\,dx'\right] = \left[\frac{d}{dx}y(x) + \alpha(x)y(x)\right]\exp\left[\int_{x_0}^{x}\alpha(x')\,dx'\right].$$

Thus, multiplying by the integrating factor $\exp[\int_{x_0}^{x}\alpha(x')\,dx']$, we have

$$\frac{d}{dx}y(x)\exp\left[\int_{x_0}^{x}\alpha(x')\,dx'\right] = S(x)\exp\left[\int_{x_0}^{x}\alpha(x')\,dx'\right].$$

Integrating between x_0 and x, we have

$$y(x) = y(x_0)\exp\left[-\int_{x_0}^{x}\alpha(x')\,dx'\right] + \int_{x_0}^{x}dx'S(x')\exp\left[-\int_{x'}^{x}\alpha(x'')\,dx''\right].$$

If α is a constant, then

$$y(x) = y(x_0)\exp[-\alpha(x - x_0)] + \int_{x_0}^{x}dx'S(x')\exp[-\alpha(x - x'')].$$

APPENDIX B

Binomial Sampling Charts

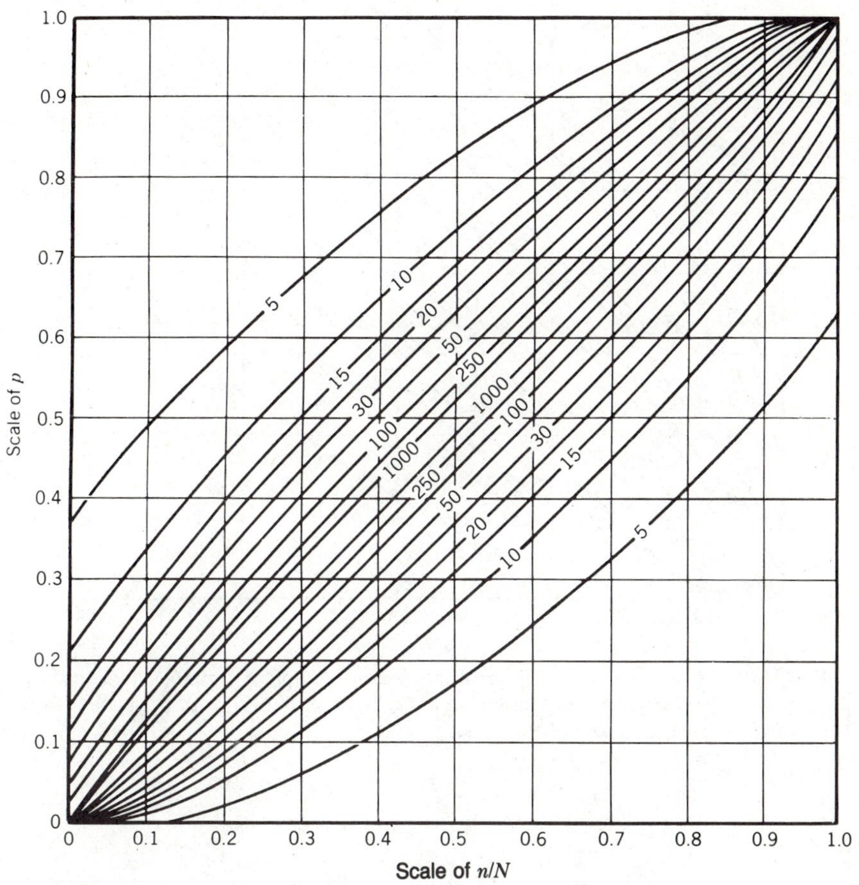

FIGURE B.1 An 80% confidence interval for binomial sampling. (From W. J. Dixon and F. J. Massey, Jr., *Introduction to Statistical Analysis,* 2nd ed., © 1957, with permission from McGraw-Hill Book Company, New York.)

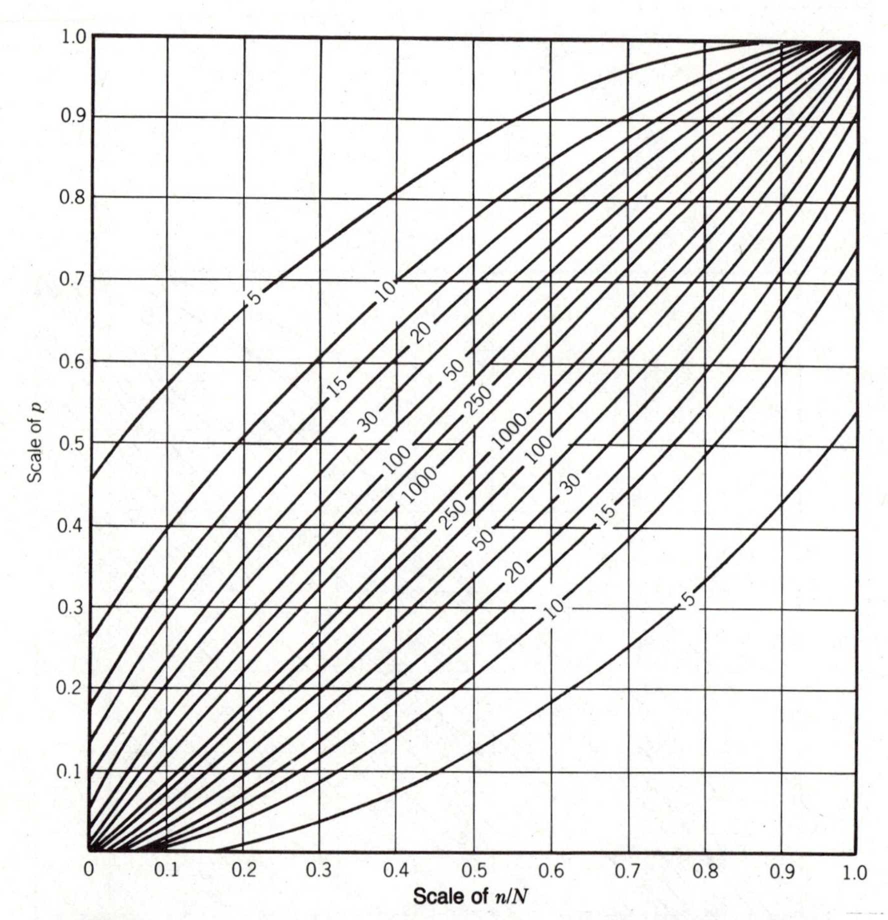

FIGURE B.2 A 90% confidence interval for binomial sampling. (From W. J. Dixon and F. J. Massey, Jr., *Introduction to Statistical Analysis,* 2nd ed., © 1957, with permission from McGraw-Hill Book Company, New York.)

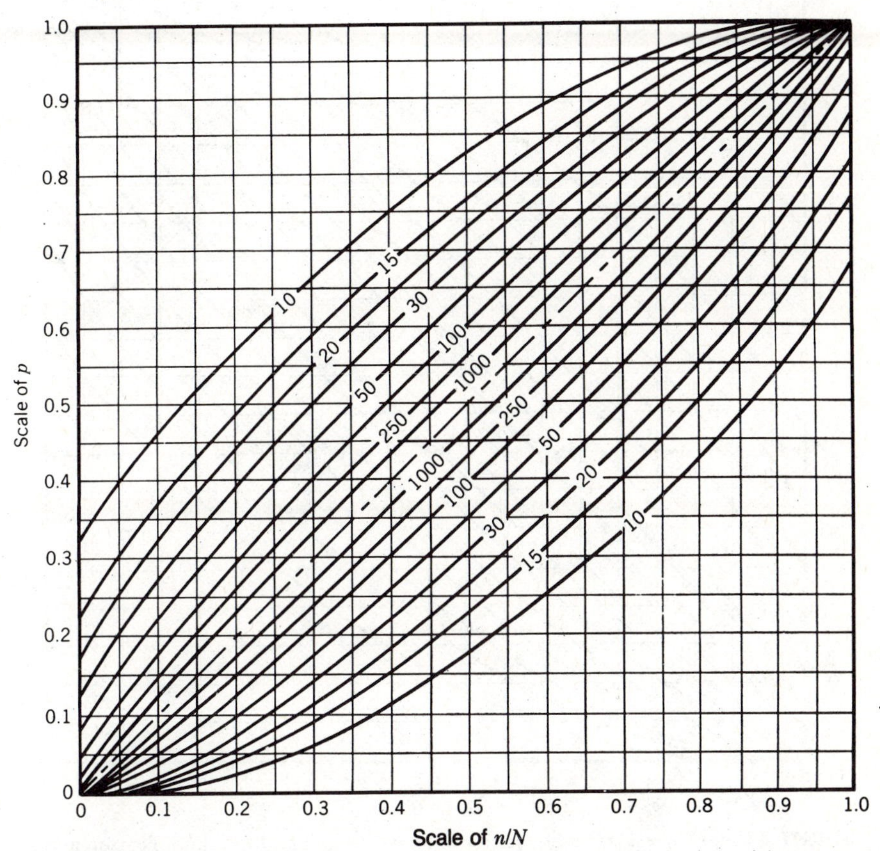

FIGURE B.3 A 95% confidence interval for binomial sampling. [From E. S. Pearson and C. J. Clopper, "The Use of Confidence or Fiducial Limits Illustrated in the Case of the Binomial," *Biometrika,* **26,** 404 (1934). With permission of Biometrika.]

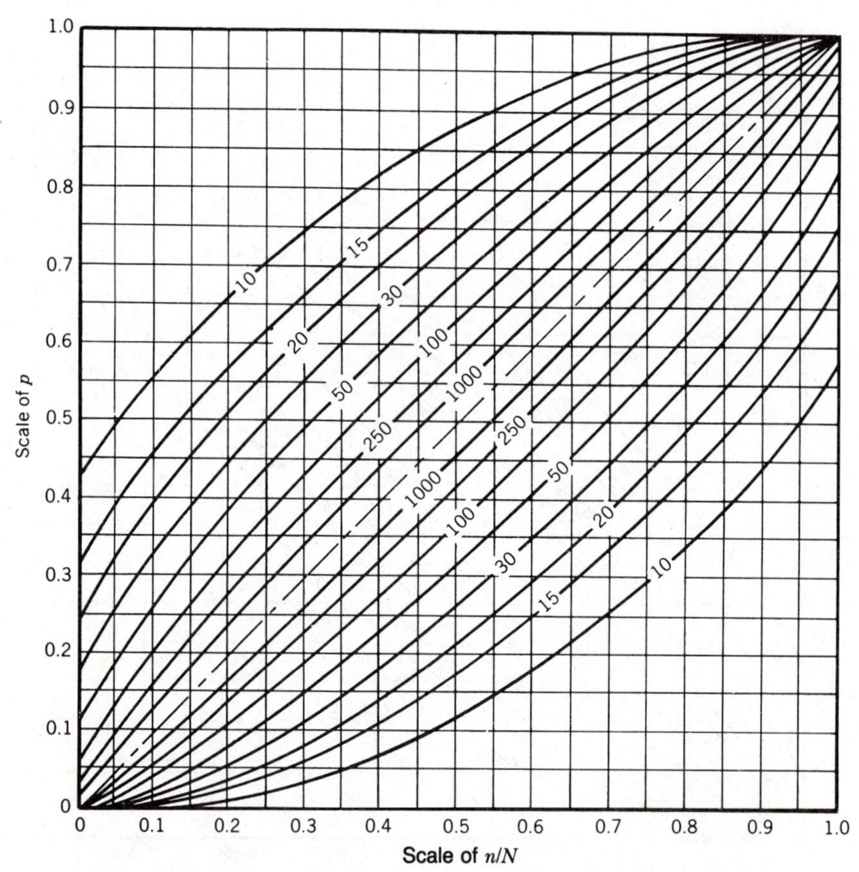

FIGURE B.4 A 99% confidence interval for binomial sampling. [From E. S. Pearson and C. J. Clopper, "The Use of Confidence or Fiducial Limits Illustrated in the Case of the Binomial," *Biometrika,* **26,** 404 (1934). With permission of Biometrika.]

APPENDIX C

Probability Tables

TABLE C.1 Standard Normal Cumulative Distribution Function $\Phi(u)$

u	.00	.01	.02	.03	.04	.05	.06	.07	.08	.09
− .0	.5000	.4960	.4920	.4880	.4840	.4801	.4761	.4721	.4681	.4641
− .1	.4602	.4562	.4522	.4483	.4443	.4404	.4364	.4325	.4286	.4247
− .2	.4207	.4168	.4129	.4090	.4052	.4013	.3974	.3936	.3897	.3859
− .3	.3821	.3783	.3745	.3707	.3669	.3632	.3594	.3557	.3520	.3483
− .4	.3446	.3409	.3372	.3336	.3300	.3264	.3228	.3192	.3156	.3121
− .5	.3085	.3050	.3015	.2981	.2946	.2912	.2877	.2843	.2810	.2776
− .6	.2743	.2709	.2676	.2643	.2611	.2578	.2546	.2514	.2483	.2451
− .7	.2420	.2389	.2358	.2327	.2297	.2266	.2236	.2206	.2177	.2148
− .8	.2119	.2090	.2061	.2033	.2005	.1977	.1949	.1922	.1894	.1867
− .9	.1841	.1814	.1788	.1762	.1736	.1711	.1685	.1660	.1635	.1611
−1.0	.1587	.1562	.1539	.1515	.1492	.1469	.1446	.1423	.1401	.1379
−1.1	.1357	.1335	.1314	.1292	.1271	.1251	.1230	.1210	.1190	.1170
−1.2	.1151	.1131	.1112	.1093	.1075	.1056	.1038	.1020	.1003	.09853
−1.3	.09680	.09510	.09342	.09176	.09012	.08851	.08691	.08534	.08379	.08226
−1.4	.08076	.07927	.07780	.07636	.07493	.07353	.07215	.07078	.06944	.06811
−1.5	.06681	.06552	.06426	.06301	.06178	.06057	.05938	.05821	.05705	.05592
−1.6	.05480	.05370	.05262	.05155	.05050	.04947	.04846	.04746	.04648	.04551
−1.7	.04457	.04363	.04272	.04182	.04093	.04006	.03920	.03836	.03754	.03673
−1.8	.03593	.03515	.03438	.03362	.03288	.03216	.03144	.03074	.03005	.02938
−1.9	.02872	.02807	.02743	.02680	.02619	.02559	.02500	.02442	.02385	.02330
−2.0	.02275	.02222	.02169	.02118	.02068	.02018	.01970	.01923	.01876	.01831
−2.1	.01786	.01743	.01700	.01659	.01618	.01578	.01539	.01500	.01463	.01426
−2.2	.01390	.01355	.01321	.01287	.01255	.01222	.01191	.01160	.01130	.01101
−2.3	.01072	.01044	.01017	$.0^{2}9903$	$.0^{2}9642$	$.0^{2}9387$	$.0^{2}9137$	$.0^{2}8894$	$.0^{2}8656$	$.0^{2}8424$
−2.4	$.0^{2}8198$	$.0^{2}7976$	$.0^{2}7760$	$.0^{2}7549$	$.0^{2}7344$	$.0^{2}7143$	$.0^{2}6947$	$.0^{2}6756$	$.0^{2}6569$	$.0^{2}6387$
−2.5	$.0^{2}6210$	$.0^{2}6037$	$.0^{2}5868$	$.0^{2}5703$	$.0^{2}5543$	$.0^{2}5386$	$.0^{2}5234$	$.0^{2}5085$	$.0^{2}4940$	$.0^{2}4799$
−2.6	$.0^{2}4661$	$.0^{2}4527$	$.0^{2}4396$	$.0^{2}4269$	$.0^{2}4145$	$.0^{2}4025$	$.0^{2}3907$	$.0^{2}3793$	$.0^{2}3681$	$.0^{2}3573$
−2.7	$.0^{2}3467$	$.0^{2}3364$	$.0^{2}3264$	$.0^{2}3167$	$.0^{2}3072$	$.0^{2}2980$	$.0^{2}2890$	$.0^{2}2803$	$.0^{2}2718$	$.0^{2}2635$
−2.8	$.0^{2}2555$	$.0^{2}2477$	$.0^{2}2401$	$.0^{2}2327$	$.0^{2}2256$	$.0^{2}2186$	$.0^{2}2118$	$.0^{2}2052$	$.0^{2}1988$	$.0^{2}1926$
−2.9	$.0^{2}1866$	$.0^{2}1807$	$.0^{2}1750$	$.0^{2}1695$	$.0^{2}1641$	$.0^{2}1589$	$.0^{2}1538$	$.0^{2}1489$	$.0^{2}1441$	$.0^{2}1395$
−3.0	$.0^{2}1350$	$.0^{2}1306$	$.0^{2}1264$	$.0^{2}1223$	$.0^{2}1183$	$.0^{2}1144$	$.0^{2}1107$	$.0^{2}1070$	$.0^{2}1035$	$.0^{2}1001$
−3.1	$.0^{3}9676$	$.0^{3}9354$	$.0^{3}9043$	$.0^{3}8740$	$.0^{3}8447$	$.0^{3}8164$	$.0^{3}7888$	$.0^{3}7622$	$.0^{3}7364$	$.0^{3}7114$
−3.2	$.0^{3}6871$	$.0^{3}6637$	$.0^{3}6410$	$.0^{3}6190$	$.0^{3}5976$	$.0^{3}5770$	$.0^{3}5571$	$.0^{3}5377$	$.0^{3}5190$	$.0^{3}5009$
−3.3	$.0^{3}4834$	$.0^{3}4665$	$.0^{3}4501$	$.0^{3}4342$	$.0^{3}4189$	$.0^{3}4041$	$.0^{3}3897$	$.0^{3}3758$	$.0^{3}3624$	$.0^{3}3495$
−3.4	$.0^{3}3369$	$.0^{3}3248$	$.0^{3}3131$	$.0^{3}3018$	$.0^{3}2909$	$.0^{3}2803$	$.0^{3}2701$	$.0^{3}2602$	$.0^{3}2507$	$.0^{3}2415$
−3.5	$.0^{3}2326$	$.0^{3}2241$	$.0^{3}2158$	$.0^{3}2078$	$.0^{3}2001$	$.0^{3}1926$	$.0^{3}1854$	$.0^{3}1785$	$.0^{3}1718$	$.0^{3}1653$
−3.6	$.0^{3}1591$	$.0^{3}1531$	$.0^{3}1473$	$.0^{3}1417$	$.0^{3}1363$	$.0^{3}1311$	$.0^{3}1261$	$.0^{3}1213$	$.0^{3}1166$	$.0^{3}1121$
−3.7	$.0^{3}1078$	$.0^{3}1036$	$.0^{4}9961$	$.0^{4}9574$	$.0^{4}9201$	$.0^{4}8842$	$.0^{4}8496$	$.0^{4}8162$	$.0^{4}7841$	$.0^{4}7532$
−3.8	$.0^{4}7235$	$.0^{4}6948$	$.0^{4}6673$	$.0^{4}6407$	$.0^{4}6152$	$.0^{4}5906$	$.0^{4}5669$	$.0^{4}5442$	$.0^{4}5223$	$.0^{4}5012$
−3.9	$.0^{4}4810$	$.0^{4}4615$	$.0^{4}4427$	$.0^{4}4247$	$.0^{4}4074$	$.0^{4}3908$	$.0^{4}3747$	$.0^{4}3594$	$.0^{4}3446$	$.0^{4}3304$
−4.0	$.0^{4}3167$	$.0^{4}3036$	$.0^{4}2910$	$.0^{4}2789$	$.0^{4}2673$	$.0^{4}2561$	$.0^{4}2454$	$.0^{4}2351$	$.0^{4}2252$	$.0^{4}2157$
−4.1	$.0^{4}2066$	$.0^{4}1978$	$.0^{4}1894$	$.0^{4}1814$	$.0^{4}1737$	$.0^{4}1662$	$.0^{4}1591$	$.0^{4}1523$	$.0^{4}1458$	$.0^{4}1395$
−4.2	$.0^{4}1335$	$.0^{4}1277$	$.0^{4}1222$	$.0^{4}1168$	$.0^{4}1118$	$.0^{4}1069$	$.0^{4}1022$	$.0^{5}9774$	$.0^{5}9345$	$.0^{5}8934$
−4.3	$.0^{5}8540$	$.0^{5}8163$	$.0^{5}7801$	$.0^{5}7455$	$.0^{5}7124$	$.0^{5}6807$	$.0^{5}6503$	$.0^{5}6212$	$.0^{5}5934$	$.0^{5}5668$
−4.4	$.0^{5}5413$	$.0^{5}5169$	$.0^{5}4935$	$.0^{5}4712$	$.0^{5}4498$	$.0^{5}4294$	$.0^{5}4098$	$.0^{5}3911$	$.0^{5}3732$	$.0^{5}3561$
−4.5	$.0^{5}3398$	$.0^{5}3241$	$.0^{5}3092$	$.0^{5}2949$	$.0^{5}2813$	$.0^{5}2682$	$.0^{5}2558$	$.0^{5}2439$	$.0^{5}2325$	$.0^{5}2216$
−4.6	$.0^{5}2112$	$.0^{5}2013$	$.0^{5}1919$	$.0^{5}1828$	$.0^{5}1742$	$.0^{5}1660$	$.0^{5}1581$	$.0^{5}1506$	$.0^{5}1434$	$.0^{5}1366$
−4.7	$.0^{5}1301$	$.0^{5}1239$	$.0^{5}1179$	$.0^{5}1123$	$.0^{5}1069$	$.0^{5}1017$	$.0^{6}9680$	$.0^{6}9211$	$.0^{6}8765$	$.0^{6}8339$
−4.8	$.0^{6}7933$	$.0^{6}7547$	$.0^{6}7178$	$.0^{6}6827$	$.0^{6}6492$	$.0^{6}6173$	$.0^{6}5869$	$.0^{6}5580$	$.0^{6}5304$	$.0^{6}5042$
−4.9	$.0^{6}4792$	$.0^{6}4554$	$.0^{6}4327$	$.0^{6}4111$	$.0^{6}3906$	$.0^{6}3711$	$.0^{6}3525$	$.0^{6}3348$	$.0^{6}3179$	$.0^{6}3019$

TABLE C.1 Standard Normal Cumulative Distribution Function $\Phi(u)$ (*Continued*)

u	.00	.01	.02	.03	.04	.05	.06	.07	.08	.09
.0	.5000	.5040	.5080	.5120	.5160	.5199	.5239	.5279	.5319	.5359
.1	.5398	.5438	.5478	.5517	.5557	.5596	.5636	.5675	.5714	.5753
.2	.5793	.5832	.5871	.5910	.5948	.5987	.6026	.6064	.6103	.6141
.3	.6179	.6217	.6255	.6293	.6331	.6368	.6406	.6443	.6480	.6517
.4	.6554	.6591	.6628	.6664	.6700	.6736	.6772	.6808	.6844	.6879
.5	.6915	.6950	.6985	.7019	.7054	.7088	.7123	.7157	.7190	.7224
.6	.7257	.7291	.7324	.7357	.7389	.7422	.7454	.7486	.7517	.7549
.7	.7580	.7611	.7642	.7673	.7703	.7734	.7764	.7794	.7823	.7852
.8	.7881	.7910	.7939	.7967	.7995	.8023	.8051	.8078	.8106	.8133
.9	.8159	.8186	.8212	.8238	.8264	.8289	.8315	.8340	.8365	.8389
1.0	.8413	.8438	.8461	.8485	.8508	.8531	.8554	.8577	.8599	.8621
1.1	.8643	.8665	.8686	.8708	.8729	.8749	.8770	.8790	.8810	.8830
1.2	.8849	.8869	.8888	.8907	.8925	.8944	.8962	.8980	.8997	.90147
1.3	.90320	.90490	.90658	.90824	.90988	.91149	.91309	.91466	.91621	.91774
1.4	.91924	.92073	.92220	.92364	.92507	.92647	.92785	.92922	.93056	.93189
1.5	.93319	.93448	.93574	.93699	.93822	.93943	.94062	.94179	.94295	.94408
1.6	.94520	.94630	.94738	.94845	.94950	.95053	.95154	.95254	.95352	.95449
1.7	.95543	.95637	.95728	.95818	.95907	.95994	.96080	.96164	.96246	.96327
1.8	.96407	.96485	.96562	.96638	.96712	.96784	.96856	.96926	.96995	.97062
1.9	.97128	.97193	.97257	.97320	.97381	.97441	.97500	.97558	.97615	.97670
2.0	.97725	.97778	.97831	.97882	.97932	.97982	.98030	.98077	.98124	.98169
2.1	.98214	.98257	.98300	.98341	.98382	.98422	.98461	.98500	.98537	.98574
2.2	.98610	.98645	.98679	.98713	.98745	.98778	.98809	.98840	.98870	.98899
2.3	.98928	.98956	.98983	$.9^{2}0097$	$.9^{2}0358$	$.9^{2}0613$	$.9^{2}0863$	$.9^{2}1106$	$.9^{2}1344$	$.9^{2}1576$
2.4	$.9^{2}1802$	$.9^{2}2024$	$.9^{2}2240$	$.9^{2}2451$	$.9^{2}2656$	$.9^{2}2857$	$.9^{2}3053$	$.9^{2}3244$	$.9^{2}3431$	$.9^{2}3613$
2.5	$.9^{2}3790$	$.9^{2}3963$	$.9^{2}4132$	$.9^{2}4297$	$.9^{2}4457$	$.9^{2}4614$	$.9^{2}4766$	$.9^{2}4915$	$.9^{2}5060$	$.9^{2}5201$
2.6	$.9^{2}5339$	$.9^{2}5473$	$.9^{2}5604$	$.9^{2}5731$	$.9^{2}5855$	$.9^{2}5975$	$.9^{2}6093$	$.9^{2}6207$	$.9^{2}6319$	$.9^{2}6427$
2.7	$.9^{2}6533$	$.9^{2}6636$	$.9^{2}6736$	$.9^{2}6833$	$.9^{2}6928$	$.9^{2}7020$	$.9^{2}7110$	$.9^{2}7197$	$.9^{2}7282$	$.9^{2}7365$
2.8	$.9^{2}7445$	$.9^{2}7523$	$.9^{2}7599$	$.9^{2}7673$	$.9^{2}7744$	$.9^{2}7814$	$.9^{2}7882$	$.9^{2}7948$	$.9^{2}8012$	$.9^{2}8074$
2.9	$.9^{2}8134$	$.9^{2}8193$	$.9^{2}8250$	$.9^{2}8305$	$.9^{2}8359$	$.9^{2}8411$	$.9^{2}8462$	$.9^{2}8511$	$.9^{2}8559$	$.9^{2}8605$
3.0	$.9^{2}8650$	$.9^{2}8694$	$.9^{2}8736$	$.9^{2}8777$	$.9^{2}8817$	$.9^{2}8856$	$.9^{2}8893$	$.9^{2}8930$	$.9^{2}8965$	$.9^{2}8999$
3.1	$.9^{3}0324$	$.9^{3}0646$	$.9^{3}0957$	$.9^{3}1260$	$.9^{3}1553$	$.9^{3}1836$	$.9^{3}2112$	$.9^{3}2378$	$.9^{3}2636$	$.9^{3}2886$
3.2	$.9^{3}3129$	$.9^{3}3363$	$.9^{3}3590$	$.9^{3}3810$	$.9^{3}4024$	$.9^{3}4230$	$.9^{3}4429$	$.9^{3}4623$	$.9^{3}4810$	$.9^{3}4991$
3.3	$.9^{3}5166$	$.9^{3}5335$	$.9^{3}5499$	$.9^{3}5658$	$.9^{3}5811$	$.9^{3}5959$	$.9^{3}6103$	$.9^{3}6242$	$.9^{3}6376$	$.9^{3}6505$
3.4	$.9^{3}6631$	$.9^{3}6752$	$.9^{3}6869$	$.9^{3}6982$	$.9^{3}7091$	$.9^{3}7197$	$.9^{3}7299$	$.9^{3}7398$	$.9^{3}7493$	$.9^{3}7585$
3.5	$.9^{3}7674$	$.9^{3}7759$	$.9^{3}7842$	$.9^{3}7922$	$.9^{3}7999$	$.9^{3}8074$	$.9^{3}8146$	$.9^{3}8215$	$.9^{3}8282$	$.9^{3}8347$
3.6	$.9^{3}8409$	$.9^{3}8469$	$.9^{3}8527$	$.9^{3}8583$	$.9^{3}8637$	$.9^{3}8689$	$.9^{3}8739$	$.9^{3}8787$	$.9^{3}8834$	$.9^{3}8879$
3.7	$.9^{3}8922$	$.9^{3}8964$	$.9^{4}0039$	$.9^{4}0426$	$.9^{4}0799$	$.9^{4}1158$	$.9^{4}1504$	$.9^{4}1838$	$.9^{4}2159$	$.9^{4}2468$
3.8	$.9^{4}2765$	$.9^{4}3052$	$.9^{4}3327$	$.9^{4}3593$	$.9^{4}3848$	$.9^{4}4094$	$.9^{4}4331$	$.9^{4}4558$	$.9^{4}4777$	$.9^{4}4988$
3.9	$.9^{4}5190$	$.9^{4}5385$	$.9^{4}5573$	$.9^{4}5753$	$.9^{4}5926$	$.9^{4}6092$	$.9^{4}6253$	$.9^{4}6406$	$.9^{4}6554$	$.9^{4}6696$
4.0	$.9^{4}6833$	$.9^{4}6964$	$.9^{4}7090$	$.9^{4}7211$	$.9^{4}7327$	$.9^{4}7439$	$.9^{4}7546$	$.9^{4}7649$	$.9^{4}7748$	$.9^{4}7843$
4.1	$.9^{4}7934$	$.9^{4}8022$	$.9^{4}8106$	$.9^{4}8186$	$.9^{4}8263$	$.9^{4}8338$	$.9^{4}8409$	$.9^{4}8477$	$.9^{4}8542$	$.9^{4}8605$
4.2	$.9^{4}8665$	$.9^{4}8723$	$.9^{4}8778$	$.9^{4}8832$	$.9^{4}8882$	$.9^{4}8931$	$.9^{4}8978$	$.9^{5}0226$	$.9^{5}0655$	$.9^{5}1066$
4.3	$.9^{5}1460$	$.9^{5}1837$	$.9^{5}2199$	$.9^{5}2545$	$.9^{5}2876$	$.9^{5}3193$	$.9^{5}3497$	$.9^{5}3788$	$.9^{5}4066$	$.9^{5}4332$
4.4	$.9^{5}4587$	$.9^{5}4831$	$.9^{5}5065$	$.9^{5}5288$	$.9^{5}5502$	$.9^{5}5706$	$.9^{5}5902$	$.9^{5}6089$	$.9^{5}6268$	$.9^{5}6439$
4.5	$.9^{5}6602$	$.9^{5}6759$	$.9^{5}6908$	$.9^{4}7051$	$.9^{5}7187$	$.9^{5}7318$	$.9^{5}7442$	$.9^{5}7561$	$.9^{5}7675$	$.9^{5}7784$
4.6	$.9^{5}7888$	$.9^{5}7987$	$.9^{5}8081$	$.9^{5}8172$	$.9^{5}8258$	$.9^{5}8340$	$.9^{5}8419$	$.9^{5}8494$	$.9^{5}8566$	$.9^{5}8634$
4.7	$.9^{5}8699$	$.9^{5}8761$	$.9^{5}8821$	$.9^{5}8877$	$.9^{5}8931$	$.9^{5}8983$	$.9^{6}0320$	$.9^{6}0789$	$.9^{6}1235$	$.9^{6}1661$
4.8	$.9^{6}2067$	$.9^{6}2453$	$.9^{6}2822$	$.9^{6}3173$	$.9^{6}3508$	$.9^{6}3827$	$.9^{6}4131$	$.9^{6}4420$	$.9^{6}4696$	$.9^{6}4958$
4.9	$.9^{6}5208$	$.9^{6}5446$	$.9^{6}5673$	$.9^{6}5889$	$.9^{6}6094$	$.9^{6}6289$	$.9^{6}6475$	$.9^{6}6652$	$.9^{6}6821$	$.9^{6}6981$

From A. Hald, *Statistical Tables and Formulas,* Wiley, New York, 1952, Table II. Reproduced by permission. See also W. Nelson, *Applied Life Data Analysis,* Wiley, New York, 1982.

TABLE C.2 Values of the Standardized Type I Extreme-Value Distribution (largest value) ($F_w(w) = \exp(-e^{-w})$)

w	CDF	PDF	w	CDF	PDF	w	CDF	PDF
−3.0		0.00000 00	−0.50	0.19229 56	0.31704 19	3.5	0.97025 40	0.02929 91
−2.9	0.00000 00	0.00000 02	−0.45	0.20839 66	0.32638 10	3.6	0.97304 62	0.02658 72
−2.8	0.00000 01	0.00000 12	−0.40	0.22496 18	0.33560 36	3.7	0.97557 96	0.02411 98
−2.7	0.00000 03	0.00000 51	−0.35	0.24193 95	0.34332 85	3.8	0.97787 76	0.02187 59
−2.6	0.00000 14	0.00001 91	−0.30	0.25927 69	0.34998 72	3.9	0.97996 16	0.01983 63
			−0.25	0.27692 03	0.35557 27	4.0	0.98185 11	0.01798 32
−2.5	0.00000 51	0.00006 24	−0.20	0.29481 63	0.36008 95	4.2	0.98511 63	0.01477 24
−2.40	0.00001 63	0.00017 99	−0.15	0.31291 17	0.36355 15	4.4	0.98779 77	0.01212 75
−2.35	0.00002 79	0.00029 29	−0.10	0.33115 43	0.36598 21	4.6	0.98999 85	0.00995 13
−2.30	0.00004 66	0.00046 47	−0.05	0.34949 32	0.36741 21	4.8	0.99180 40	0.00816 23
−2.25	0.00007 58	0.00071 89	0.0	0.36787 94	0.36787 94	5.0	0.99328 47	0.00669 27
−2.20	0.00012 04	0.00108 63	0.1	0.40460 77	0.36610 42	5.2	0.99449 86	0.00548 62
−2.15	0.00018 69	0.00160 46	0.2	0.44099 10	0.36105 29	5.4	0.99549 36	0.00449 62
−2.10	0.00028 41	0.00232 00	0.3	0.47672 37	0.35316 56	5.6	0.99630 90	0.00368 42
−2.05	0.00042 31	0.00328 66	0.4	0.51154 48	0.34289 88	5.8	0.99697 70	0.00301 84
−2.00	0.00061 80	0.00456 63	0.5	0.54523 92	0.33070 43	6.0	0.99752 43	0.00247 26
−1.95	0.00088 61	0.00622 81	0.6	0.57763 58	0.31701 33	6.2	0.99797 26	0.00202 53
−1.90	0.00124 84	0.00834 67	0.7	0.60860 53	0.30222 45	6.4	0.99833 98	0.00165 88
−1.85	0.00172 97	0.01100 04	0.8	0.63805 62	0.28669 71	6.6	0.99864 06	0.00135 85
−1.80	0.00235 87	0.01426 93	0.9	0.66593 07	0.27074 72	6.8	0.99888 68	0.00111 25
−1.75	0.00316 82	0.01823 15	1.0	0.69220 06	0.25464 64	7.0	0.99908 85	0.00091 11
−1.70	0.00419 46	0.02296 12	1.1	0.71686 26	0.23862 28	7.2	0.99925 37	0.00074 60
−1.65	0.00547 82	0.02852 48	1.2	0.73993 41	0.22286 39	7.4	0.99938 89	0.00061 09
−1.60	0.00706 20	0.03497 81	1.3	0.76144 92	0.20751 91	7.6	0.99949 97	0.00050 02
−1.55	0.00899 15	0.04236 34	1.4	0.78145 56	0.19270 46	7.8	0.99959 03	0.00040 96
−1.50	0.01131 43	0.05070 71	1.5	0.80001 07	0.17850 65			
−1.45	0.01407 84	0.06001 78	1.6	0.81717 95	0.16498 57	8.0	0.99966 46	0.00033 54
−1.40	0.01733 20	0.07028 48	1.7	0.83303 17	0.15218 12	8.5	0.99979 66	0.00020 34
−1.35	0.02112 23	0.08147 77	1.8	0.84764 03	0.14011 40	9.0	0.99987 66	0.00012 34
−1.30	0.02549 44	0.09354 65	1.9	0.86107 93	0.12879 04	9.5	0.99992 51	0.00007 48
−1.25	0.03049 04	0.10642 20	2.0	0.87342 30	0.11820 50	10.0	0.99995 46	0.00004 54
−1.20	0.03614 86	0.12001 76	2.1	0.88474 45	0.10834 26	10.5	0.99997 25	0.00002 75
−1.15	0.04250 25	0.13423 10	2.2	0.89511 49	0.09918 16	11.0	0.99998 33	0.00001 67
−1.10	0.04958 01	0.14894 68	2.3	0.90460 32	0.09069 45	11.5	0.99998 99	0.00001 01
−1.05	0.05740 34	0.16403 90	2.4	0.91327 53	0.08285 05	12.0	0.99999 39	0.00000 61
−1.00	0.06598 80	0.17937 41	2.5	0.92119 37	0.07561 62	12.5	0.99999 63	0.00000 37
−0.95	0.07534 26	0.19481 41	2.6	0.92841 77	0.06895 69	13.0	0.99999 77	0.00000 23
−0.90	0.08546 89	0.21021 95	2.7	0.93500 30	0.06283 74	13.5	0.99999 86	0.00000 14
−0.85	0.09636 17	0.22545 23	2.8	0.94100 20	0.05722 24	14.0	0.99999 92	0.00000 08
−0.80	0.10800 90	0.24037 84	2.9	0.94646 32	0.05207 75	14.5	0.99999 95	0.00000 05
−0.75	0.12039 23	0.25487 04	3.0	0.95143 20	0.04736 90	15.0	0.99999 97	0.00000 03
−0.70	0.13348 68	0.26880 94	3.1	0.95595 04	0.04306 48	15.5	0.99999 98	0.00000 02
−0.65	0.14726 22	0.28208 67	3.2	0.96005 74	0.03913 41	16.0	0.99999 99	0.00000 01
−0.60	0.16168 28	0.29460 53	3.3	0.96378 87	0.03554 76	16.5	0.99999 99	0.00000 01
−0.55	0.17670 86	0.30628 08	3.4	0.96717 75	0.03227 79	17.0	1.00000 00	0.00000 00

Source: National Bureau of Standards, "Probability Tables for the Analysis of Extreme Value Data," Applied Math Series 22, Washington, D.C., 1953.

APPENDIX D

Probability Graph Papers

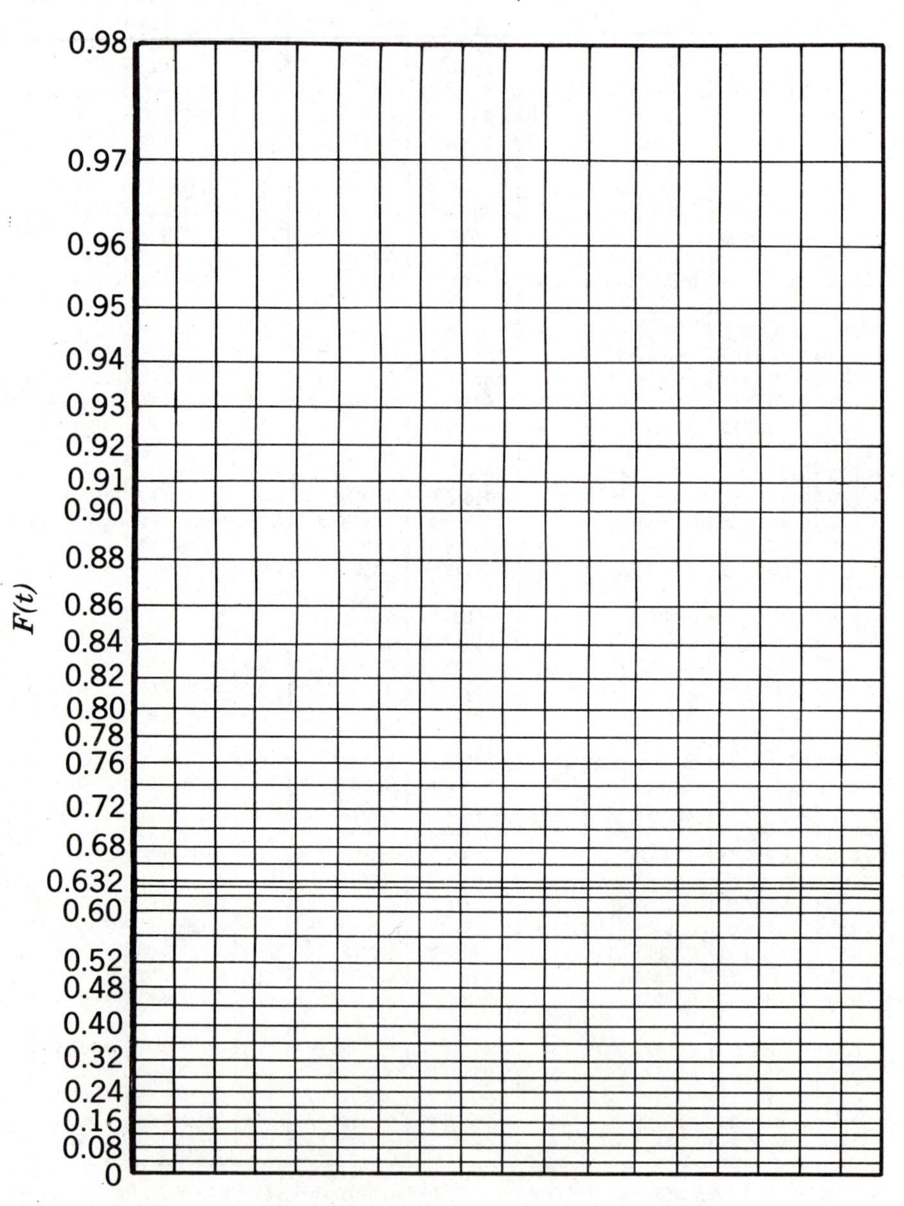

FIGURE D.1 Exponential distribution probability paper.

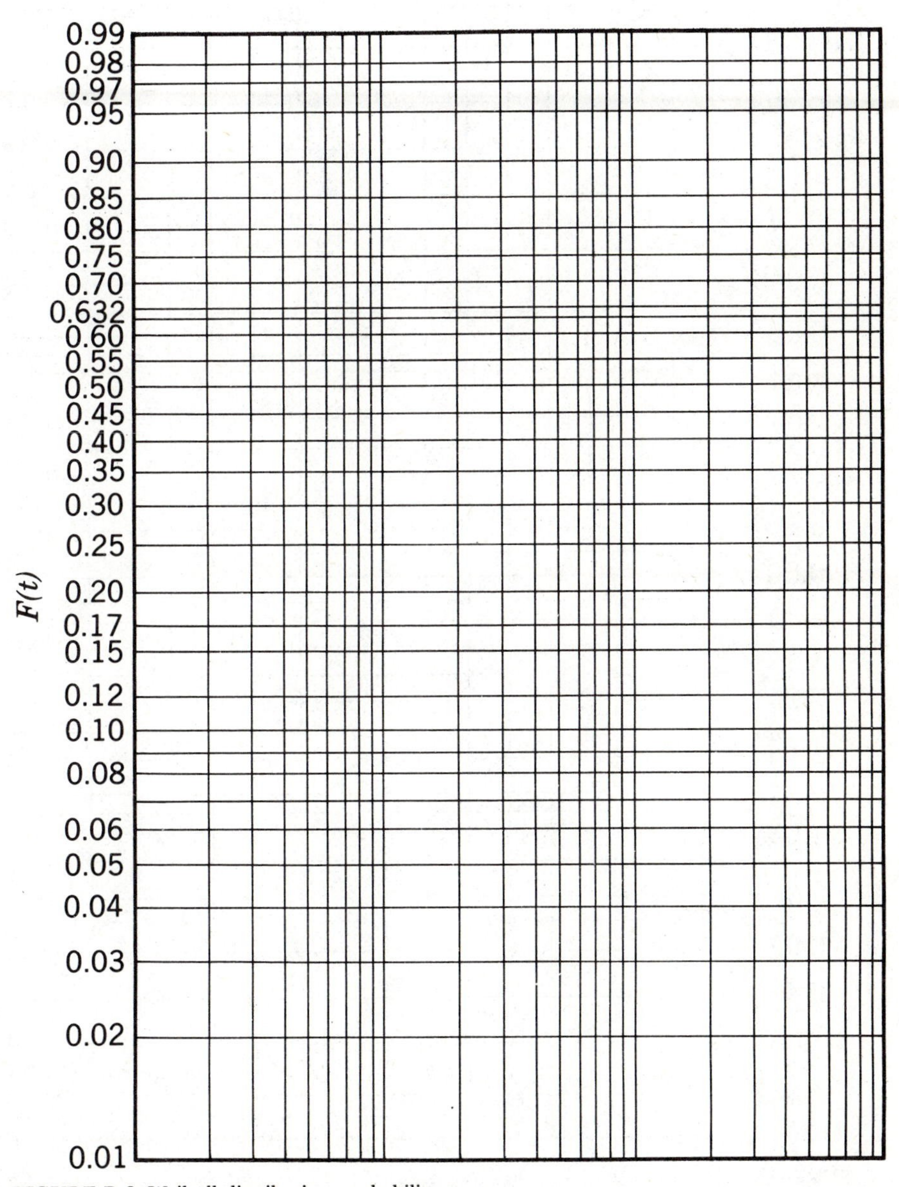

FIGURE D.2 Weibull distribution probability paper.

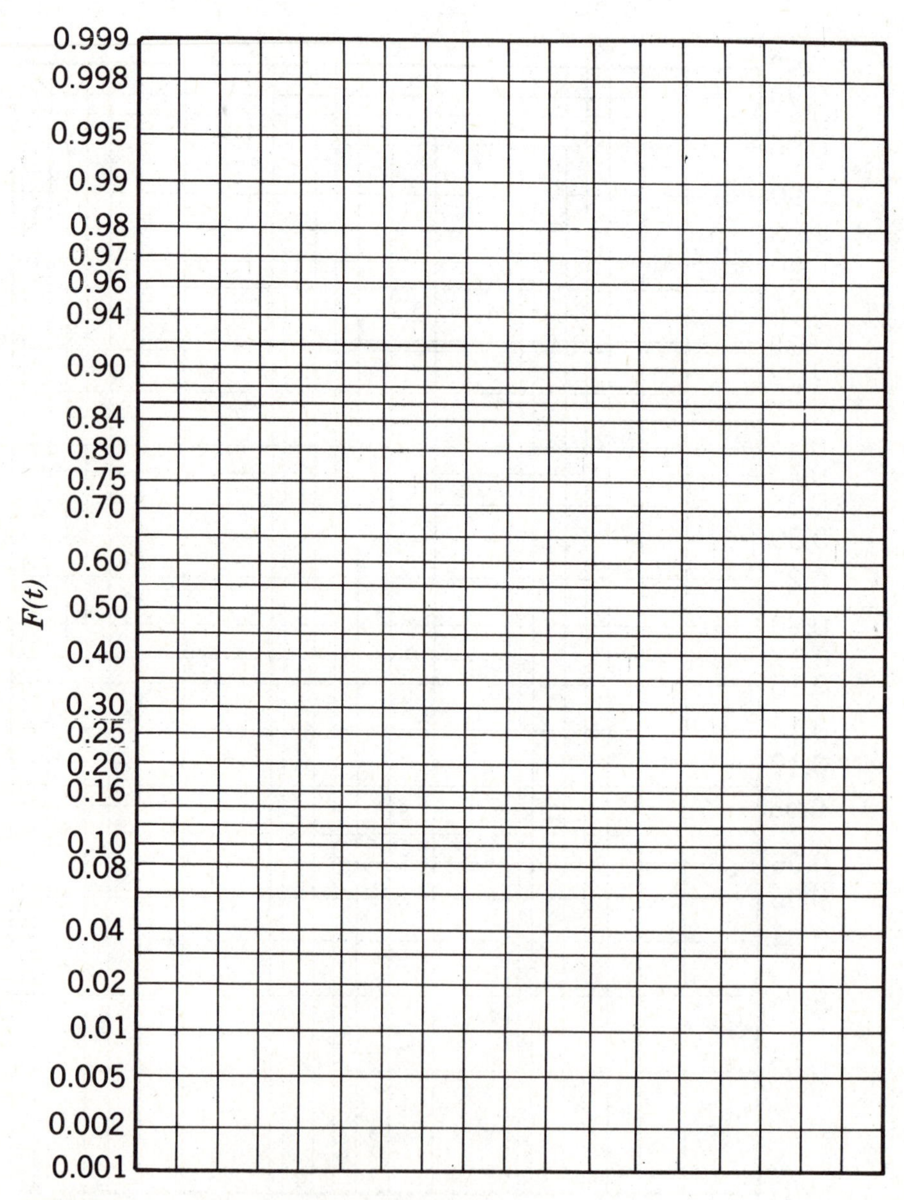

FIGURE D.3 Normal distribution probability paper.

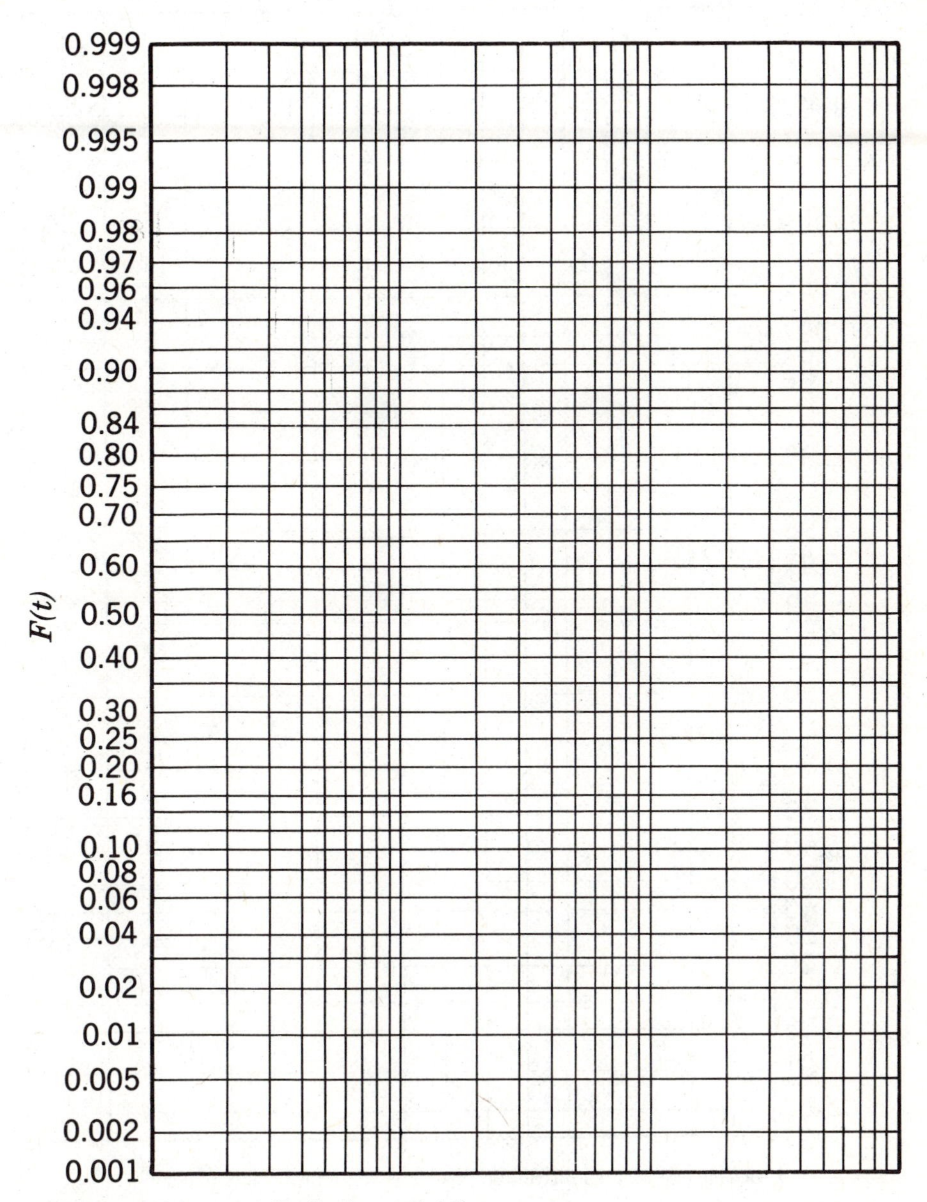

FIGURE D.4 Lognormal distribution probability paper.

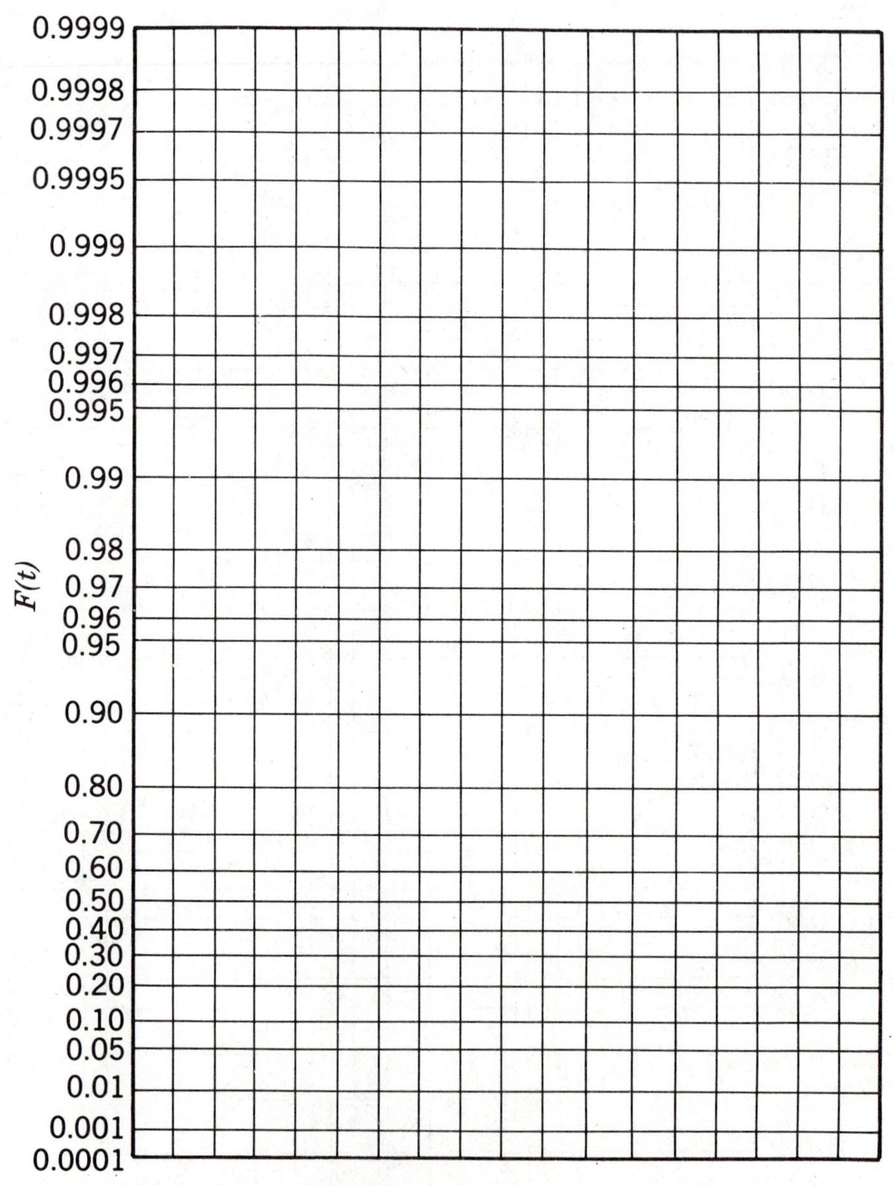

FIGURE D.5 Double-exponential distribution extreme-value probability paper.

Answers to Selected Exercises

CHAPTER 2

2.2 (*a*) no, (*b*) no, (*c*) 0.409, (*d*) 0.563.
2.3 (*a*) 0.5, (*b*) 0.25, (*c*) 0.625, (*d*) 0.5.
2.5 (*a*) $P\{X\} = 0.04$, (*b*) $P\{X_1|X_2\} = 0.25$.
2.7 $R_{DL} = 0.9048$.
2.9 (*a*) $C = 1/14$.
(*b*) $F(1) = 1/14$, $F(2) = 5/14$, $F(3) = 1$.
(*c*) $\mu = 2.57$, $\sigma = 2.10$.
2.11 $\mu = 1.53$, $\sigma^2 = 1.97$.
2.13 (*a*) 10, (*b*) 36, (*c*) 792, (*d*) 20.
2.15 $P_{NEW} = 0.0036$
2.16 (*a*) 0.058, (*b*) 6.6×10^{-5}, (*c*) 0.058.
2.18 (*a*) 0.594, (*b*) 0.0166.
2.20 (*a*) 0.353, (*b*) 3.
2.21 230 consecutive starts.
2.22 74 units.
2.23 (*a*) 2×10^{-4}, (*b*) 0.061, (*c*) 0.678.
2.26 415 units to test; no more than 18 failures to pass.
2.27 (*a*) $0.1 \sim 0.34$, (*b*) 100 units.

CHAPTER 3

3.1 $b = 6$, $\mu = 0.5$, $\sigma = 0.22$.
3.2 (*a*) $a = 18 \times 10^6$ hr^3, $b = 3000$ hr.
(*b*) 0.64, (*c*) 77 hr.
3.3 (*a*) $f(x) = 0.04xe^{-0.2x}$.
(*b*) $\mu = 10$, $\sigma^2 = 50$, (*c*) 0.0278.
3.4 (*a*) 1 μm, (*b*) 80.8%, (*c*) 0.720 μm.
3.6 (*a*) $(e^{-x/\gamma} - 1)/(e^{-\tau/\gamma} - 1)$, (*b*) 0.168.
3.7 (*a*) no answer, (*b*) 8.32 cm, (*c*) 9.76 cm, (*d*) no answer.

3.8 skewness $= \dfrac{\langle x^3\rangle - 3\langle x^2\rangle\langle x\rangle + 2\langle x\rangle^3}{(\langle x^2\rangle - \langle x\rangle^2)^{3/2}}$.

3.10 (*a*) $f_y(y) = \dfrac{1}{b-a}\dfrac{1}{B}\left(\dfrac{y-a}{b-a}\right)^{r-1}\left(1 - \dfrac{y-a}{b-a}\right)^{t-r-1}$.

(*b*) $\mu_y = (b-a)\dfrac{r}{t} + a$.

3.13 (*a*) PDF $= 2xe^{-y}e^{-x^2}$, (*b*) 0.0855, (*c*) yes.

3.14 (*a*) $f_x(x) = 2/3(2x + 0.5)$,

$f_y(y) = \frac{2}{3}(1 + y)$.

(*b*) $f(x|y) = (2x + y)/(1 + y)$,

$f(y|x) = (2x + y)/(2x + 0.5)$.

(*c*) no.

3.15 $\rho_{xy} = -0.0818$.

3.22 (*a*) $n = 5.58$, (*b*) $n = 1.57$.

CHAPTER 4

4.1 (*a*) $16/(t+4)^2$, (*b*) $2/(t+4)$, (*c*) 4.

4.2 (*a*)130 hr, (*b*) 256 hr, (*c*) 155 hr, (*d*) 513 hr.

4.3 (*a*) 0.966, (*b*) 0.980, (*c*) 0.975, (*d*) 0.990.

4.4 (*a*) 0.905, (*b*) 0.9275.

4.7 (*a*) 1.63, (*b*) 0.224.

4.8 47 days.

4.9 $\lambda = 0.105/\text{hr}$.

4.12 MTTF $= \sqrt{\pi}\,\theta/2$.

4.13 28%.

4.14 (*a*) 123 hr, (*b*) 6.3%, (*c*) 86%.

4.16 (*a*) 3.98 yr, (*b*) 3.14 yr.

4.22 2.5%.

4.24 (*a*) 0.2856, (*b*) 0.1315, (*c*) 1.25.

4.27 (*a*) 70.2 failures/year, (*b*) nine flashlights.

CHAPTER 5

5.6 $\hat{R}(t_i) = \dfrac{N + 0.7 - i}{N + 0.4}$,

$\hat{f}(t) = \dfrac{1}{(t_{i+1} - t_i)(N + 0.4)}$,

$\hat{\lambda}(t) = \dfrac{1}{(t_{i+1} - t_i)(N + 0.7 - i)}$.

5.9 increasing with time.

5.10 $m \approx 2.4$, $\theta \approx 12$.

5.12 $m \approx 2.5$, $\theta \approx 130$ hr.

5.13 (*a*) no answer, (*b*) $\mu \approx 7000$ hr, $\sigma \approx 3000$ hr, (*c*) 48%.

5.16 (*a*) no answer, (*b*) $t_0 = 43$ hr, $s = 1.67$,

5.17 (*a*) no answer, (*b*) ~493 hr.

5.18 (*a*) constant, (*b*) 8.1 months.

5.19 MTTF = 9.76,

CHAPTER 6

6.1 (*a*) 1.39×10^{-3}, (*b*) 721 V, (*c*) 2161 V.

6.4 $r = 1 + \dfrac{1}{a\gamma}(e^{-2a\gamma} - e^{-a\gamma})$.

6.7 (*a*) $f = (e^{2l} - e)/(e^2 - 1)$, (*b*) $f = (e^{10l} - e^5)/(e^{10} - 1)$

6.8 15.7 Nm.

6.9 $\geqslant$ 11 strands.

6.11 $\tilde{R} = 0.2090$.

6.14 $c_0/l_0 = 4.64$.

6.15 (*a*) 1.24×10^{-6}, (*b*) 0.037, (*c*) 0.311.

6.17 10^{-15}.

6.20 (*a*) 2.5×10^{-4}, (*b*) 5×10^{-5}.

6.21 (*a*) 0.18, (*b*) 0.06, (*c*) 2.40 yr.

6.23 (*a*) 87 cycles, (*b*) 1.25 million cycles.

6.24 0.670.

CHAPTER 7

7.1 $R' = 0.9289$.

7.2 6 units.

7.3 (*a*) 0.827, (*b*) 0.683, (*c*) 0.696.

7.5 (*a*) $1/4\lambda^2$, (*b*) $5/4\lambda^2$, (*c*) parallel larger.

7.8 (*a*) $R = e^{-3\lambda_1 t}$, (*b*) $R = 1 - (1 - e^{-\lambda_1 t})^3$,
(*c*) $R = 2e^{-2\lambda_1 t} - e^{-3\lambda_1 t}$, (*d*) plot.

7.11 (*a*) $ze^{-(t/\theta)^m} - e^{-2(t/\theta)^m}$ (*b*) $1 - (t/\theta)^{2m}$.

7.15 part *b*.

7.16 $\dfrac{5 - 2\beta}{\lambda(2 - \beta)}$.

7.17 (*a*) $\frac{2}{3}$ $MTTF_0$, (*b*) $\frac{11}{6}$ $MTTF_0$.

7.19 (*a*) $2R^2 - R^4$, (*b*) $(2R - R^2)^2$.

7.21 (*a*) 0.9867, (*b*) 0.9952.

7.22 3.2×10^{-8}.

7.27 $1 - 2\lambda t$.

CHAPTER 8

8.1 (*a*) 0.885, (*b*) every 6300 hr, (*c*) every 4275 hr.

8.6 (*a*) 0.7225, (*b*) 0.8825, (*c*) 0.7188.

8.8 (*a*) 4.04θ, (*b*) 455%.

8.13 (*a*) 0.9315, (*b*) 20.4 hr.

8.14 (*a*) 0.897, (*b*) $\lambda = 0.013$/hr, $\mu = 0.111$/hr, (*c*) 2% difference.

8.15 0.980.

8.16 65.5 days.

8.17 (2.2×10^{-4})/day.

8.18 (*a*) 0.968, (*b*) 0.946, (*c*) every 18.6 days.

8.19 every 1980 hr.

8.22 (*a*) 0.9594, (*b*) every 87.5 days.

CHAPTER 9

9.4 (*a*) $1 - \lambda(2\lambda^* - \lambda)t^2$, (*b*) 1.56.

9.8 standby: $2/\lambda^2$, active parallel: $5/4\lambda^2$.

9.11 (*a*) shared-load system, (*b*) 1.063.

9.14 (*a*) $2(1 + \lambda t)e^{-\lambda t} - (1 + \lambda t)^2 e^{-2\lambda t}$.
(*b*) $1 - \frac{1}{4}\lambda^4 t^4$, active parallel: $1 - \lambda^4 t^4$.

9.16 1.2×10^{-3}.

9.22 0.9902.

9.24 (*a*) 0.9961, (*b*) yes.

CHAPTER 10

10.12 0.12800, 0.12385, 0.12387.

10.13 (*a*) M_1: 0.382, M_2: 0.637.
(*b*) A: 0.382, B: 0.382, C: 0.637.

Index